Animal Migration,
Orientation,
and Navigation

This is a volume in
PHYSIOLOGICAL ECOLOGY
A series of monographs, texts, and treatises

A complete list of the books in this series appears at the end of the volume

Animal Migration, Orientation, and Navigation

Edited by

Sidney A. Gauthreaux, Jr.

Department of Zoology
College of Sciences
Clemson University
Clemson, South Carolina

1980

ACADEMIC PRESS
A Subsidiary of Harcourt Brace Jovanovich, Publishers
New York   London   Toronto   Sydney   San Francisco

ACADEMIC PRESS, INC.
111 Fifth Avenue, New York, New York 10003

United Kingdom Edition published by
ACADEMIC PRESS, INC. (LONDON) LTD.
24/28 Oval Road, London NW1 7DX

Library of Congress Cataloging in Publication Data
Main entry under title:

Animal migration, orientation, and navigation.

(Physiological ecology series)
Includes bibliographies and index.
1. Animal migration. 2. Animal orientation.
3. Animal navigation. I. Gauthreaux, Sidney A.
II. Series.
QL754.A56   591.52'5   80–1680
ISBN 0–12–277750–6

PRINTED IN THE UNITED STATES OF AMERICA

80 81 82 83   9 8 7 6 5 4 3 2 1

Contents

5 Mechanisms of Orientation, Navigation, and Homing
Kenneth P. Able

List of Contributors

Numbers in parentheses indicate the pages on which authors' contributions begin.

Kenneth P. Able (283), Department of Biological Sciences, State University of New York at Albany, Albany, New York 12222

Charles R. Blem (175), Academic Division, Department of Biology, Virginia Commonwealth University, Richmond, Virginia 23284

Hugh Dingle (1), Program in Evolutionary Ecology and Behavior, Department of Zoology, University of Iowa, Iowa City, Iowa 52242

*Albert J. Fivizzani** (225), Department of Zoology and Physiology, Louisiana State University, Baton Rouge, Louisiana 70803

Sidney A. Gauthreaux, Jr. (103), Department of Zoology, College of Sciences, Clemson University, Clemson, South Carolina 29631

Albert H. Meier (225), Department of Zoology and Physiology, Louisiana State University, Baton Rouge, Louisiana 70803

* Present address: Department of Biology, University of North Dakota, Grand Forks, North Dakota 58202

Preface

From the outset, the organizational plan for this volume was to be different from those published previously. Most books on animal migration have chapters devoted to a particular taxonomic group, and within a chapter various aspects of migration are discussed. In this volume the chapters are arranged by major aspects of animal migration, and within a chapter information from diverse taxonomic groups is included. In this way similarities and differences are more apparent.

Seven aspects of animal migration are included: evolution of migration, climatic and meteorological influences, biological clocks and phenology, bioenergetics, physiological control, sensory systems, and orientation and navigation. The biological clocks and phenological aspects are integrated in the chapter on climatic influences and the sensory aspects in the chapter on orientation and navigation.

In Chapter 1 on the evolution of migration, Hugh Dingle provides a summary of the migration strategies of animals in the context of a time, space continuum. His efforts demonstrate clearly the importance of migration to the study of population biology and evolutionary ecology. Continuing this approach in Chapter 2, I examine the influence of long- and short-term climatic cycles on the spectrum of migratory patterns found in nature, from biogeographical migrations to the diel migrations of planktonic organisms. Charles Blem covers the bioenergetics of animal migrations in Chapter 3 in terms of the cost of transport, the energetic requirements of various migration strategies, and the energy stores of the migrants. His efforts to integrate his topic with those covered in other chapters is very successful. In Chapter 4 Albert Meier and Albert Fivizzani discuss the physiological basis of animal migration, paying particular attention to the recent endocrinological findings on the timing and energetic aspects of various migration strategies. They also cover some of the developments that concern the physiological basis of migratory orientation. In Chapter 5, the last in the volume, Kenn Able does a thorough job of examining the array of mechanisms used in direction finding by migrating animals and the sensory systems that are essential to the orientation mechanisms. Although these authors had little or no collaboration during the preparation of their chapters, the final results of their efforts mesh together surprisingly well.

This volume would not have been completed without the kind, yet forceful encouragement of the series editor, T. T. Kozlowski. During the initial phases of planning his advice was indispensable. Much the same can be said for the staff of Academic Press, whose patience toward the end deserves some form of special recognition. My gratitude to my wife Kay and my children David and Renee is hard to express. They were very understanding of my long hours at work and my frequent distractions from family affairs. My final thanks go to my colleagues who so graciously agreed to contribute a chapter to this volume. Without their efforts, this volume would not have been possible.

Sidney A. Gauthreaux, Jr.

1 Ecology and Evolution of Migration

HUGH DINGLE

Animal Migration, Orientation, and Navigation
Copyright © 1980 by Academic Press, Inc.
All rights of reproduction in any form reserved.
ISBN 0-12-277750-6

1

I. The Phenomenon of Migration and Some Problems of Definition

Since antiquity people have been aware of animal migrations. Frequently this was because migrants profoundly influenced livelihood. For example, locust plagues are mentioned in the Bible (Genesis and Proverbs), northern groups have depended on the seasonal appearance of marine (Eskimos) or terrestrial (Amerindians, Lapps) mammals, coastal Amerindians have fished for salmon, and African tribes have followed the wet and dry season movements of large ungulates. Yet knowledge of migrations, especially of birds and butterflies, was not always based on practical considerations. The beauty of these organisms and their spectacular movements have excited both curiosity and artistic responses. Bird migrations were known to both Aristotle and Homer and the flights of birds and butterflies are featured in English poetry from the Elizabethans to the present.

Because of these readily observed and spectacular movements both popular and scientific views of animal migration have incorporated long distance return movements. Thus return movements were established early on by vertebrate biologists as the criterion for "true migration" (Landsborough Thomson, 1926; Heape, 1931). Landsborough Thomson described true migrations as "changes of habitat, periodically recurring and alternating in direction, which tend to secure optimum environmental conditions at all times" (1926, p.3). Because return movements were unknown except in rare instances, insects and most mammals were specifically excluded. Later authors (Heape, 1931; Dorst, 1962; Harden Jones, 1968; Orr, 1970) have adopted similar definitions often emphasizing a distinction between migration and other movements. These latter were referred to as "emigration," "nomadism," "passive dispersal," and the like (Heape, 1931; Williams, 1958).

More recent study of a number of organisms, especially insects and zooplankton, has revealed movement patterns which, although they do not involve round trips, have the same function as so called true migrations. That is, they allow exploitation of different habitats as life history requirements alter or as environments change seasonally or successionally. Often these movements are "on the whole a rather quiet, humdrum process . . . taking place all the time as a result of the normal life of the animals" (Elton, 1927), but some are spectacularly dramatic like the movements of migratory locusts. Movements of awkward, fragile organisms like aphids or plankton were considered even by entomologists or marine biologists to involve passive transport by winds or currents (Williams, 1958; Hardy, 1958).

Even apparently weak and fragile organisms, however, are active participants in the migration process. An excellent example is the black bean aphid (*Aphis fabae*) (Kennedy, 1958 et seq.). Like others of its kind, this species has a complex life history involving alternation of winter sexual and summer asexual generations on different host plants (Dixon, 1973). Kennedy focused on the asexual summer generation which arises from the wingless offspring of spring migrants to young bean plants (*Vicia faba*). These parthenogenic females produce wingless young viviparously until crowding occurs or the host plant senesces; in these cases winged offspring are produced which migrate to a new host, settle, lose the wings, histolyze the wing muscles, and produce wingless young parthenogenically.

The entire sequence of flight and settling involves a discrete series of events. The aphid climbs to the top of the plant and takes off, being attracted at this time to blue light (Kennedy et al., 1961). After a period of flight, the individual becomes sensitive to yellow wavelengths reflected from leaf surfaces and descends to search for a suitable host plant. If one is found, settling and larvaposition ensue. By experimentally manipulating host plants encountered by free-flying aphids in a laboratory flight chamber, Kennedy demonstrated that stimuli evoking flight inhibit settling and vice versa. The two behaviors also have complex effects on each other. Depending on the respective excitatory states, one can get antagonistic induction or rebound where inhibiting say flight by settling stimuli results in even greater amplitude when flight stimuli are again presented, or antagonistic depression with the opposite effect. Once airborne, direction of flight is determined by the wind, but duration is under control of the aphid. In many insects, e.g., locusts (Rainey, 1976, 1978), there is behavioral orientation specifically so that migrants are carried downwind. In sum, rather than being haphazard vagabonds at the mercy of the vicissitudes of weather, aphids and other insects have evolved highly specialized behavior to become airborne and to settle and colonize when migration is complete.

So too, have responses to specific stimuli evolved in the larvae of benthic marine organisms to ensure entry into the plankton. An example occurs in the larvae of the stomatopod crustacean *Gonodactylus bredini* (Dingle, 1969). These animals live in rock cavities in the marine littoral. The eggs are laid in the cavities, and the females show considerable maternal behavior (Dingle and Caldwell, 1972). The larvae when hatched remain with the female for three stadia; at this time they are negatively phototactic and strongly thigmotactic, clinging to the sides of the chamber. Virtually at the instant of molt to the fourth stadium, they cease to be thigmotactic and become positively phototactic, responses that take them out of the cavity and into the plankton. At the time of the molt to the postlarva, they again become thigmotactic and settle out to seek shelter in the interstices of the substrate. A similar sequence of behavior is common to other marine invertebrates with planktonic larvae. In a survey of 141 species Thorson (1964) found that the larvae of 82% respond photopositively and migrate to the surface early in life and that most become photonegative when they metamorphose to bottom-dwelling adults.

Some seasonal movements of birds also do not fit classical concepts of true migration. Perhaps the best known example is the red-billed Quelea (*Quelea quelea*) in East Africa (Ward, 1971). These birds migrate at the beginning of the wet season to areas where the rains began some weeks earlier; distance, timing, and direction of movement are dependent on the timing of the rains and the movement of the rain front. During the return from this "early-rains migration" aggregations of individuals stop to breed at suitable locations. Breeding is often repeated, and a given female may produce successive broods in the same breeding season at sites far apart. This "breeding migration" with "itinerant breeding" is apparently an adaptation to an extreme wet–dry seasonal cycle and may occur in other bird species as well (Serventy, 1971).

What is apparent from all these organisms is that, no matter how undramatic or dramatic migrations may be, they involve specialized behavior. This behavior differs from locomotory behavior occurring during the course of other life history events. Thus many birds exhibit *Zugunruhe* or migratory restlessness during the night at a time when they would usually sleep, aphids produce wings and respond to blue wavelengths of light, spiders balloon on long threads, and benthic marine organisms become photopositive in contrast to later photonegativity. Even those species which drift on air or water currents have some means of getting into those currents and staying there. Our working definition of migration is then as follows: *Migration is specialized behavior especially evolved for the displacement of the individual in space.* This is somewhat broader than the definition I have used before for insects (Dingle, 1972, 1974) which followed that of Kennedy (1961a *et seq.*), although in keeping with Kennedy's studies, I emphasize that specialized behavior also involves specialized physiology (cf. e.g., Meier and Fivizzani, Chapter 4).

The definition used here is very similar to R. R. Baker's (1978) definition of migration as "the act of moving from one spatial unit to another." Baker, recognizing the enormous variability in migratory behavior, attempts a virtually all-inclusive definition. I think, however, that an important point is missed: this is simply that accidental or unintentional movements, included in Baker's definition, cannot be acted on by natural selection. Specialized behavior can, and that is why I emphasize it here. Baker is quite right in emphatically pointing out that the observed variability within and among species provides the raw material for natural selection and is critical to the understanding of the evolution of migration. In any event, no definition of behavior which is part of a continuum can be totally exclusive (although contra Baker there is a point at which quantitative differences are major enough to be considered qualitative, e.g., language), and rather than in attempts at refinement of definition, effort should go toward understanding the underlying biology.

In trying to comprehend that biology, I shall deal here largely with ultimate causes (J. R. Baker, 1938). Proximate and ultimate have often been confused in studies of migration, and in any case far more effort has been devoted to the former, as in studies of orientation, than to the latter. The reasons are obvious; ultimate causes involving natural selection and adaptation are far more difficult to discern. But questions of not only "when" and "how," but also "why" are central to biology, and the why of migration will be the central theme of this chapter.

I shall examine migration in different groups of organisms briefly surveying the movements that occur and then examining the various theories proposed for their evolution. I then consider theoretical work on the adaptiveness of migration finally closing with a comment on the goodness of fit between theory and fact and suggesting needed data and experiments to fill the existing often distressingly large lacunae in our knowledge and understanding. In a chapter such as this my choice of topics and examples is necessarily eclectic. Therefore, it is likely that some organisms or theories will be discussed only briefly.

II. The Vertical Migration of Plankton

There are many unsolved puzzles of pelagic natural history, but one seems perhaps more baffling than any other: that of vertical migration.

(Sir Alister Hardy, *The Open Sea: The World of Plankton*, p. 199)

A. The Phenomenon of Vertical Migration

The existence of vertical migration in the plankton was known from scattered observations early in the last century, but the first extensive evidence came from the *Challenger* expedition of 1872–1875 (Hardy, 1958). The *Challenger* data made it obvious that the depth distribution of many planktonic species changed dramatically over a daily cycle. Characteristically, the greatest concentration occurred nearer the surface during the night and at greater depths during the day. These observations have been confirmed many times over, and it is now evident that vertical migration has evolved independently in virtually every major group of planktonic animals (Hardy, 1958). More recently it has also been found that the sonic "deep scattering layer" is usually composed of pelagic organisms whose movements account for its daily rising and sinking (Mauchline and Fisher, 1969; Omori, 1974).

The work of Russell (1927) and of Clarke (1934), Nicholls (1933), and others (Waterman *et al.*, 1939) extended the earlier observations and began to define variation in migratory behavior. Seasonal differences were noted in the timing of the daily cycle with the peak abundance at the surface occurring earlier during the long winter nights (Fig. 1). Different stages of the life history migrated to a greater or lesser extent. For example, in the copepod *Calanus finmarchicus* early copepodite stages display little vertical movement while adults move over several meters (Nicholls, 1933; also Fig. 1), and in the epipelagic decapod shrimp *Sergia lucens* immature individuals remain at the surface until the postlarva whereupon migration begins and its magnitude increases through the adult (Omori, 1974). Differences between the sexes are also apparent (Nicholls, 1933; Clarke, 1934; Enright and Honegger, 1977), and there may be considerable individual variation. For example, individuals of the freshwater copepod *Diaptomus gracilis* captured nearer the surface tended to aggregate near the top when placed in a submerged tube in daylight while those captured near 20 m depth tended to move downward (Cushing, 1955).

Several observations suggest that light is the major proximate factor influencing vertical migration with the organisms rising in dim light and sinking in bright. This had been suspected in the 1930s. For example, Waterman *et al.* (1939) showed by plotting the daily cycle of light penetration into the sea that movements of several species in the western North Atlantic followed this pattern. They also demonstrated that at least some species descended to shallower depths on cloudy days. Experimental confirmation was provided by Harris (1953) who mimicked vertical migration in *Daphnia magna* by varying light intensity above a population swimming in a suspension of india ink equivalent in light absorption to a 20-m column of lake water.

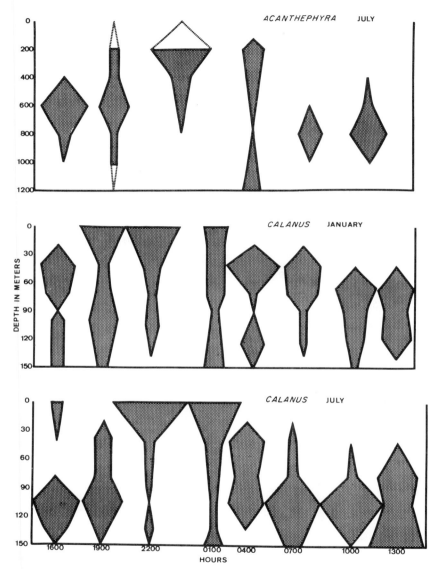

Fig. 1. Examples of daily vertical migration in marine crustaceans. Top: *Acanthephyra*, a sergestid shrimp, in the North Atlantic (after Waterman *et al.*, 1939). Middle and bottom: the copepod *Calanus finmarchicus* off Scotland in winter and summer (after Nicholls, 1933). The width of the "kites" indicates relative densities.

TABLE I

Swimming Speeds of Vertically Migrating Copepods in Meters per Hour[a]

Species	Rate	Habitat
Metridia pacifica	31–91 (best estimates)	Marine
M. lucens	21–50 (minimum estimates)	Marine
Calanus finmarchicus	12–16 (minimum estimates)	Marine
Pleuromamma abdominalis	100	Marine
Eudiaptomus gracilis	4–5	Fresh water
Cyclops strenuus	11 (maximum estimate)	Fresh water

[a] Based on calculations in Enright (1977a).

The *Daphnia* were required to move over a distance of only 24 cm, but in nature the distances traversed in vertical migration can be impressive especially for small organisms. Pelagic shrimp may ascend and descend distances of 800 m or more in a single daily cycle (Waterman *et al.*, 1939; Omori, 1974) and movements of over 100 m have been recorded for copepods (Nicholls, 1933; Clarke, 1934; Enright, 1977a). To cover the distance in the allotted time, these planktonic organisms must achieve considerable swimming speeds. By noting the time it took populations of *Metridia pacifica* to cross a delimited zone, Enright (1977a) was able to make field estimates of swimming rates. He estimates that these plankters are capable of speeds of 31–91 m hr^{-1} and perhaps higher. These and other estimates calculated by Enright from the literature are given in Table I. Although calculated from short bursts in a laboratory "plankton wheel," values produced by Hardy and Bainbridge (1954) for copepods ranged up to 66 m hr^{-1}, and were likely underestimates because of the unnatural conditions (Enright, 1977a). Plankters thus have the capability for extensive daily vertical migrations. But the major question, as implied by Hardy (1958) in the opening quotation above, is: What is the adaptive significance?

B. THE ADAPTIVE SIGNIFICANCE OF VERTICAL MIGRATION

1. Light and Optimal Depth

Prompted by observed correlations between zooplankter movements and changing intensities of light, early workers sought explanations of migration in terms of levels of illumination (for reviews, see Cushing, 1951; Raymont, 1963). Michael (1911) seems to have been the first to suggest that a migrant, the chaetognath *Sagitta bipunctata*, followed its optimum light intensity. The theory was elaborated by Rose (1925) in terms of a phototaxis which could keep individuals in a zone of optimum intensity. This zone would move up at night and down during the day accounting for movements of the zooplankton. Russell (1927) and Cushing (1951) supported the idea in their reviews. A modification proposed by Harris (1953; Harris and Wolfe, 1955) suggested that vertical migration was a by-product of the nature of the visual systems concerned with the animals attempting to remain in areas of unvarying light levels, thus rising at night and descending during the day. In terms of adaptive significance the theory, whether in its original or modified form, is highly dissatisfying (McLaren, 1963; Enright 1977b). First it is totally unclear what is "optimal" and indeed this seems to be nowhere indicated in the relevant literature. Second, there is no apparent mechanism by which natural selection can act on the behavior since there is no way to ascertain the disadvantages of being outside the optimal light zone.

An updated version of the theory, proposed by Hairston (1976), does specify selective advantage. He studied the effects of light on high and low carotenoid strains of the freshwater copepod *Diaptomus nevadensis*. Both strains suffered higher mortality when exposed to visible blue light (450 nm) than when kept in the dark, but some protection was offered by the red pigments. An advantage to lower depths during the day would be the avoidance of the deleterious effects of blue light which can penetrate to 20 m or more in clear water. The problem here is that the adults of many copepods migrate while the immature stages do not, and it is difficult to conceive of only adults suffering adverse effects from light. Also differences in pigmentation do not in general occur between migrant and nonmigrant individuals, strains, or species. Any general validity for Hairston's theory therefore seems doubtful (Enright, 1977b).

2. Predation

An intuitively appealing theory regarding the adaptiveness of vertical migration is that it is a response to predation. Most major predators of zooplankters such as whales and pelagic fish feed near the surface during the day. Further, the available evidence suggests that they are visual predators. Thus the selective advantage of moving to greater depths during the day is obvious: predation pressure is less where visibility is low.

Assessments of the influence of predation appear in many recent discussions of the adaptive significance of vertical migration, but there is little direct evidence one way or the other. A notable exception is the study of Zaret and Suffern (1976) on two freshwater zooplankters, the copepod *Diaptomus gatunensis* in Gatun Lake, Panama, and the cladoceran *Daphnia galeata mendotae* in Connecticut. *Diaptomus gatunensis* is the dominant and largest zooplankter in Gatun Lake, and the adults particularly exhibit pronounced diurnal vertical migration. Because of its size and abundance it should be a common prey of the chief fish planktivore

Melaniris chagresi, yet analyses of fish stomachs indicate it is rarely taken. Laboratory and field observations demonstrate that *M. chagresi* is a visual surface feeding predator quite capable of feeding on *D. gatunensis*, and, in fact, on the occasional calm moonlit nights it does so. The evidence thus suggests that the vertical migration of *D. gatunensis* is effective in reducing predation as the fish do not feed in deeper water and usually cannot see the zooplankters at night.

In the Connecticut study the feeding efficiency of the chief surface-feeding planktivorous fish *Notemigonous crysoleucus* on adult *Daphnia galeata mendotae* increased markedly at higher light intensities. *Daphnia* shows strong vertical migration reaching a peak at the surface at 2200 hr when light intensities inhibit predation. Two further observations also support the predation hypothesis. First, *N. crysoleucus* is an introduced species in the study pond, and since its introduction the vertical migration of *D. galeata mendotae* has apparently increased. Second, in the winter when fish predation is reduced, the *Daphnia* population shows only slight migration. In these species of *Daphnia* and *Diaptomus*, therefore, there is strong suggestive evidence that predation is a major selective force influencing vertical migration.

3. Feeding

Pelagic food chains start with the phytoplankton, and investigators have considered these primary producers when pondering the adaptive significance of vertical migration. In one of the first models to suggest a relation between vertical movements and phytoplankton density, Hardy (1958) suggested that by moving up and down in the sea zooplankters could take advantage of the different rates and depths at which horizontal currents move. By the simple expedient of changing depth, a zooplankter could move from a slower to a faster current and increase its horizontal displacement severalfold. The hypothesis originally attempted to explain differing horizontal densities of zooplankton by "animal exclusion" (Hardy and Gunther, 1935) in which dense phytoplankton was supposed to produce some antibiotic metabolite which influenced zooplankton to leave the area; this they could do by changing depth and current speed. This mechanism could account for the frequently observed negative correlation between zooplankton and phytoplankton densities. Bainbridge (1953) reversed the emphasis by suggesting instead that rather than dispersing, zooplankton used different currents to search for dense phytoplankton in which they then concentrated and fed. The inverse densities of the two sorts of plankton are then explained as "overgrazing."

The evidence is not compelling for either version of the Hardy–Bainbridge hypothesis. There are also several problems with the model (McLaren, 1963). First, it does not explain the daily regularity of vertical migration since there seems to be no good reason why dispersal rates should be altered cyclically especially under irregular selection. Second, migration should vary with food abundance increasing with scarcity and vice versa; there is no evidence that it does so. Finally, patchiness is frequently absent in freshwater phytoplankton because of mixing in the euphotic zone, yet vertical migration is still conspicuously present.

Two independently derived but similar optimization hypotheses relating to migrations were advanced by McAllister (1969) and Kerfoot (1970, 1972). McAllister's statement of the notion is more explicit if less well developed. Basically the idea is that vertically migrating zooplankters may better utilize phytoplankton productivity by allowing unimpeded photosynthesis during daylight hours when the migrants are deep, followed by rapid feeding in the evening before nocturnal respiration is lost to the ecosystem. Although not patently explicit, the notion of "management" of the resource by optimal utilization is still implicit in Kerfoot's more sophisticated development of the hypothesis (McLaren, 1974; Enright, 1977b). Kerfoot takes as his point of departure Harris's (1953) statement that migration was a natural by-product of the use of light for orientation (see above). With a somewhat elaborate series of assumptions and computations, Kerfoot calculates plankton productivity over different ranges of light intensity (dividing the total range into 10 μW cm^{-2} blocks, e.g., one block was 10^{-2} to 10^{-3} μW cm^{-2}). Then if light-orienting consumers are adapted to different light ranges, if the ecosystem is at equilibrium, and if each portion of the consumer population shows a stable age distribution, the consumer biomass within each light range is approximately proportional to the productivity within that range, thus optimizing the transfer of phytoplankton production to zooplankton. To the question of why zooplankters do not simply concentrate in the area of richest productivity (Miller et al., 1972), Kerfoot (1972) responds that high productivity does not mean more food because populations expand to reach equilibrium implying that zooplankters distribute themselves to ensure sustained, equilibrium food supplies (McLaren, 1974; Enright, 1977b). The increased time spent at the surface in winter when productivity is low allows more time to feed and hence stabilization of annual food input.

Kerfoot has been criticized for his assumptions and for manipulating computations and selecting data to fit the model (Miller et al., 1972; McLaren, 1974), but the most profound problem involves the implicit notion of management of resources (Enright, 1977b). This would seem to require group selection. Simply put, it is not at all obvious why selective advantage would not accrue to the glutton rather than the cooperative gourmet, i.e., why Darwinian selection acting on individuals would not be primary.

The demographic model of McLaren (1974), a revision of an earlier

(1963) metabolic model, is clearly based on Darwinian selection. The hypothesis incorporates well-established experimental observations that crustacean growth is slower in colder (and hence deeper) water with the result that in some species adults reach a larger size and produce more eggs in the presence of excess food. An additional assumption is that grazer populations exist at approximately equilibrium abundance. Using empirical data on fecundity and development rates from the copepod *Pseudocalanus minutus*, McLaren analyzed the demographic consequences of vertical migration in thermally stratified waters. In seasonal life cycles, the increased fecundity resulting from delayed growth caused by spending part of each day in deeper colder water would confer selective advantage above that conferred by a shorter generation time (Cole, 1954; Lewontin, 1965) at higher temperatures. This is not always true in continuous life cycles, but here again advantages do accrue to increased fecundity if the ratio of mortality of early (prior to stage III copepodite) to late life-cycle stages if sufficiently high. The ratio pre- to post-stage III has little influence on adult size. Indeed a further common observation is that in many copepods early stages do not migrate while later ones do.

Firmly rooted as it is in empirical data and individual selection, McLaren's model is appealing. It nicely applies to any zooplankter which vertically migrates in a thermally stratified body of water and in which size and fecundity are negatively correlated with temperature. The model does, however, have two conspicuous weaknesses (Enright, 1977b). First, the assumption of regularly available excess food is somewhat questionable since it has been observed that zooplankton and phytoplankton densities are often inversely related (Hardy and Gunther, 1935; Hardy, 1958). Second, and perhaps more important, it does not account for the daily cyclicity of vertical migration as any proportionate time spent at cooler temperatures will fulfill the model's requirements. McLaren (1963) also recognized this difficulty when he incorporated the predation hypothesis (see above) into his original conclusions. Thus it would appear that at the least the McLaren model cannot stand alone. The criticism of Mauchline and Fisher (1969) that McLaren failed to take into account the energetic cost of vertical travel now seems irrelevant since this cost is apparently trivial (Vlymen, 1970; Enright, 1977a).

Probably the most complete model of the adaptiveness of vertical migration is the metabolic model of Enright (1977b). It also has the virtue of having been subjected to an experimental test (Enright and Honegger, 1977). Enright, following Miller *et al.* (1972), asks what selective advantage accrues to vertical migrators that would compensate grazers for loss of food intake during daylight. His model, like McLaren's (1974), also takes temperature differences into account by assuming that a considerable metabolic saving can be realized by resting at depth during the day both because the energy required to combat sinking is less (water viscosity is inversely related to temperature) and because of reduced basal metabolic rate. A second assumption is that the photosynthetic rate of the phytoplankton is sufficient to cause an appreciable increase in algal biomass during the day; empirical evidence is supportive and suggests such increases can be of the order of 50% (Enright, 1969). A third assumption is that following an extended period of nonfeeding, e.g., during daylight, the initial grazing rate of planktonic herbivores is higher than the steady-state rate and then declines in a few hours; ingestion rates of the copepod *Calanus pacificus* do follow this pattern. Finally, the model is also a competition model that assumes that zooplankton behavior has been influenced by competition among grazers for food because these grazers remove a major fraction of the phytoplankton standing crop daily.

   Calculations of energy input and cost for migratory zooplankters suggest that the net energy gain by migrating would be greater than that for feeding. Lowering energetic costs at colder depths during the day followed by a burst of feeding activity in the evening would thus more than compensate for the daily loss in energy resulting from departing the photic zone. This model makes a clear prediction about migratory behavior which distinguishes it from the predation model. This prediction is that if maximizing net energy gain is more important selectively than avoiding predators, then grazers ought to take advantage of daily increases in phytoplankton production by migrating to the surface to begin feeding approximately 1–2 hr *before* sunset. To avoid visually searching predators migrants do best to avoid the surface until after sunset. A further suggestion is that since reduction of metabolic rates in nonfeeding as opposed to feeding activity is so critical, such physiological characteristics may distinguish migrants from nonmigrants. If the ability to reduce metabolism were slow to develop, this could account for observed differences between nonmigrant subadults and migratory adults of the same species. Such physiological differences have yet to be measured.

   The prediction of presunset upward movement was, however, tested by Enright and Honegger (1977). They ran 3-day field sampling programs in late April, late May, and early July off the coast of Southern California taking duplicate plankton samples at 30-min intervals during late afternoon and evening around the time of sunset. The results indicated considerable variation in vertical migration as a function of both season and stage in the life history. They assessed the amount of migration by determining the proportion of the population of the most abundant copepod, *Calanus helgolandicus (pacificus)*, in the warmer waters above the bottom of the thermocline (about the first 20 m).

   The prediction of presunset migration was borne out in all three of the May samples for adults of both sexes and for stage V copepodites. In April stage V's showed conspicuous presunset upward movement, males showed a small amount, and females virtually none. Finally, in July essentially no copepods appeared in the upper stratum prior to sunset. The May migrations occurred when the copepod population was most numerous and the proportion of immatures highest suggesting that predation was lowest at this time. Enright and Honegger surmise that predation may be important in July when the population declined to about a third of the May level. The obvious presunset migration in April juveniles may be a consequence of the relatively greater importance of feeding and growth at this stage. Suffice it to say, as Enright and Honegger point out (cf. also Zaret and Suffern, 1976), zooplankters can exercise more than one behavioral option, with the probability of taking one or the other alternative calibrated by natural selection. Also more than one selective agent can be operating. Simplistic notions of selection are thus inappropriate (Zaret and Suffern, 1976).

Finally, Brinton (1967) analyzed vertical migration in euphausiids occurring off the California coast and noted that those species and genera clearly displaying vertical migration are the least specialized morphologically. All migrating species possess a spherical eye and uniform, setose, comblike thoracic legs that apparently sweep or strain the water for phytoplankton. The predation and feeding models discussed are likely to be applicable to this group. In contrast non-migratory species and genera possess a bilobed eye and a pair of extra long thoracic legs which are brushlike, bristled or spearlike, subchelate, or even chelate. Such limbs and eyes evidently adaptive for predation and selective feeding. Thus there is apparent coevolution between degree of migratory behavior and feeding and other morphology. This subject has been little explored in the plankton but seems to merit more attention than it has so far received.

4. Other Models

Various other models have been proposed to account for vertical migration, and two seem worth a brief mention. Extending Hardy's drift hypothesis, David (1961) suggests that population shifts bring about outbreeding and genetic recombination in turn preventing the development of new specialized species at selective disadvantage. Aside from potential problems because the model seems tinged with notions of group selection, McLaren (1963) points out that among certain copepods, the one with the least migration is the least polytypic contra predictions of the theory. David himself admitted that the theory was "regrettably speculative" with little evidence to support it.

Then, of course, there is the Wynne-Edwards (1962) hypothesis. According to this zooplankters, like other organisms, congregate to undertake "epideictic" displays which allow them to assess population density and exercise appropriate control over reproduction. The daily cyclicity is seen as a timing device to bring concentrations of organisms together. Again the usual difficulties with group selection apply, besides which there is absolutely no evidence of the potential for social assessment in zooplankters (McLaren, 1963). Wynne-Edwards's arguments now seem to have been quite thoroughly laid to rest and should probably be spared further exhumation.

5. Conclusions on Adaptive Significance

No single hypothesis of adaptive significance is consistent with all the known facts concerning vertical migration. This is hardly surprising since it is likely that different selective forces influence different species or even life history stages within a species in different ways (Mauchline and Fisher, 1969; Enright, 1977b). Any single factor theory is bound to be simplistic (Zaret and Suffern, 1976). The strongest supporting evidence favors avoidance of visual predators, at least by grazing herbivores, and metabolic efficiency resulting from residing at cool depths during the day and a burst of feeding at the height of phytoplankton density. This evidence, however, comes from only a few species of copepods and cladocerans.

For the remainder and vast majority of vertically migrating planktonic organisms, there are many data that are interesting, some that are suggestive, and none that is conclusive. Because of the technical difficulties of working in the ocean and other bodies of water, progress in unlikely to be rapid, but the iterative combining of theory and test (Enright, 1977b) seems the logical way to proceed where possible. Yet for many zooplankters, we still need basic natural history information before proceeding further with theory. Most of the literature deals with crustaceans; the numerous planktonic species in other phyla are still unexplored. So too most efforts have been directed toward grazing herbivores, and planktonic predators represent a great void. Where something is known of the biology, a logical way to proceed would be a close examination of related migrant and non-migrant species or stages within a species rather than whole assemblages at a community level where a number of selective interactions, difficult to sort out, may be operating. Natural selection still acts primarily on individuals, and it is to individuals we must look for an understanding of adaptive significance and the evolutionary process.

III. The Dispersal of Planktonic Larvae

Of the known species of shallow water marine bottom invertebrates some 80% or approximately 90,000 species reproduce by means of pelagic larvae (Thorson, 1964). The great majority of the species for which we have data possess larvae that are initially photopositive, thus showing specialized behavior that will take them to the surface of the sea (Thorson, 1964). Two characteristics of ocean surface waters seem worth considering in attempting to understand the significance of the larval phase; these are the productivity of the phytoplankton in the photic zone and the greater average velocity of the surface currents which can disperse larvae over wide areas (Scheltema, 1971). The advantages of a planktonic larval stage could accrue because of the ability to exploit an otherwise unavailable food resource or to take advantage of currents for dispersal. That food can not be the sole selective agent is suggested by the fact that many lecithotrophic (i.e., dependent on stored yolk for nutrition) larvae are still strongly photopositive. This suggests additional advantages; the most frequently mentioned is dispersal. The advantage must be considerable because of the extraordinarily high mortality among pelagic larvae (Thorson, 1964; Menge, 1975).

The question of the short-term (circa 10–20 generations) advantage of large-scale dispersal is considered in detail by Strathmann (1974) in terms of individual selection of sibling larvae. He notes first that there may be advantages to long larval life independent of dispersal such as permitting the adults to produce smaller eggs (Thorson, 1950) or maximize settlers per unit of reproductive effort (Vance, 1973). A number of possible factors such as avoidance of crowding or deteriorating local conditions do not seem to influence dispersal because planktonic larvae are produced even in their absence. Factors that do favor dispersal include variations in success of settling, subsequent benthic and spawning mortality, and survival of early larval stages released in an area or, in other words, a coarse grained environment with quality of patches varying independently over time (Levins, 1968; Horn and MacArthur, 1972). Spreading one's offspring evenly over varying favorable and unfavorable habitats should confer higher average fitness over time. The relative success of dispersing offspring over several areas compared to nondispersers in the home area is the ratio of the product of their successess in the two situations; clearly if this is greater than unity (with dispersal as the numerator), it will pay to disperse. It is also interesting to note here William's (1975) postulate that survival of offspring in heterogeneous environments is one of the selective forces promoting the evolution of sexual reproduction, since many pelagic larvae are the product of sexual reproduction in forms which spread vegetatively once they have colonized (e.g., many coelenterates). Strathmann also notes that dispersal of siblings can be increased by increasing the length of the pelagic period (but only up to a point), spreading release times of the larvae by the parent, spreading the vertical distribution of the larvae (to drift on currents moving at different speeds), and producing larvae with variable pelagic

periods. Unfortunately there are virtually no data on the relative importance of these factors.

   In view of the widespread occurrence of pelagic larvae, it is perhaps
instructive to examine situations in which such larvae do not occur to in-
vestigate selective factors which may mitigate against dispersal. The most
obvious is in fresh water. Thus the marine lobsters produce large numbers
of pelagic young, but freshwater crayfish produce relatively few young that
are extensively brooded by the female and that have evolved a specialized
behavior of attaching to her swimmerets even when they are capable of
swimming freely. Another interesting contrast occurs between semiterrestrial forms that return to the sea to release young such as the coconut
crab *Birgus latro* and the West Indian land crab *Gecarcinus lateralis* and
various other semiterrestrial tropical species that occur along streams or in
wet forests. The former produce relatively large numbers of small eggs
(although usually larger than the eggs of fully marine form) while the latter
frequently brood very few young to a large size relative to the adult (Bliss,
1968, and personal observations). Physical factors in fresh waters in
general and in stream habitats in particular are probably strong selective
forces favoring brooding over the broadcast of pelagic larvae. Thus the
evolution of dispersal is not independent of other considerations influencing reproductive mode.

   Menge's (1972, 1974, 1975) studies on two species of intertidal sea stars
make this point very well. These two species, *Leptasterias hexactis* and the
large *Pisaster ochraceus*, occur together on the Northwest Pacific Coast of
North America. Menge's studies suggest that the small body size of *L. hexactis* is the result of competition with *P. ochraceus*. *Pisaster ochraceus*
broadcasts its numerous young, but *L. hexactis* broods relatively few
young. Menge suggests that the small species could not replace itself in its
short expected lifespan by broadcasting because of the high pelagic mortality; therefore brooding is a coadaptive consequence of competition induced small body size. Although information is scanty, data from some
other marine organisms also suggest brooding as a consequence of small
size, and the issue certainly deserves further investigation not least because
of its profound implications for dispersal. Among other factors influencing
reproductive patterns and hence dispersal, Menge (1975) lists (1) food
availability for both adults and offspring, (2) planktonic mortality rates,
(3) interactions between species, (4) latitudinal variance in these, and (5)
various physical constraints. The potential complexity of influences and interactions demand consideration of the ecological and evolutionary effects
of biological interactions and physical factors plus interaction between
planktonic and benthic stages of life cycles. It is all too apparent that ac-
quiring the necessary data and theory will not be easy.

IV. The Migrations of Demersal and Terrestrial Crustaceans

Many species of shrimps, crabs, and lobsters migrate with the commonest movements occurring inshore–offshore on a seasonal basis or in different stages of the life history (Bainbridge, 1961; Allen, 1966). In various shrimp, young stages move into estuaries to mature presumably to take advantage of high nutrient concentrations (Bainbridge, 1961) and possibly to avoid predators. As individuals mature, they leave the estuaries for the open ocean. The age at which they do so varies with temperature; in subtropic and temperate species falling temperatures result in younger and smaller individuals leaving the estuary (Allen, 1966; Pullen and Trent, 1969; Ruello, 1975). Movement to warmer waters offshore is apparently necessary for continued growth and development. Species that may reach maturity in estuaries still migrate offshore to breed, and these movements may develop into extensive alongshore migrations (Allen, 1966). The most spectacular case of this is the king prawn, *Peneus plebejus*, of the east coast of Australia (Ruello, 1975). This species can develop in estuaries southward into the temperature zone, but newly hatched larvae evidently require warmer temperatures. Northward migrations by the adults into the tropics to breed apparently occur in large schools and can cover 960 km, the longest known movement of any crustacean.

A number of species of crabs also mature in estuaries and move offshore to breed (Hill, 1975; Norse, 1975). In tropical and subtropical portunids the primary selective agent seems to be salinity as the eggs hatch prematurely when salinities are too low (Hill, 1975; Norse, 1977). As in shrimp, migration into estuaries probably evolved to take advantage of rich nutrients, and where several species occur, they partition the salinity gradient by interspecific agonistic behavior (Norse, 1975). Salinity may not be the only factor, however, as portunids seem to require soft substrates in which to burrow to facilitate egg attachment to the pleopods. *Euphylax dovii*, for example, has adapted to an offshore pelagic existence, but females still require soft substrates for egg attachment (Norse and Fox-Norse, 1977). This results in inshore migration exclusively by the females at the time of breeding.

Various spiny lobsters (Palinuridae) also make extensive migrations. Generally, although not always, these are associated with breeding (Allen, 1966). These breeding movements are usually from deeper to shallower water inshore although there may be movement from deeper lagoons to shallower offshore reefs in Bermuda (Sutcliffe, 1953). *Panulirus argus* in the Caribbean migrates into deeper water in the autumn, frequently in long queues (Herrnkind and Cummings, 1964; Herrnkind and McLean, 1971). The adaptive significance of these migrations is, however, not understood.

As with other demersal crustaceans migration in this group seems to involve favorable breeding sites or possibly favorable thermal environments.

Many terrestrial crabs and hermit crabs must migrate back to the sea to release their young. These movements, although often spectacular and well known to native peoples, seem to have attracted little attention from biologists (Bainbridge, 1961; DeWilde, 1973). An exception is the hermit crab *Coenobita clypeatus* in Curacao (DeWilde, 1973). These crabs live far inland obtaining water from wells. From July to October females migrate to the sea, often in large groups and over distances of several kilometers. A similar seasonal migration occurs in the large terrestrial land crab *Cardisoma guanhumi* in Florida (Gifford, 1962) and Venezuela (Taisson, 1974). For *Coenobita rugosus* on the coast of Somalia the selective agents are food and water (Vannini, 1976). The crabs generally spend the day in the sand dunes, but come out at night to feed. In the dry season they must replenish water in the shell so migration is to the beach at low tide. In the wet season with plenty of water available nightly migration is away from the sea and into the bush where food is more abundant; some individuals may spend the whole wet season in the bush. Migrations to the sea to release eggs also occur.

Similar studies of other terrestrial crustaceans should yield useful comparative information on migrations and insights into its adaptive value and evolution. It might be particularly useful to explore the coevolutionary relation between egg size and number and migrations. Terrestrial crabs returning to the sea to spawn produce thousands of small eggs like their wholly marine relatives. In contrast, wholly terrestrial crabs or crabs and crayfish breeding in fresh water produce few large eggs (Bliss, 1968, gives several examples). Why some groups made the transition and others did not is an intriguing question deserving more attention than it now receives. (See also discussion of planktonic dispersal above.)

V. Migration of Insects and Other Terrestrial Arthropods

The east wind brought the locusts . . . they covered the face of the whole earth . . . and there remained not any green thing . . .

Book of Exodus

A. The Behavior of Migrants

Migration is a widespread phenomenon in terrestrial arthropods and has evolved independently in group after group (Johnson, 1969; Dingle, 1972, 1974). The vast majority of insects migrate by flight as adults, but several have evolved phoresy (displacement by attachment to another organism); Clausen (1976) discusses species in 16 families transported by this method. Phoresy is also a common means of transport among pseudoscorpions (Weygoldt, 1969), and aerial dispersal by ballooning spiders is well known (Duffy, 1956; Southwood, 1962). Some insects, e.g., locust hoppers, move long distances by walking or crawling. Whatever the means of locomotion, however, migration in terrestrial arthropods involves specialized behavior.

The most complete analysis of migration as specialized behavior is that on the aphid *Aphis fabae* (Kennedy, 1963 et seq.) as discussed above, where it was noted that flight and settling interact in the manner of antagonistic reflexes. During migration settling and related responses are strongly inhibited so that the aphid fails to respond to stimuli that would ordinarily attract it. Kennedy in fact advocates the suppression of "vegetative" responses, i.e., feeding, growth, and reproduction, as one of the criteria by which insect migration can be identified. The milkweed bug, *Oncopeltus fasciatus*, also shows suppression of feeding and reproductive activity during migration, further enhanced by a circadian rhythm which separates them in the daily cycle (Caldwell and Rankin, 1974). As flight proceeds, the migrants become increasingly sensitive to stimuli which eventually arrest flight. This is revealed in aphids by the increasing sensitivity to yellow light and unsuitable hosts. The bark beetles *Dendroctonus pseudotsugae* and *Trypodendrum lineatum* after prolonged flight display a lowered landing threshold in response to olfactory stimuli from the host trees (Bennett and Borden, 1971).

Age and reproductive condition also influence migratory flight. Characteristically migration occurs after the cuticle has hardened (Johnson, 1969; Dingle, 1972, 1974), and in females prior to oviposition. Johnson (1969) refers to the interaction between migration and reproduction as a prereproductive "oogenesis-flight syndrome"; migration in some species also occurs between bouts of reproductive activity (Kennedy, 1961a, 1975; Baker, 1968). Prior to flight the appropriate muscle and enzyme systems are also undergoing a period of maturation (Johnson, 1976). The partitioning of metabolic investment between migration and reproduction depends, at least in the species so far examined, on titers of juvenile hormone which in *Oncopeltus fasciatus* promotes flight at intermediate levels and reproduction, with suppression of flight, at high levels (Rankin, 1974, 1978; Rankin and Riddiford, 1977). In other species juvenile hormone may suppress flight by promoting histolysis of wing muscles (Stegwee et al., 1963; Davis, 1975).

Insect migration thus involves not only specialized behavior but also specialized physiology. It also varies considerably both within and among species and is influenced by a complex of interacting environmental and genetic factors. The evolution of migration is determined by the action of natural selection on those interactions. In discussing this evolution, I shall begin with those species showing complete macroptery (wing development for flight), then proceed from species which histolyze the wing muscles during sedentary periods to species polymorphic for wing development, and finally discuss species that have lost the power of flight altogether.

B. Variation within and among Species

1. Variation in Flight Behavior in Macropters

Because of their relatively small size and the ease with which many species can be cultured in the laboratory, insects offer several advantages of the study of migration. In particular the experimental analysis of flight behavior and related life history phenomena becomes possible (Dingle, 1974). The standard technique for studying migration has been that of tethered flight where some measure of flight duration or strength is used as a presumptive index for migratory capability. Several species ranging in size from locusts to aphids have been studied in this manner. As an example, when milkweed bugs, *Lygaeus* and *Oncopeltus* spp., are glued to an applicator stick at the prothorax and lifted free of the substrate, they will usually fly without further stimulation often for several hours (Fig. 2; see also Fig. 3). That this tethered flight does indeed index migration is suggested by the fact that (1) immigrant bugs in early summer, (2) bugs in the autumn just prior to emigration, and (3) bugs entering a milkweed patch from which all bugs had been previously removed all display greatly enhanced flight durations when tethered (Caldwell, 1969, 1974; Blakley, 1977). Thus bugs that are in all probability migrants fly considerably longer when tethered than the average bug.

A major intrinsic factor influencing tethered flight duration is age. Some examples are shown in Fig. 2. The milkweed bugs *Oncopeltus fasciatus* on long photoperiods and *Lygaeus kalmii* display maximum flight durations 8–10 days following adult eclosion when cuticle deposition is complete and prior to reproduction (Dingle, 1965 et seq.; Caldwell, 1969, 1974). The frit fly *Oscinella frit* (Rygg, 1966) and the mosquito *Aedes aegypti* (Rowley and Graham, 1968) reach full flight capacity somewhat earlier. Riegert (1962) tested three species of grasshopper, *Camnula pellucida*, *Melanoplus sanguinipes* (= *M. bilituratus*), and *M. bivittatus* and found maximum flight between 7 and 21 days with little flight either before or after that time. In the locust *Locusta migratoria* maturation of the flight system is

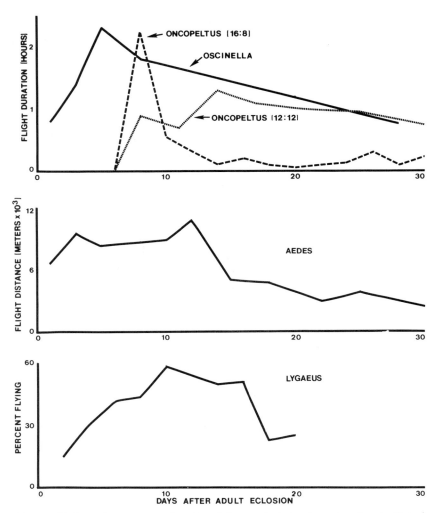

Fig. 2. Flight performance as a function of days posteclosion in various insects. Data for *Oncopeltus fasciatus* from Dingle (1968), for *Oscinella* from Rygg (1966), for *Aedes* from Rowley and Graham (1968), and for *Lygaeus kalmii* from Caldwell (1969).

revealed by a rise in wing-beat frequency for about the first 30 days of adult life (Kutsch, 1973). Sex differences in mean flight duration (Riegert, 1962) or proportion of the test population displaying long flights (Dingle, 1965 et seq.) have also been noted.

Flight activity is also phenotypically modifiable by factors such as photoperiod, temperature, sexual activity, and food availablity. *On-*

copeltus fasciatus from temperate latitudes diapause on short days and extend the ages over which long flights occur (Fig. 2). (The relation between migration and diapause is discussed further below.) Temperature influences flight directly and via its effect on metabolic rates. For example, individuals of the coccinellid beetle *Coleomegilla maculata* captured at a hibernation site displayed long flights earlier at higher temperatures (Solbreck, 1974, 1978). In *O. fasciatus* raising the ambient temperature from 23° to 27°C reduced the proportion of migration to less than 10% of the population; warmer temperatures lead to earlier reproduction, so once bugs found a thermally favorable environment, they would tend to stay there (Dingle, 1968). Food deprivation for a few days may also increase the proportion of migrants (Rankin, 1978); longer periods of starvation depress flight. Low quality food, e.g., stems and flowers rather than seeds (Ralph, 1976), also promotes flight and is probably a significant factor in the northward migration of *O. fasciatus* in the spring (Caldwell, 1974). Fresh leaves containing the plant growth hormone gibberellin A_3 promote reproduction and suppress migration in locusts (*Schistocerca*) (Ellis et al., 1965). Depriving milkweed bugs of sexual partners also enhances flight activity and extends the peroid of migration. The adaptive advantage of leaving an area short of food and sexual partners is obvious.

A species in which food quality and temperature appear to be the major determinants of migration is the tortricid moth *Epiphyas postvittana* in Australia (Danthanarayana, 1976). Large individuals predominate during the cool wet months of the year while during the warm dry season individuals are smaller. Ratios of body length to wing length indicate that lower wing loadings of smaller individuals render them more efficient flyers. Danthanarayana suggests that the smaller body size is an adaptation for dispersal in the face of environmental adversity. The situation is analogous to migratory *gregaria* phase locusts.

Crowding can profoundly influence migration in migratory locusts (*Locusta, Nomadacris, Schistocerca*). The subject of phase polymorphism in locusts has been extensively reviewed (Kennedy, 1961; Uvarov, 1966, 1977; Albrecht, 1967) so a brief summary will suffice here. Density influences morphology, reproduction, and migratory behavior. With respect to morphology, both body form and color are affected. For example, the ratio of elytron (wing) length to hind femur length in *Schistocera gregaria* is about 2 in fully solitary populations and about 2.3 in fully swarming ones. The relatively long elytron of *gregaria* might be associated with smaller wing loading and hence greater efficiency of flight. Sex differences in body proportions are also greater in *solitaria* than *gregaria*. Phase differences in color are even more impressive with *gregaria* nymphs showing a bold pattern of black melanin on yellow carotenoid pigments while *solitaria* are uniformly pale or green. As adults the egg production of *solitaria* females exceeds that of *gregaria* females (Kennedy, 1961a,b). Phase variation is also under genetic influences, but selected lines are all subject to density effects.

The more impressive differences between the phases, however, are in behavior. First, *gregaria* are intensely gregarious actively seeking contact with other individuals. Second, they are migratory to an extraordinary degree throughout their lives with nymphs or "hoppers" "marching" across the landscape in large groups while the adults form large milling swarms carried by the prevailing winds (Rainey, 1976, 1978). Swarm cohesion is maintained by the intense gregariousness of individual locusts. Phase *solitaria* also migrates but not in swarms and by night rather than by day. The winds bring *gregaria* to areas of rains where they consume fresh vegetation and reproduce (Rainey, 1976, 1978). Kennedy (1961b, 1962) views phase polymorphism as an adaptation to fluctuations between a "soft" habitat where resources are abundant and concentrated (high fecundity *solitaria*) and a "hard" habitat where resources are erratic and highly dispersed (gregarious migratory *gregaria*). The influence of density on migration in other insects is reviewed by Johnson (1969).

Migratory capability can also vary geographically. The tethered flight of four populations of *Oncopeltus fasciatus*, from Iowa, Florida, Puerto Rico, and Guadeloupe, was analyzed according to four criteria: mean flight duration, proportion of the population showing long flights (operationally defined as 30 min or longer), longest single flight per individual, and threshold for flight (Dingle, 1978). The rank order of these populations with respect to all four of these criteria was Iowa > Florida > Puerto Rico > Guadeloupe; the distribution of flights in the Iowa and Puerto Rico bugs is shown in Fig.3. Thus migratory flight decreased as populations were progressively more geographically isolated.

What is apparent from all studies is that whereas different environmental factors can profoundly influence flight, variances between individuals are still high (Johnson, 1976). Two examples are indicated in Fig. 3 comparing populations of *Oncopeltus fasciatus*. Although means differ significantly (Dingle, 1978), both distributions are highly skewed with most individuals flying for only a few seconds, or not at all in spite of the fact that all are macropterous and no other morphological variation is apparent. Similarly skewed distributions have been observed in a variety of other insects (Johnson, 1976). Such intrinsic variation in flight performance is potentially of great importance for the evolution and adaptive significance of migration.

Regrettably, however, there have been few studies of genetic variances on flight although the available data suggest the genetic variances may be high. Dingle (1968) applied intense selection for long flight to *Oncopeltus*

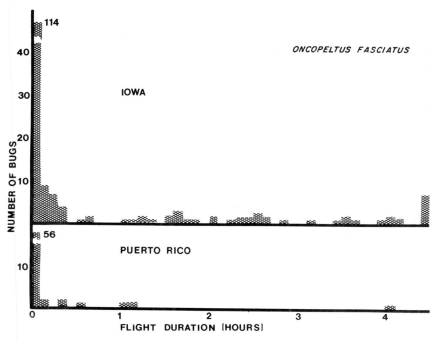

Fig. 3. Frequency histograms of flight duration for *Oncopeltus fasciatus* populations from Iowa and Puerto Rico.

fasciatus and raised the proportion of bugs making flight of 30 minutes or longer from approximately 25% to over 60% of the population in one generation. Rankin (1978) caused a delay of several days in age of maximum flight activity by selecting for late flying bugs. Caldwell and Hegmann (1969) estimated the heritability (h^2) for flight duration in *Lygaeus kalmii* to be about .2 when offspring were regressed on male parent and about .4 when regressed on female suggesting significant maternal effects. Since heritability is the ratio of additive genetic variance to total phenotypic variance, it indicates the potential effectiveness of natural selection (Fisher, 1958; Falconer, 1960; Dingle 1974). The maternal effects suggest that flight is a character closely associated with fitness (Fisher, 1958; Falconer, 1960). The polymorphism for size, fecundity, flight and tolerance of environmental stress seen in populations of spruce budworms (*Choristoneura fumiferana*) and tent caterpillars (*Malacosoma disstria* and *M. pluviale*) (Wellington, 1960, 1964) is largely accounted for by a model that assumes that the characters are transmitted by the X chromosome (Campbell, 1962, 1966). This seems to be the only study which has examined genetic covariances between flight and other relevant characters, yet such covariances are clearly of critical importance if we are to understand the evolution of migration in the appropriate context of life history variation. The whole area of genetic variances and covariances influencing migration is in obvious need of greater future research effort.

Finally, insight into the evolution and adaptive significance of migration should come from studies of closely related species. Some instances with respect to alary polymorphism are discussed below, but there seem to be surprisingly few studies of species comprising only macropters. Dingle (1978) has studied several species of *Oncopeltus* from tropical and temperate habitats and found that there is little long duration flight in the relatively aseasonal species in contrast to *O. fasciatus* from the highly seasonal temperate mainland. A similar contrast is apparent between the butterflies *Danaus plexippus*, which migrates to and within the temperature zone, and *D. gilippus* which is nonmigratory and largely tropical (Urquhart, 1960). The European *Lygaeus equestris* (Solbreck, 1972, 1976) and the North American *L. kalmii* (Caldwell, 1969, 1974) show similar patterns of relatively short distance flight to diapause sites in fall and away from such sites in the spring. Other genera with migratory and nonmigratory species include *Pieris* butterflies (Baker, 1968), and several genera of grasshoppers (*Uvarov*, 1966); other examples are given in Johnson (1969) and Williams (1958). Most of these genera would undoubtedly repay further study.

2. Wing Muscle Histolysis and Development

Breakdown of flight muscles during adult life occurs in more than 50 species of insect in over 20 families in 8 orders (Johnson, 1976) suggesting that it has evolved independently many times. In some groups such as aphids histolysis may occur with shedding of the wings or brachyptery. In cases where degeneration does not involve wing modification, it occurs in a variety of situations. Thus in beetles it may occur before as in *Leptinotarsa decemlineata* (de Kort, 1969), of after diapause as in *Sitona regensteinensis* (Danthanarayana, 1970). In many species histolysis is irreversible, e.g., *Dysdercus* bugs (Edwards, 1969), but in others such as the bettles *Leptinotarsa*, *Ips* (Bhakthan et al., 1971), and *Dendroctonus* (Reid, 1958) regeneration followed by further flight can occur.

Extensive comparative information on the ecology of histolysis in closely related species comes from studies of cotton stainer bugs (*Dysdercus*). These bugs colonize various malvaceous host plants especially trees with large dehiscent fruits such as *Adansonia, Thespesia, Sterculia, Ceiba*, and *Bombax* (Dingle and Arora, 1973; Fuseini and Kumar, 1975; Derr, 1977). So long as sufficient moisture is present to permit the bugs to feed (Derr, 1977), immigrant females histolyze the wing muscles and begin ovipositing so that large populations build up. When host trees are not fruiting, adults scatter and seek refuge on the forest floor (Derr, 1977), or maintain low population densities on forbs such as *Hibiscus* and *Sida* (Fuseini and Kumar, 1975). Under conditions of starvation in the absence of either food or moisture, the wing muscles of females do not histolyze, and the bugs emigrate to new hosts or scatter until the next fruiting season. Males do not histolyze the wing muscles and may move from one host tree to another as food or females become scarce (Fuseini and Kumar, 1975).

The relation between feeding and flight in three African species, *D. fasciatus, D. nigrofasciatus,* and *D. superstitiosus,* was studied in the laboratory by Dingle and Arora (1973). Males and starved females of all three species retain the flight muscles and show considerable flight which ceases as a result of histolysis when females are fed. Fed females of *D. fasciatus* never fly, but rather histolyze the wing muscles and begin forming oocytes almost immediately (Odhiambo and Arora, 1973). Some fed females of *D. nigrofasciatus* and *D. superstitiosus,* on the other hand, do show a brief period of flight around 4 days posteclosion before histolysis and oocyte development. All three species are facultative migrants using starvation as a releasing cue, but there are differences in tactical details. *Dysdercus fasciatus* feeds primarily on baobab (*Adansonia*)fruits and faces an alternating "feast or famine" environment. It is the most opportunistic, shutting down the flight system completely when dehiscing fruits are available, but showing the strongest flight when they are not. The other two species feed on a variety of herbaceous annuals and perennials where food is not usually superabundant. The variation in the timing of wing muscle histolysis allows some colonization of new host plants on the continuous basis. *Dysdercus voelkeri* in Ghana is similar to *D. fasciatus* in its migration–reproduction pattern (Fuseini and Kumar, 1975) as is *D. bimaculatus* in Central America (Derr, 1977). *Dysdercus cardinalis* is apparently more similar to the other two species. Feeding in all species causes muscle histolysis and in *D. intermedius* mating does also even in starved females (Edwards, 1969). The ecology of these species with respect to the variation present would be worth further study.

A relation between habitat and flight muscle development is also evident in the aquatic Corixidae (Hemiptera). Several species of *Sigara* were studied in southern England by Young (1965). The wing muscles of these species fail to develop in populations occurring in permanent lakes and streams. In temporary habitats the muscles develop normally to produce migrants. Muscle development is evidently cued by higher temperatures and low food levels. For example, portions of a *Sigara scotti* population were cut off from the main body of a lake in small pools which eventually dried up. The temperatures in the pools reached 31°C as opposed to a maximum of 24°C in the lake; flight muscles developed in pool individuals while those in the lake remained flightless. The selective advantage of such a mechanism is clear. The advantage of flightlessness is not so obvious, but flightless individuals survived longer and laid more eggs under conditions of food deprivation suggesting greater efficiency in the aquatic habitat. Further studies on the adaptive advantage of flightlessness are clearly needed. In still more permanent habitats corixids may show wing polymorphism as well as variation in flight muscle development. Patterns of flight in Finnish (Pajunen, 1970) and North American (Scudder, 1964) corixids also seem to involve migration in temporary habitats and flightlessness in more permanent ones.

The adaptive significance of variations in wing muscle histolysis and development appear to be associated with the diversion of energy and materials to other purposes, primarily reproduction (Young, 1965), but possibly also to maintaining basic metabolic needs during long periods of diapause as in *Leptinotarsa* (de Kort, 1969). Such diversion, however, requires a long-term commitment to a particular location, especially if histolysis is irreversible, so there must be sufficient resources at the site in question. Exploitation of resources that may be scattered but highly abundant where they occur seems to be characteristic of insects that use this tactic whether or not accompanied by deciduous wings (see Table 10.1 of Johnson, 1976). Thus resources may be suitable pond habitats as in corixids (cf. Johnson, 1976), trees with great quantities of dehiscing fruit as in *Dysdercus*, or trees of appropriate age and physiological state as in the bark beetles *Ips* and *Dendroctonus*. Appropriate sites allow *Dysdercus*, for example, to build up populations of thousands of individuals (Janzen, 1972; Fuseini and Kumar, 1975; Derr, 1977) as is also the case in aphids, ants, and termites where histolysis is accompanied by shedding of the wings.

3. Wing Length Variation

Leafhoppers and planthoppers (Homoptera) display a spectrum of variation with respect to alary polymorphism and capacity for migration. Many species migrate long distances on weather fronts such as *Empoasca fabae* and *Macrosteles fascifrons* in North America (Pienkowski and Medler, 1964; Chiykowski and Chapman, 1965) and *Sogatella furcifera* and *Nilaparvata lugens* from the Asian mainland to Japan (Kisimoto, 1971). Other species such as the salt marsh fulgoroid *Aphelonema simplex* (Denno, 1976) are almost wholly brachypterous. An interesting intermediate situation occurs in three cicadellid leafhoppers of the genus *Cicadulina* (Rose, 1972). Populations of these species can be divided into short- and long-distance flyers on the basis of tethered flight. These flight characteristics are heritable and are correlated with a polymorphism in body size. Absolute wing lengths do not vary but body lengths do, so that short-bodied individuals have relatively longer wings and display longer flights. *Cicadulina* occupies perennial grasses lush in the wet season, but supplying little nourishment in the dry. The short-bodied form predominates in field populations in the dry season and serves as a means of dispersal from unfavorable habitats.

Salt marsh fulgoroids also vary in alary polymorphism (Denno, 1976). Species occupying relatively permanent high marsh sites and feeding on the grass *Spatina patens*,such as *Aphelonema simplex*, were composed of high proportions of brachypterous individuals. In contrast, the intertidal grass *Spartina alterniflora* was subject to wide environmental fluctuations and displays variable growth form. The extremely abundant and host-specific delphacid *Prokelisia marginata* was about equally macropterous and brachypterous, and the brachyptery was less extreme. At high population densities the proportion of macropters increases in all species. The relatively high proportions of macropters in *P. marginata* allow this species to exploit the temporally variable intertidal grass habitat by continual shifting of the population in space. Crowding also stimulates macropter formation in the long distance migrant *Nilaparvata lugens* and other plant hoppers and serves as a general indicator of habitat deterioration; photoperiod is also used as a cue in some forms (Kisimoto, 1956, 1972).

Aphids also display alary polymorphism. These insects have complex life cycles alternating between sexual and parthenogenetic generations (Dixon, 1973) either on a single host plant (monoecious), or alternating between hosts (dioecious). Species such as the monoecious sycamore aphid, *Drepanosiphum platanoides* (Dixon, 1971, 1973), emerge from the egg in the spring and produce a succession of winged parthenogenetic generations which may migrate to colonize other trees if the original host becomes heavily infested. In the autumn wingless females and winged males are produced, and these sexual females lay overwintering eggs. In many dioecious species individuals emerging in the spring produce wingless parthenogenetic females. At different times during the summer these females produced winged offspring which migrate to new host plants; an example is the migration of *Aphis fabae* discussed above (Kennedy, 1961 et seq.). In the autumn sexuals are produced which return to the original host where females lay overwintering zygotic eggs.

The production of winged individuals, whether sexual or asexual, is a function of a variety of environmental cues. Common stimuli for parthenogenetic alates appear to be crowding and changes in the host plant (Lees, 1966; Dixon, 1973). *Megoura viciae*, for example, produces winged forms under the influence of contact stimuli from crowding or when the bean leaves on which they are feeding senesce (Lees, 1966). In many species the primary stimuli for the production of sexual forms is short daylength (Lees, 1966). Where winters are mild, as in southern England, different clones of *Myzus persicae* vary in their photoperiodic responses; some clones remain asexual, others produce all sexual offspring, and some are intermediate producing some males or intermediate females (Blackman, 1971). In mild years some aphotoperiodic asexual clones survive the winter, the number varying from year to year and place to place. Offspring of crosses between males of a male producing clone and females of a fully sexual clone displayed what appeared to be Mendelian segregation suggesting the evolution of a relatively simple genetic polymorphism. Clearly ability to produce alates and disperse with crowding or host plant deterioration is of selective advantage. The extremely short generation time of aphids also makes alary polymorphism a viable strategy. The reason for the production of sexual forms is less obvious. The most reasonable hypothesis is that of G. C. Williams (1975). In a lottery analogy he suggests that when natural selection is severe as in the aphid overwintering generation, sexual reproduction results in genetically diverse offspring. The different "tickets" stand a greater chance of producing "winners" in the selective lottery than the redundant tickets produced by asexual reproduction. Combining migration with sexual reproduction in dioecious species permits seeking the optimal host for the overwintering egg.

Many beetles also display alary polymorphism especially the Curculionidae, Carabidae, and various aquatic families (Jackson, 1928, 1956; Darlington, 1943, 1971; Lindroth, 1945, 1949; Carter, 1976). Entirely winged species occur where habitats fluctuate and wingless forms where habitats are stable (Lindroth, 1945, 1949; Southwood, 1962). Alary polymorphism with occasional dispersal by winged forms is found in intermediate habitats (Carter, 1976). Isolated habitats such as mountain tops and islands also contain a high proportion of wingless or flightless species (Darlington, 1943, 1971).

The relation between habitat and wing polymorphism is also evident in the aquatic water striders (Gerridae, Hemiptera). The extensive studies of Vepsäläinen (1973, *et seq.*) and others make this one of the best understood groups of species. Populations of the different species show an array of wing development strategies (Table II) varying from fully winged (*Gerris rufoscutellus*) to totally wingless (*G. najas*) determined by relative isolation, habitat characteristics, and length and climate of the breeding season interacting with breadth of the niche and population density (Vepsäläinen, 1978). Thus populations of *G. rufoscutellus* and *G. thoracicus* in highly temporary habitats are monomorphically long-winged while populations of *G. lacustris* and *G. najas* in isolated lakes or pools are totally wing-

TABLE II

Habitats and Alary Polymorphism Strategies in Gerrids[a]

Habitat	Strategy	Example
Temporary ponds	Monomorphic long-winged, univoltine	*Gerris rufoscutellus in* Finland
Ponds and gravel pits subject to drying in long warm summers	Monomorphic long-winged, multivoltine	*G. thoracicus* in Central Europe
Small, temporary, highly unproductive rock pools	Monomorphic long-winged, long life, interreproductive flights, refuging in brackish water; "supertramp" strategy	*G. thoracicus* in Tvärminne Archipelago of southern Finland
Temporary ponds, streams, ditches; cool summers	Seasonal dimorphism (polyphenism); long-winged winter (diapause), short-winged summer generations	*G. thoracicus* and *G. odontogaster* in Finland
Temporary ponds, streams, ditches; long warm summers	Long-winged diapause (overwintering) generation of especially strong flyers; summer generation dimorphic (polyphenic)	*G. odontogaster* in Central Europe
Ponds, medium sized lakes and streams; cool summers	Seasonal dimorphism (polyphenism); long-winged winter (diapause), short-winged summer generations	*G. argentatus* and *G. paludum* in Finland
Ponds, medium sized lakes and streams; long, warm summers	Long-winged winter (diapause); dimorphic (polyphenic) summer generations	*G. argentatus* and *G. paludum* in Central Europe
Rivers, lakes; cool summers	Genetically dimorphic winter generation; short-winged summer generation	*G. lacustris* in Finland
Ponds, medium sized lakes, ditches, rivers	Seasonal dimorphism (genetic and polyphemic); long-winged winter (diapause), short-winged summer generation	*G. lacustris* in parts of Finland and Europe
Rivers, lakes; long warm summers	Genetically dimorphic summer and winter generations	*G. lacustris* in Central Europe
Large lakes and rivers; cool summers	Genetically dimorphic, univoltine	*G. lateralis* in Finland
Large lakes and rivers; long warm summers	Genetically dimorphic, probably bivoltine	*G. asper* in Central Europe
Streams; warm summers	Genetically dimorphic	*G. najas* in Poland
Isolated permanent ponds; cool summers	Monomorphic short-winged, univoltine	*G. lacustris* in southern Finland
Isolated pools of permanent streams; cool summers	Monomorphic short-winged, univoltine	*G. najas* in Finland
Isolated bogs	Monomorphic short-winged	*G. sphagnetorum* in Finland

[a] Data from Vepsäläinen, 1973, 1974a,b, 1978). Arrangement is an approximate progression from completely long-winged populations to completely short-winged.

less. The catholic *G. lacustris* has populations with environmentally induced dimorphism (polyphenism of Shapiro, 1976), genetic dimorphism
(polymorphism), and genetic short-winged monomorphism; *G. thoracicus*
is almost as diverse. The gerrid system is maximally flexible as both long-
and short-winged morphs are known from at least some populations of all
the species studied (Vepsäläinen, 1978). Habitats can exhibit morphism
cycles as populations go extinct, are reestablished by long-winged im-
migrants, and finally evolve to the appropriate morph equilibrium under
the influence of the selective regime characteristic of the habitat (Järvinen
and Vepsäläinen, 1976).

The genetic basis for the alary polmorphism in gerrids involves
Mendelian segregation with short wings dominant to long wings
(Vepsäläinen, 1974b, 1978). Relative thresholds determine the degree of
brachyptery and the sensitivity to environmental stimuli influencing wing
length. Thus increasing daylengths promote short-wingedness and decreasing long-wingedness, but at high temperatures long-winged individuals can
be produced even at increasing photoperiods. The frequency of long wings
will be determined by genotype with the homozygote dominant short wing
requiring higher temperatures than the heterozygote which in turn will require higher than the homozygous recessive long wing. Relative thresholds
within genotypes are presumably the result of modifiers selected under
specific environmental regimes. Polyphenically dimorphic populations thus
have lower temperature thresholds than genetically dimorphic ones. The
dominance of the short-winged allele is accounted for by a model which
assumes that as habitats become more isolated, the cost of long-wingedness
increases, resulting in the evolution of a dominant allele for short wings
(Vepsäläinen, 1978).

Wing variation and flightlessness are also known in the Lepidoptera. In
many sedentary species females are flightless, the males retaining the ability
to fly presumably to seek mates (Hackman, 1966). In the gypsy moth,
Lymantria dispar, marginal populations in Europe, North America
(descended from European imports), and Japan are characterized by
flightless females and are as a consequence nonmigratory. In contrast,
females from populations from the center of the distribution in the
U.S.S.R. can fly, and migrations do occur (Mikkola, 1971). Environmental
fluctuation due to drought could select for migration in central Eurasia
while lack of new habitats to colonize may select against it in marginal
areas. Much further research on these populations is needed. As in other insect groups, brachypterous species are known from isolated habitats such
as mountaintops and islands. Wingless Lepidoptera seem to be especially
well represented on islands in the sub-Antarctic (Gressit, 1970; Mikkola,

1971; Carlquist, 1974) where the extreme environments strongly select against flight.

A different situation occurs in *Orgyia thyellina* (Lymantriidae) which displays seasonal polyphenism for brachyptery (Kimura and Masaki, 1977). Long days result in winged females while short days produce wingless females laying diapause eggs. Males are winged under both conditions. The selective advantage to brachyptery apparently results from the choice of protected sites for cocooning by last instar short day larvae. The emerging wingless females attract males by pheromones and lay their eggs on the cocoon where they are protected. Eggs laid in more exposed sites, as occurs in the summer generation, suffer more severe winter mortality.

Finally, an interesting case of alary polymorphism occurs in the European bug *Pyrrhocoris apterus* (Honěk, 1976a,b). The frequency of macropters in populations in Central and Southeastern Europe was never more than 23% with no geographic trends observed. Long-day photoperiods increase the proportion of macropters, and selection for macroptery increases the frequency to around 75% in seven generations suggesting high additive genetic variance (Falconer, 1960; Dingle, 1974). What is curious about *P. apterus* is that the macropters examined by Honěk are also flightless. Honěk (1976b) suggests the trait may be adaptively neutral. However, in addition to problems of demonstrating lack of function and other difficulties with the neutrality theory (Lewontin, 1974), there appear to be insufficient data to draw conclusions at present. Further more extensive comparative studies of population variation in this species should be both useful and interesting.

C. MIGRATION AND LIFE HISTORIES

Migration as well as life table statistics must be considered if life history strategies are to be understood (Dingle, 1972, 1974; Solbreck, 1978). Along with diapause, migration can serve as an escape mechanism. The insect thus faces choices with regard to reproduction; it can reproduce "here" and "now" (neither migration nor diapause), "here" but "later" (diapause), "now" but "elsewhere" (migration), or "later" and "elsewhere" (both diapause and migration) in Solbreck's (1978) terminology. Diapause and migration are physiologically similar since both involve temporary suppression of reproduction (Kennedy, 1961a), and it is therefore not surprising that the two interact in important ways in the lives of individual insects. The most straightforward cases of this interaction are migrations to and from diapause sites when these are located some distance from habitats where feeding and reproduction occur. Various species of the bugs *Aelia* and *Eurygaster* (Brown, 1965) and coccinellid beetles (Hagen, 1962) migrate to high altitude diapause sites where large aggregations may form, and then remigrate to valleys the following season. In the bug *Lygaeus equestris* diapausing aggregations form in sheltered buildings or rocky outcrops away from the host plants (Solbreck, 1972) and in *L. kalmii* (Caldwell, 1969) or the coccinellid *Coleomegilla maculata* (Solbreck, 1974) beneath suitable leaf litter.

Diapause can also be involved with migration more directly. In *Oncopeltus fasciatus* the delay in reproduction resulting from diapause increases the duration of the migratory period in the adult (Fig. 2, top). In temperate North America this allows escape from winter since individuals cannot survive extremes of frost. This diapause is ordinarily induced by short days although starvation can also cause it (Dingle, 1978).

Selection and parent–offspring regression indicate that the proportion of additive genetic variance for photoperiodic response in Iowa *O. fasciatus* is around 70% (Dingle *et al.*, 1977). Since the response determines age at first reproduction, a character closely associated with fitness, we would predict a much lower value as expected of an intensely selected fitness character (Fisher, 1958; Dingle, 1974). The selection pressures on *O. fasciatus*, however, regularly reverse direction apparently maintaining high proportions of additive genetic variance. In the temperate zone as *O. fasciatus* moves north in the spring, selective advantage accrues to those individuals reproducing earliest on the abundant milkweed crop. As days shorten, reproductive risks increase because of the inability to overwinter. The extent of the risk varies with latitude and habitat and is indexed by photoperiod and food abundance. In this case advantage accrues more and more to those entering diapause and migrating, and they will eventually have an absolute selective advantage. The genetic mechanism available to *O. fasciatus* to accomplish the association between diapause, migration, and environment thus serves as a "genetic rheostat" with continuous adjustment of age at first reproduction to meet environmental exigencies. Diapause and migration, jointly cued by photoperiod, are important elements in the evolution of the *O. fasciatus* life history strategy. High additive genetic variances for diapause have been found in a number of other insects (reviewed by Dingle *et al.*, 1977; Hoy, 1978) and in at least one case, the pitcher plant mosquito *Wyeomyia smithii*, seem to be maintained by reversing selection pressures (Istock, 1978). Further investigation of the relation between diapause and migration is clearly in order.

But whether following diapause or not, migration in most species involves colonization of new habitats with profound life history implications. In colonizers selection favors those individuals who reproduce early with high fecundity (Cole, 1954; Lewontin, 1965; Dingle, 1972, 1974).

Migratory species thus tend to be "r-strategists" rather than "K-strategists" in current life history terminology (e.g., Stearns, 1976) and to show high intrinsic rates of increase (r). Because determinations of r require the assumption of a stable age distribution, they are not directly applicable to migrant colonizers whose populations are not so distributed; they do, however, provided a useful index of potential for increase. Three examples of r values calculated for closely related species displaying differing migratory tendencies are given in Table III. They suggest that migrants do possess a higher population growth potential than nonmigrants; life history data from other migratory and nonmigratory species show similar trends (e.g., Southwood, 1977).

On the other hand, an apparent paradox emerges when we examine r values or fecundity for migratory and nonmigratory populations of the same species (Kennedy, 1961a, 1975). Three examples are given in Table IV; in each case r is higher for nonmigrants which is the opposite of what one would predict from the r–K spectrum. Similarly, fecundity is reduced in migrant morphs of species such as locusts (Kennedy, 1961a). The paradox is resolved, however, if we examine the temporally dynamic nature of the reproductive strategies (Nichols et al., 1976). In each case, the insect must weight the "costs" of various trade-offs available to it (e.g., whether to reproduce "here" or "elsewhere"). Clearly if current habitats are deteriorating, there is strong selection against offspring, and there is selective advantage in favor of delayed reproduction (the oogenesis–flight syndrome) to reproduce elsewhere (Dingle, 1968). This is what happens in the species in Table IV. *Oncopeltus fasciatus* diapauses and migrates in short

TABLE III

Values of r Calculated under Similar Conditions for Pairs of Species Showing High and Low Migratory Tendencies

Tendency	r
High migratory	
Dysdercus fasciatus[a]	0.094
Oncopeltus fasciatus[b]	0.044
Tribolium castaneum[c]	0.128
Low migratory	
D. superstitiosus	0.062
O. unifasciatellus	0.034
T. confusum	0.100

[a] H. Dingle and G. K. Arora (unpublished).
[b] Landahl and Root (1969).
[c] Ziegler (1976).

TABLE IV

Values of r Calculated for Migratory and Nonmigratory Populations of the Same Species

Species	r (migratory)	r (nonmigratory)
Oncopeltus fasciatus[a]	0.037	0.050
Neacoryphus bicrucis[b]	0.109–0.116	0.123
Aphis fabae[c]	0.91–1.27	0.99–1.52

[a] Dingle (1968).
[b] Solbreck (1978).
[c] Dixon and Wratten (1971).

days to escape winter, N. bicrucis migrates when food is short, and A. fabae forms alates when crowded or the host plant senesces. Interestingly, alate A. fabae mature embryos rapidly so that a burst of reproduction occurs at colonization, "restoring" some of the decreased r (Dixon and Wratten, 1971). The higher fecundity of A. fabae apterae is in part due to the larger size and longevity of these morphs (Taylor, 1975)., but in two other aphids, Sitobion avanae and Metopolophium dirhodum, apterae are consistently more fecund than alatae of equivalent weight (Wratten, 1977). In Drosophila melanogaster long periods of flight result in reduced egg output, but the "cost" of migration is mitigated by increased body size (Roff, 1977).

Temporally and morph variable reproductive strategies are illustrated by the bird cherry-oat aphid Rhopalosiphum padi (Dixon, 1976). In the spring, winged emigrants leave the primary host (bird cherry) to colonize abundant fresh growing grasses; these emigrants are large and highly fecund. During the summer apterous exules walk from plant to plant spreading the population. In late summer crowding produces alate exules. These have lower fecundity and are smaller than emigrants, apparently adaptations to colonizing harsher environments of drying grass. In the autumn these produce less fecund alate gynoparae and males which return to bird cherry. The gynoparae produce small apterous oviparae which mate with the males to produce overwintering eggs. After return to bird cherry, there is only a short time before winter and oviparae must be produced quickly. The small size and low numbers of oviparae are apparently adaptations for rapid development and early reproduction. The production of more smaller oviparae may also promote efficient scanning of the habitat for appropriate winter egg oviposition sites.

The relation between demographic factors and colonizing ability is also incorporated in the concept of reproductive value (Fisher, 1958; Dingle, 1972, 1974). This is a measure of an individual's contribution to future population growth and is analogous to compound interest. The oogenesis–flight syndrome means that migrant females carry their future offspring with them so that at the time of colonization reproductive value is high as one would expect in a good colonizer (Dingle, 1965, 1974). Special adaptations such as the carrying of almost mature young followed by a burst of reproduction at the end of migration as occurs in *Aphis fabae* (Dixon and Wratten, 1971) causes further enhancement of reproductive value at the time of colonization (e.g., Dingle, 1965).

D. Conclusions: The Adaptive Significance of Migration in Terrestrial Arthropods

Because of their small size, short generation times, tractability for both laboratory and field analysis, and considerable economic importance, we perhaps know more about the migrations of insects than of any other group. That there is enormous variation in migration strategies, keyed to specific life history adaptations, is obvious. Yet running through this diversity is a common thread of adaptation to shifting environments. It was pointed out a long time ago by Jackson (1928) and later by Brown (1951) that migrants were likely to be species occupying ephemeral habitats. This has been amply confirmed since (e.g., Vepsäläinen, 1978; also Table II). Southwood (1962) summarized abundant data from insects and Arachnoidea indicating that migrants are characteristic of "temporary" habitats (early successional or seasonally transient) such as old fields and ephemeral ponds. As old habitats deteriorate new ones are colonized. More recently Southwood (summarized 1977) has expressed this in terms of H/τ where H = length of time the habitat remains favorable and τ = generation time. When this ratio approaches unity a migration strategy is favored (note that τ can be varied by diapause). The degree to which the emigration response is genotypically or phenotypically influenced may depend on the reliability of environmental cues with unreliable cues favoring genotypic stereotypy and reliable ones phenotypic plasticity (Ziegler, 1976).

The ability of migrants to colonize new habitats adds a new dimension to the dynamics of populations (Taylor and Taylor, 1977). The shifting environmental mosaic creates lacunae which eventually recover their hospitality and can be repopulated by migrants able to reach them. Population fluctuations thus occur in space as well as time, and individuals are replaced by immigration as well as by births. Populations may therefore stabilize below carrying capacity in which case we would expect "*r*-strategies" in migratory species. For species facing environmental uncertainty, we would expect considerable "bet-hedging" especially where juvenile mortality is likely to be high (Stearns, 1976). Spatial strategies result from selective balance between making use of available resources and dispersal to find new exploitable resources. The interplay of these pressures determines the movement of individuals and the spatial pattern of the population. As the foregoing examples have shown, evolutionary calibration and fine tuning have produced a diversity of strategies on that common theme.

VI. Migration in Fishes

The migrations of fishes can be approximately divided into four main types: (1) migration of predominantly marine species to freshwater to breed (anadromous), (2) migration of predominantly freshwater species to the ocean to breed (catadromous), (3) movements wholly within fresh water (potomadromous), and (4) movements entirely within the sea (oceanodromous) (Myers, 1949; Harden Jones, 1968). The term anadromous is also sometimes used with respect to breeding migrations from lakes into streams. These categories are not entirely exclusive and indeed many species have populations falling into more than one category. As with other groups, there is considerable flexibility in migratory strategies, but specialized behavior is apparent (e.g., Hoar, 1958).

A. Oceanodromous and Potomadromous Migrations

The movement patterns of the oceanodromous herring (*Clupea harengus*), cod (*Gadus morhua*), and plaice (*Pleuronectes platessa*) have been extensively described by Harden Jones (1968); summary is given in Fig. 4. Considerable variation within species is evident with different populations, for example, spawning in spring or fall (e.g., herring populations spawning in spring off Norway and in fall on the Dogger Bank) and on various spawning grounds. The separation of spawning, nursery, and feeding areas allows exploitation of more resources and possibly results in the well-known abundance of these migratory species (Nikolsky, 1963). Indeed this very abundance leads to their exploitation by commercial fisheries and hence to much of our knowledge of migratory patterns. Selection has presumably favored particular feeding and wintering areas for appropriate reasons of resource abundance and temperature. Spawning, however, must also be situated and timed so that eggs and larvae have access to suitable nursery areas where and when productivity is high; usually this means that spawning occurs upcurrent from the nurseries. The difficulties of sampling and marking pelagic fish and of ascertaining details of ecology and demography mean that our understanding of adaptive strategies in

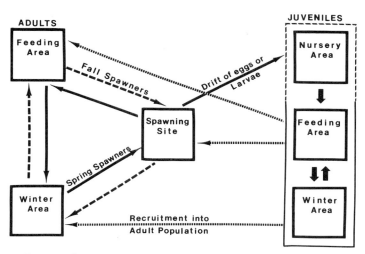

Fig. 4. Patterns of movement in oceanodromous fish migrants summarized from data in
Harden Jones (1968). Left, adult movements indicated as clockwise for fall spawners and
counterclockwise for spring spawners. Right, juveniles may move from nursery to feeding and
winter areas and may be recruited into the adult population at different times from different
areas depending on species or populations.

oceanodromous migration is superficial at best. In the case of noncommer-
cial species, our ignorance is profound.

Potomadromous migrations are usually seasonal whether they occur in
the tropics or the temperate zone. In Lake Mendota, Wisconsin, the white
bass (*Roccus chrysops*) migrates to spatially limited spawning grounds in
May and June when water temperatures range from 16° to 24°C (Hasler,
1969). Concentration at spawning sites evidently is a consequence of the requirement of specific depth and bottom requirements for nest construction
and associated behavior in these serranids. Karr and Gorman (1975)
observed upstream spring movement of the creek chub (*Semotilus
atromaculatus*), the white sucker (*Catostomus commersonii*), and the
northern pike (*Esox lucius*) to spawning areas of a small tributary of the
Maumee River in northeastern Indiana; these movements are apparently
fairly typical of stream species. Selection for such migration could result
from the necessity for particular substrate and current characteristics for
maximum egg hatch, from reduced predation on eggs and young in shallow
headwaters, from higher production of food for young fish, and from exclusion of adults from headwaters by winter ice or very likely all four. *Esox*
may also exploit migrating *Catostomus*. In Texas the Mexican tetra *Astyanax mexicanus* has apparently extended its range northward by evolv-

ing seasonal migration to exploit the warmest sections of streams (Edwards, 1977). Small streams such as those studied by Karr and Gorman and Edwards are logical sites for comparative studies of the adaptiveness and coevolution of migratory and reproductive behaviors, but unfortunately such considerations are rare in the fish literature (cf. also similar comment in Wootton, 1976, p. 128). The relative influences of different selective forces therefore remain largely subjects of speculation only.

The movements of fishes in tropical rivers are discussed by Lowe-McConnell (1975). In seasonal rivers in South America, Africa, and Asia patterns are similar suggesting parallel selection pressures. Migration upstream occurs with rising waters at the onset of the rains and proceeds upriver until the fish can spread out over the inundated flood plain. Here reproduction takes place somewhat before peak water level. The peak is the time of maximum food production for both adults and young and hence is the main feeding time and the period of rapid growth. As the dry season progresses, the rivers contract often to a series of shallow channels and pools at lowest water levels. At this time environmental pressures such as deoxygenation, low food production, and heavy predation are at their extreme. Fishes may still move up and down the available channels, however. The spawning movements to the floodplains may be a mechanism for minimizing predation since predators, usually larger fish, are apt to be more restricted to deeper water. Nelson (1964) postulates that floodplain breeding may have selected for internal fertilization in glandulocaudine fishes. Mating in these species takes place in pools in the dry season when the population is concentrated, but predation is severe. With the annual floods the top-living glandulocaudines can move out to the shallows and adundant food whereas predators require deep water and are unable to penetrate this zone. Most of the migrant fishes displaying these seasonal cycles have short life cycles (one to two years) with populations characterized by high proportions of young of the year and high turnover rates. Fish faunas of equatorial forest rivers show similar but less extreme migration–reproduction cycles as a consequence of more modest fluctuations in water level.

Spawning migrations either to areas near shore or up tributary streams are also known in many fishes of tropical lakes. With respect to African lakes, Jackson (1961) argues that the relevant selection involves the need for cover so that young fish can escape predation especially by the large tiger fish *Hydrocymus vittatus*. An example is the cichlid *Tilapia macrochir* in Lake Mweru which breeds in swamp vegetation; smaller fish remain in this cover, but the larger ones can apparently escape predation by virtue of their size (*H. vittatus* swallows prey whole) and therefore live in open water. Jackson further argues that migration into rivers is an extension of lakeshore movement also evolving as a result of predation pressure and the need for protection of young in specific substrates. In contrast Whitehead (1959) and Corbet (1961) argue that river spawning species were originally riverine and moved into lakes as an extension of downstream feeding movements (known from other populations and species). Jackson was challenged by Fryer and Greenwood (Fryer, 1965) who supported Whitehead and Corbet, and indeed most evidence indicates that river spawners were originally riverine species. Fryer's argument that migration evolves to "ensure dispersal over the whole colonizable river course" rather than for protection of young is, however, unconvincing since there is no indication of what constitutes "colonizable" or indeed what this means for a more or less constantly moving population. There is also no suggestion as to how the "ancestral habit" of river spawning is maintained by selection. Unfortunately while there is a certain logic in all the arguments, the question must remain an open one in the face of a paucity of relevant data.

B. CATADROMOUS MIGRATION

The best-known catadromous migrant is the eel, *Anguilla*, usually considered two species, the European (*A. anguilla*) and American (*A. rostrata*) eels based on vertebral counts (Harden Jones, 1968). The movements of freshwater adults back to the sea to breed and the leptocephalus larvae in Atlantic currents is described by Harden Jones; he also considers the controversy over whether one or two species are involved. Since the breeding site of eels is unknown (it is rather vaguely placed in the depths of the Sargasso Sea), speculation on the selective forces driving the system is largely futile. Therefore I shall avoid it here and say only that the long life cycle with the freshwater phase lasting several years remains a fascinating problem in evolutionary biology. Genetic differentiation of eel populations in different rivers does occur (Williams *et al.*, 1973), but the absence of any knowledge as to whether larvae return to the streams from which their parents emerged, even though that seems unlikely, makes speculation concerning selection in the streams premature. Therefore, the eel problem is left for what it is, an intriguing mystery.

C. ANADROMOUS MIGRATION

1. Salmon and Trout

Because of their importance to both commercial and sport fishing, the most extensively studied anadromous migrants are the salmonids. The three major genera are the trout and Atlantic salmon (*Salmo*), the Pacific salmon (*Oncorhynchus*), and the lake trout and char (*Salvelinus*). Species and populations vary from confinement to freshwater streams or lakes to migrations to the sea almost immediately after hatching with most of the adult life spent in the open ocean. In both Atlantic and Pacific salmon extensive oceanic migrations also take place apparently to exploit a number of feeding grounds. Homing to the stream of birth using chemical cues is now well documented (Hasler, 1960; Harden Jones, 1968) and is exploited to stock streams where salmon were previously absent, e.g., coho (*O. kisutch*) in the Great Lakes of North America (Scholz *et al.*, 1976). A major difference between *Salmo* and *Oncorhynchus* is that the former are repeat spawners (*iteroparous*) whereas the latter spawn only once and then die (*semelparous*). Variations in migratory pattern are extensively described by Harden Jones (1968), Mills (1971), and Moyle (1976).

The Salmonidae apparently evolved as a family adapted to cold nutrient-poor fresh waters of glaciated areas (Moyle, 1976). The evolution of anadromous migration is seen as a secondary adaptation for opportunistic feeding resulting from the transient nature and frequent isolation of the waters in which the group evolved. The behavioral flexibility of the salmonids allowed them to colonize most accessible streams and coldwater lakes of the Northern Hemisphere, and homing capability and frequent geographical isolation has resulted in the evolution of morphologically distinct populations, sometimes classified as separate species, but usually readily hybridizing when contact is restored either naturally or by man. A case in point is *Salmo gairdnerii* which occurs as the rainbow trout in headwater streams and some lakes, the anadromous steelhead in the lower reaches of coastal streams, and the Kamloops rainbow in the lakes of British Columbia (Anonymous, 1969). The often confused taxonomy of salmonids reflects their recent and probably rapid evolution.

The different migratory strategies of various species involves the evolution of specialized behaviors. This is illustrated by Hoar's (1958) study of migratory behavior in four species of *Oncorhynchus*. The chum (*O. keta*) and pink (*O. gorbuscha*) are anadromous to such an extent that the young migrate to the sea immediately after hatching. The coho (*O. kisutch*) is anadromous but spends one and sometimes two years in streams before migrating. Finally, the sockeye (*O. nerka*) is the only North American *Oncorhynchus* which is entirely landlocked although it assumes a lake habitat prior to returning to the streams to breed. In the oceanic or lake-dwelling phase of the life history, all four are pelagic schooling species in contrast to the solitary bottom dwelling trout and char which are territorial. The pink is the most highly evolved anadromous form and enters the sea as fry rather than as a transformed smolt (the next juvenile stage). Almost immediately after hatching these fry show oriented nocturnal downstream swimming with little tendency to hide under stones. The only slightly less highly evolved chum is similar but shows somewhat less intense nocturnal downstream swimming, less intense schooling, and a greater tendency to hide when startled. The freshwater lake sockeye fry shows schooling and also downstream migration, but older fry and underyearlings may orient upstream; the smolt transformation is retained. Finally, the stream dwelling coho shows no nocturnal downstream orientation, strong hiding reactions, rapid responses to current changes, and is cryptically colored. Differences among the species are summarized in Table V.

Hoar (1958) envisions the evolution of the anadromous migratory habitat as occurring in four steps. First, a type ancestral to the sockeye, pink, and chum (but more like the sockeye) separated from troutlike salmonids. Second, some species migrated into estuaries as fry but the smolt transformation was retained. The sockeye line developed when some populations were blocked from entering estuaries. The third step involved development of earlier resistance to hypertonic sea water and an abbreviated presmolt period leading to eventual loss of the smolt transformation in pink and chum and their separation from the coho. Finally, there was a split between chum and pink with the pink retaining minimum capabilities for freshwater life.

The range of variation found within the genus *Oncorhynchus* can almost be found within a single species of *Salmo* such as the rainbow-steelhead *S. gairdnerii* (Northcote, 1969). Northcote (1969) examined *S. gairdnerii* spawning in both inlet and outlet streams of various North American lakes. The problem here is that the fish apparently must respond to opposite sets of stimuli, whether migrating into or out of the lake, depending on whether the home stream enters into or empties from the lake. As might be expected no single mechanism is universally responsible for the observed patterns;

TABLE V

Comparisons of Behavior and Development in Young of Four Species of *Oncorhynchus*

Coho	Sockeye	Chum	Pink
River resident	Lake resident	Fry migrate to sea	Fry migrate to sea
Marked hiding behavior	Hiding	Some hiding	Little hiding
Territorial	Somewhat territorial	Schooling	Strong schooling
Smolt transformation	Smolt transformation	Smolt transformation absent	Smolt transformation absent
Little nocturnal activity	Nocturnal	Nocturnal	Nocturnal

rather, various populations have developed response systems dependent to differing degrees on photoperiod, water temperature, and innate control. In some cases environmental control predominates with, for example, higher temperatures in outlet streams inducing upstream migration while cooler water in inlets leads to downstream movement; the lake is reached in either case. In other lakes genetically controlled shifts in thresholds of response may be operating. Finally, in headwater streams above waterfalls, movement within the stream may be largely genetic with little environmental influence. The control of migration is thus determined by exigencies of the habitat occupied. Transplant experiments indicate that responses can change dramatically in a few decades. Northcote, concerned that sufficient genetic variation may not be available for such a rapid response to selection, postulates that the new response may have been latent in the transplanted population, but not expressed until entering the new habitat. In view of the ample genetic variation now known for many species and populations, however (e.g., Krebs *et al.*, 1973; Dingle *et al.*, 1977), this explanation seems unnecessary. Indeed because of the constantly fluctuating selection pressures even within habitats (Dingle *et al.*, 1977), such variation is to be expected in *Salmo*. There is thus no reason why strong selection, which is definitely present, cannot shift the responses in a relatively few generations.

2. The Shad

The American shad, *Alosa sapidissima* (Clupeidae), resembles many salmon in its extensive oceanic migrations and movements up rivers to breed. Its biology has been extensively studied by Leggett and associates (Leggett, 1972, 1976, 1977a; Leggett and Whitney, 1972). The oceanic migration occurs from New Brunswick to Florida on the east coast of North America and over a similar range on the west coast where the species has been introduced. Tagging experiments indicate that ocean migration rates can be at least 21 km/day (Leggett, 1977a) and movement is at still higher speeds in fresh water (Dodson and Leggett, 1973, 1974). In both oceanic migration and upstream spawning runs there is a strong correlation between movements and water temperature. Peak spawning runs take place when temperatures are near 18.5°C with the result that in southern rivers like the St. Johns River in Florida spawning occurs as early as January while in New Brunswick it may not take place until June (Leggett and Whitney, 1972; Leggett, 1976). Marine migration follows water temperatures in the 13°–18°C range. Such temperature–movement correlations have been found in many species of migrant fish (Leggett, 1977b). The reason may lie in the energetic cost of migration. Shad do not feed in fresh water, may lose up to 50% of their body weight in spawning runs, and may suffer high postspawning mortality; less than 10% are repeat spawners. Mortality is even higher late in the season with warmer water (Leggett, 1972). Migration at lower temperatures also assures optimal temperatures later for developing eggs and young (Leggett, 1976).

Shad apparently home to their natal streams by a combination of olfactory and rheotactic mechanisms (Dodson and Leggett, 1974) permitting return to the tributary of birth. Carscadden and Leggett (1975a) examined five meristic charters in shad from the St. John River, New Brunswick, and several other streams with multivariate techniques. In the St. John system 71% of the comparisons indicated significant differences between populations in different tributaries suggesting that precision homing has led to the evolution of reproductively isolated populations. In this respect shad resemble salmon. It would be interesting to know if any behavioral differences occur as well, e.g., in the timing of spawning or the choice of optimal temperatures. Life history variation does occur (Carscadden and Leggett, 1975b) and is discussed below.

3. Lampreys

Much of the research on the biology of lampreys has been done by Hardisty and Potter and the following account draws heavily on their chapters in Hardisty and Potter (1971). Anadromous species such as the sea lamprey *Petromyzon marinus* and the European *Lampetra fluviatilis* move up rivers to spawn in gravel substrates. In *P. marinus* this movement is highly synchronous and occurs in the spring while in *L. fluviatilis* a less synchronized migration begins in late summer and continues well into autumn. After spawning the adults die. The newly hatched ammocoetes larvae drift downstream and eventually settle in soft substrates where they remain as detritus feeders for several years (5 in *L. fluviatilis*). There then follow the transforming and macrophthalmia stages during which there is no feeding and which may last for several months; the cycle is completed with migration to the sea. In the sea, lampreys pursue the parasitic habit of feeding on fishes; this they do for about 2 years before returning up rivers to spawn.

Of the 31 recognized living species of lampreys, however, only 9 possess a parasitic marine phase. The other 22 are freshwater of which 6 are parasitic feeding in large river basins while the remaining 16 are entirely nonparasitic and confined to streams. Like their marine counterparts, the freshwater species move upstream to breed. The sea lamprey (*P. marinus*) has invaded the Great Lakes of North America, where it decimated the commercial fisheries, and also has entirely landlocked populations in the Finger Lakes of New York State. An interesting situation occurs with so-called "paired species," two closely related species with one a parasitic form and the other a nonparasitic entirely stream dwelling form apparently derived from it. Nonparasitic adults never feed. Ten parasitic species have given rise to from one to three nonparasitic ones; there are also two "unattached" nonparasitic forms with restricted and probably relict distributions. The evolution of the nonparasitic species has apparently resulted from paedogenesis although the long life cycle is retained. For example, the brook lamprey, *L planeri*, derived from *L. fluviatilis*, prolongs the larval phase almost two years so that life-span is similar to the parental species. Nonparasitic species migrate from larval habitats in soft bottoms back upstream to more sandy or gravelly substances to spawn. Hardisty and Potter speculate that the long larval phase is necessitated by the long evolution of lamprey life cycles. Any major deviation from programmed lifespan would involve such complex physiologic and metabolic deviations that they would be severely disruptive. It is characteristic of these nonparasitic species, however, that they are smaller than the parasitic ones as any further prolongation of the larval period for additional growth would for demographic reasons show rapidly diminishing returns for any further increase in body size or fecundity. The repeated formation of paired species in widely separated geographic areas implies similar selective forces; exactly what these are, however, remains obscure.

4. Sticklebacks

The three-spined stickleback (*Gasterosteus aculeatus*) occurs as two distinct morphs throughout its range which includes Europe and both coasts of North America (Hagen, 1967; Bell, 1976; Wootton, 1976). These morphs are not only genetically distinct morphologically, but also display two genetically programmed life history modes, one anadromous and one confined to fresh water (Hagen and McPhail, 1970; Hagen, 1973; Bell, 1976). The morphs are maintained by selection, and similar selective regimes, specifically predation, rather than gene flow account for phenetic similarities where these occur (Moodie, 1972; Hagen and Gilbertson, 1973; Bell, 1976). Although there are intermediates between the forms, positive assortative mating is estimated at 62–65% within morphs (Hay and McPhail, 1975). There is also evidently selection against intermediates (Hagen, 1967). The following summary of differences between the morphs is drawn largely from Bell (1976) and Wootton (1976), but also the other authors cited above.

The anadromous form is usually referred to as *trachurus* and the freshwater as *leiurus*. The *trachurus* fish spend most of the year at sea or in estuaries but migrate into streams in the spring to breed. They are relatively slender-bodied and generally larger than *leiurus* and are especially distinguished by the large number of lateral plates on the body (30 or more). In contrast *leiurus* is confined to fresh water, deeper-bodied, smaller, and has a modal number of lateral plates around seven, as a consequence of selection by predators. *Trachurus* fish also have more and longer gill rakers and are plankton feeders while *leiurus* with stouter, fewer gill rakers is more benthic in trophic behavior. *Trachurus* shows a distinct preference for currents, especially in nesting sites, as opposed to the more still water habits of *leiurus*. Finally, *trachurus* shows a higher fecundity. Differences between the forms are summarized in Table VI.

Not surprisingly *trachurus* fish display morphology and behavior expected of migratory fish, e.g., slender bodies and preference for currents. They also tend to be phenotypically uniform and stable, probably because of selection in more uniform environments and possibly also because of greater likelihood of gene flow preventing drift. Resident freshwater *leiurus* display a complex pattern of phenotypic divergence because of the variation in selective regimes experienced in diverse habitats, especially with respect to predation. The adaptive significance of the migratory habit is, however, not known. Wootton, (1976, p. 128) puts it as follows: "Since *leiurus* sticklebacks do not make extensive migrations between fresh and salt water, such migrations cannot be essential for the success of the stickleback. In general the functional significance of migrations is poorly understood. Even in a thorough review on fish migrations, little space is devoted to the topic (Harden Jones, 1968). This is a true reflection of the current state of knowledge of the problem." The behavioral polyphenism allows the species (or superspecies, Bell, 1976) to exploit a range of en-

TABLE VI

Differences between the Anadromous *Trachurus* and Freshwater *Leiurus* Forms of the Stickleback, *Gasterosteus aculeatus*

Character	*Trachurus*	*Leiurus*
Body type	Slender	Deep bodies
Lateral plates	Many (30+)	Few (usually ca. 7)
Gill rakers	More, long, slender	Few, short, stout
Feeding preference	Planktonic	Benthic (relatively)
Nest site preference	Current, open, sandy bottom, in *Elodea*	Quiet water, secluded, mud buttom, in *Oenanthe*
Body size	Larger	Smaller
Fucundity	More eggs	Fewer eggs
Breeding season	May–September	March–June

vironments, as Wootton indicates, but the problem is still to explain the evolution of population divergence on the basis of individual selection.

D. MIGRATION AND LIFE HISTORIES

As the data for fishes suggest, migration must be considered an integral part of life history strategies (cf. also insect migration). Developmental variation in lampreys and reproductive characteristics of sticklebacks are cases in point. The problem has been considered in detail by Schaffer and Elson (1975) with respect to the evolution of age at first spawning in the Atlantic salmon, *Salmo salar*. Theoretical considerations (Schaffer, 1974a,b; Stearns, 1976) suggested that environmental factors increasing reproductive success per unit reproductive effort select for greater reproductive effort at all ages and an earlier age at first reproduction. On the other hand, increased postbreeding survival and increased growth per unit effort should select for reduced reproductive effort at all ages and later age at first reproduction.

For salmon, Schaffer and Elson predicted that long fast rivers requiring high energy expenditure to reach the spawning grounds (i.e., increasing effort relative to reproductive success) would select for a longer feeding period at sea to provide energy for the spawning run and hence delayed breeding; conversely breeding would be earlier in short easy rivers. A second prediction was that rapid growth at sea subsequent to the age at first possible spawning (i.e., increased growth and hence future return per unit effort) should be characterized by delayed spawning relative to fish with slower growth. Highly exploited stocks, however, should be characterized by earlier breeding because commercial nets take larger (older) fish. Finally, if juvenile survival varies from year to year, selection should lead to populations in which adults first spawn at different ages (i.e., individuals should "bet-hedge" by producing offspring with high variances for age at first reproduction). This means that variance should either increase with latitude and greater variance in harshness of winters, or since in the north young spend longer in streams (up to 6 years) and are more likely to experience a harsh winter, it should increase then decrease with latitude. The latter turned out to be the case for the final prediction; all other predictions were also borne out for a series of rivers in eastern Canada.

The migration patterns in the various rivers thus involve complex coevolutionary relationships with life histories. Elson (1970) also reported evidence from selective breeding experiments indicating that age at first reproduction is indeed heritable. This plus the good agreement with theory suggests that the observed life history differences are in fact adaptations. The relatively rapid evolution of early reproduction in commercially exploited stocks is indicative of high additive genetic variance for this trait in turn strongly suggesting that salmon populations are subject to continually varying selective pressures (cf. Dingle *et al.*, 1977).

Life histories also vary in the American shad (Carscadden and Leggett, 1975b; Shoubridge, 1977). Theory predicts that if the ratio of adult to juvenile survival is high, iteroparity will be favored while semelparity is more likely if the reverse is true (Schaffer, 1974a). Environmental variance which influences adult survival will select for early age at first reproduction and increased fecundity at each breeding, while with the converse there is delayed maturity and decreased fecundity (Schaffer, 1974b; cf. also Stearns, 1976). These predictions were tested on populations of shad from both coasts of North America.

Results confirmed predictions. In the St. John's River in Florida where environmental variance was low and temperatures for juvenile survival optimal over a long period, the population is semelparous with high fecundity (412,000 eggs per spawning). This result also reflects the high energy expenditure of the spawning migration due to the relatively high water temperature. The St. John's is thus harsh for adults, but equitable for juveniles. In contrast in the Miramichi River in New Brunswick, environmental variance was high and lower temperatures reduced the cost of migration but increased the risk to juveniles. Here iteroparity was the rule with reduced fecundity (266,000 eggs per lifetime) and delayed age at first reproduction. Similar patterns were observed in Pacific rivers where temperatures were stable but relatively low. Variation within rivers was also noted in both Atlantic and Pacific streams as a function of length (cost) of the migratory journey.

Life history patterns in Pacific rivers have evolved in less than a century since shad were originally introduced. These shad came from the Hudson, Potomac, and Susquehanna Rivers where 20 to 40% of adults are repeat spawners, age at maturity is about 4 years for males and 4.6 years for females, and mean lifetime fecundity 300,000 to 350,000 eggs. In contrast Pacific rivers have from 32 to 77% repeat spawners, age at maturity varies from 3.3 to 3.8 years for males and from 4.0 to 4.5 years for females, and mean lifetime fecundities range from 321,000 to 500,000 eggs. Again this rapid evolution suggests strong local selection pressures and genetic variance for these traits. Fluctuating selection pressures in the original rivers are therefore once more indicated. Analysis of this genetic and environmental variance should yield important evolutionary insights.

The analyses of shad and Atlantic salmon life history variation also suggest a reason for the reproductive pattern of Pacific salmon (*Oncorhynchus*). The migration routes involve movements over great distances (as much as 1000 km) up rivers frequently interrupted with swift currents and waterfalls caused by steep altitudinal gradients. The high energetic cost would be expected to reduce the chances of repeat spawning, i.e., adult mortality should be high. These circumstances would be expected to select for semelparity and this is in fact the characteristic reproductive mode.

E. Conclusions: The Evolution of Fish Migration

The evolution of fish migrations seems to have been primarily driven by selection for more efficient adult feeding and growth (or greater winter survival) in one area but more successful breeding in another. Within this general theme there is a variety of individual migration patterns from simple pool to riffle movements within a stream to complex anadromous (or in the case of the eel, catadromous) migrations involving long-distance journeys both in the ocean and up rivers. What is the nature of selection for such complex patterns?

If one compares species or populations with respect to anadromous versus entirely freshwater habits, it is immediately apparent that the anadromous forms are larger. This is true even for the relatively tiny stickleback (Wootton, 1976). In lampreys the freshwater nonparasitic species derived from parasitic anadromous species are about half the size of the latter at maturity (Hardisty and Potter, 1971). Similar differences are observed between landlocked and anadromous populations of the sea lamprey *Petromyzon marinus* (Smith, 1971), and the steelhead trout (*Salmo gairdnerii*) is much larger than the entirely freshwater rainbow. An extreme example is the arctic char (*Salvelinus alpinus*). The sea-run fish normally prey on other fish and weigh 8–10 pounds (20 pounders are known) while the plankton feeders of Alpine lakes may barely attain 6 inches and weigh 3 ounces (Mills, 1971). The differences are evidently due to the high productivity of the sea compared to streams and oligotrophic lakes (although individuals feeding in lakes are larger than those confined to streams). Allen (1969) and Mundie (1969) discuss stream productivity with respect to salmon.

The advantage of greater body size in fish, assuming similar time to mature, is greater fecundity (e.g., Wootton, 1976; Shoubridge, 1977). Age at maturity may in fact vary and influence life histories (Schaffer and Elson, 1975), but does not seem to be a consequence of differences in sea versus fresh water [cf. e.g., stream and anadromous lampreys (Hardistry and Potter, 1971) or sticklebacks (Wootton, 1976)]; if anything, maturation is delayed in fresh water. Fish that were tolerant of osmotic shifts and able to penetrate first estuaries and then the sea were thus at an advantage, because of greater size and fecundity, relative to conspecifics which stayed behind. This would remain true so long as migration costs did not become too great. In this regard it is worth noting that where streams have both freshwater and anadromous populations, the former usually occur in headwaters some distance from the sea while the latter spawn in lower reaches. A case in point is the stickleback which would be an interesting species on which to test the model. The prediction is that the cost of migration to headwaters is too high for the *trachurus* form.

The above model tacitly assumes a freshwater origin. A marine origin with invasion of fresh water could occur with sufficient selective advantage to freshwater breeding. Lower productivity means fewer predators so it is possible that this was a major selective force. Others, such as oxygenation of the eggs, have also been postulated. Whatever the initial driving selection or the origins of the species in question, once the system became established it would selectively adapt to local conditions with the coevolution of the appropriate behavioral and life history strategies as demonstrated in shad (Carscadden and Leggett, 1975b; Shoubridge, 1977) and salmon (Schaffer and Elson, 1975).

VII. Migration in Amphibians and Reptiles

A. Amphibians

The invasion of terrestrial and semiterrestrial habitats combined with the necessity to return to water to breed provide the circumstances leading to migration in amphibians. Migrations are of two major sorts: a movement to water for breeding and an exodus by newly metamorphosed juveniles invading the terrestrial habitat. The most extensive studies of amphibian movements throughout the annual cycle are those of Dole (1965a, et seq.) on the leopard frog, *Rana pipiens*, in Michigan. In the spring, overwintering adults emerge from hibernation and move to ponds to breed; young frogs move to ponds somewhat later even though no breeding takes place in this age group (Dole, 1967). Dole postulates that water balance may be involved in these young frogs as well as in adults which remain in the ponds for a period after breeding. Following breeding, adults move back to terrestrial habitats where they establish home ranges (Dole, 1965a), and in late summer migration from home ranges to hibernation sites may occur. When tadpoles metamorphose into young frogs in midsummer, they leave the pond either by scattering in all directions or, on some occasions, by defined routes (Dole, 1971). The exodus is often dramatic, especially on rainy nights, with thousands of froglets departing en masse (Bovbjerg and Bovbjerg, 1964; Dole, 1971). Dole recovered some marked animals as adults 5 km from their pond of origin suggesting that these juveniles are the colonizers of new areas.

In many amphibians selection has produced individuals that home repeatedly to extremely limited areas of streams and ponds often breeding within a few feet year after year. The reasons for this extreme site fidelity are not known, but since many species breed in ephemeral ponds or streams site specificity may be a guarantee of breeding success. At any rate precise homing is found in *Ambystoma* salamanders (Sheppe, 1965; Whitford and Vinegar, 1966), newts of the genus *Taricha* (Parker, 1963; Twitty, 1966) and *Notophthalmus* (Hurlbert, 1969), in *Bufo* (Oldham, 1966), and in *Rana pipiens* (Dole, 1968). In moving to breeding areas most amphibians migrate at night, but *Notophthalmus* moves in conspicuous fashion by day. Its apparent distastefulness to predators may make it immune to predation selecting for nocturnal migration (Hurlbert, 1969). Breeding migrations have been relatively well studied, but other movements, especially those immediately postbreeding, have been largely neglected. These need further assessment before we can fully understand the entire seasonal cycle (Dole, 1967).

B. REPTILES

A number of snakes migrate to and from winter hibernacula. In Manitoba, for example, the red-sided garter snake, *Thamnophis sirtalis parietalis*, hibernates in limestone sinkholes in aggregations of several thousand individuals (Aleksiuk, 1976). These sites evidently provide considerable protection because they fill with snow; the aggregations also allow concentration of metabolic heat. Ease of mate finding may be an additional selective pressure since breeding takes place before spring dispersal. Navigation and homing are well developed, suggesting that the system is under relatively intense selection. In Utah several species of snakes use rock piles as communal hibernacula (Hirth *et al.*, 1969; Brown and Parker, 1976). The racer, *Coluber constrictor mormon* dispersed as far a 1.8 km from the den, but the geometric mean distance was a more modest 383 m (Brown and Parker, 1976). Whip snakes, *Masticophis t. taeniatus*, were captured as far away as 3.6 km. Brown and Parker (1976) interpret communal hibernacula in terms of a refuging model with a core, a dispersal limit, and an intervening area where resources are acquired (Hamilton and Watt, 1970). In all cases there is annual variation in distance and direction of dispersal which seems to be a function of current climatic conditions.

Breeding migrations have been observed in the Australian lizard *Amphibolurus ornatus* (Bradshaw, 1971) and the iguanid lizard *Iguana iguana* in Central America (Rand, 1968). In *Iguana* the migrations involve a short swim to an isolated islet which presumably provides a protected breeding site. The migrations of *Amphibolurus* are more complex. These lizards live on scattered granite outcrops in the desert with large outcrops being preferred breeding sites. In the spring young adults immigrate to large outcrops to establish territories and breed, in the process displacing juveniles born the previous year to smaller more scattered outcrops. These in turn return as adults in succeeding years. There is thus a regular emigration–immigration cycle in the preferred breeding habitat.

The best-known migrants among reptiles are various species of marine turtles. The green turtle (*Chelonia mydas*) has been most extensively studied, with some data also on the ridley (*Lepidochelys olivacea*), the leatherback (*Dermochelys coricea*), and the hawksbill (*Eretmochelys imbricata*). The longest-known voyage is a transatlantic movement of some 5900 km by a leatherback (Pritchard, 1973) and distances of 1000–1500 km, based on recoveries of tagged animals, are common (Carr, 1967, 1975; Pritchard, 1973). The general pattern involves a 2–4 year breeding cycle with egg-laying on specific beaches and long intervals spent in offshore feeding areas. Turtles breeding on the same beach may divide into two or more populations returning to widely spaced feeding areas or, conversely, feeding populations may divide to breed on geographically separated beaches (Carr, 1975). The selective driving force behind the migrations is evidently the spacing of suitable breeding beaches and feeding areas.

The reptiles show a high degree of philopatry and site tenacity (Carr, 1967, 1975) and real questions exist as to why these traits are maintained by selection. There is a correlation between pattern of oceanic currents and most populated nesting beaches on both the Atlantic and Pacific sides of Costa Rica, and Richard and Hughes (1972) suggest that currents are important in facilitating transport of turtles. Currents proceeding directly from Ascension Island in the South Atlantic to mainland Brazil may also ease orientation and travel problems for turtles nesting on that island (Carr, 1967). The worldwide tendency among green turtle populations to use small remote islands for nesting may be adaptive because of freedom from terrestrial predators (Carr, 1967).

These factors, however, do not explain why such apparently rigorous migrations are undertaken when (1) many apparently suitable beaches are bypassed and (2) other turtles in the same feeding grounds migrate to other beaches. Carr (1967, 1975; Carr and Coleman, 1974) concludes present selection pressures cannot explain these pathways and that some routes are vestiges of previously highly adaptive patterns whose survival value is now waning. He cites as examples the Ascension and Aves Island breeding populations (Aves is a small island near Guadeloupe in the Caribbean). In the case of Aves, the island was apparently larger at one time and is now shrinking as a result of subsidence. The consequence is overcrowding on the turtle beaches. With Ascension, migration to the island was intially ingrained, because of predator protection, when it was nearer the South American mainland. As Ascension receded due to continental drift and seafloor spreading, the migration routes became progressively longer. The problem, as Carr (1967) points out, is that this model demands descent of this migration route from ancestral species of *Chelonia* present in the Cretaceous. This is a very long time for the maintenance of a "vestigial" trait even given the long generation time of sea turtles. In spite of Carr's demurrers, it would seem more useful to examine present or recent selection regimes more carefully (cf. also Lack, 1968a, on birds). Suffice to say, the evolution of particular migration routes is still very much an open question.

VIII. Migrations of Birds

> We cannot escape the fact that between about 15,000 and 5000 years ago and again in the last 5000 years. . .the migration of most species and its incredibly various patterns must have been evolved. . .All such adaptations must be the product of evolution in something like the last 10,000 years, a conclusion shattering to much current evolutionary theory.
>
> R. E. Moreau (1972), p. 44.

A. THE DIVERSITY OF BIRD MIGRATION

In most birds migration involves highly specialized behavior and physiology (Berthold, 1975; Rappole and Warner, 1976) characterized by two distinct physiologic states, *Zugdisposition* and *Zugstimmung* (Groebbels, 1928). In *Zugdisposition* the bird is hyperphagic increasing food intake as much as 40% above normal and storing the excess as lipid fuel for the migratory journey. The inductive stimulus is usually photoperiod. In *Zugstimmung* the bird undertakes and maintains a migratory flight; in caged birds this state results in the well-known *Zugunruhe* or "migratory restlessness." There is considerable variation in these physiologic states within and among species, populations, individuals, and different parts of the migratory journey. Rappole and Warner (1976) suggest that migration involves trade-offs between appropriate physiology and currently favorable environmental conditions, e.g., favorable following winds (Gauthreaux, Chapter 2).

As with other groups, birds also display great variety in migratory pattern (reviewed by Salomonsen, 1955; Dorst, 1962; Orr, 1970). Long-distance movements between summer and winter ranges represent only one type of migration. At one extreme is the justly famous arctic tern (*Sterna paradisaea*) which makes a round trip from arctic and subarctic breeding grounds to wintering areas in the Antarctic (Lincoln, 1952). At the other are tropical species moving only a few miles between seasonally changing habitats (Karr, 1976). Long-distance migration may take place entirely within the temperate zone, e.g., starlings (*Sturnus vulgaris*), European (*Erithacus rubecula*) and American (*Turdus migratorius*) robins, or juncos (*Junco hyemalis*), or entirely within the tropics, e.g., *Quelea quelea* (Ward, 1971), as described at the beginning of this review, and the yellow-green vireo (*Vireo flavoviridis*) (Morton, 1977). Altitudinal migrations also occur in both temperate and tropical regions (Orr, 1970).

There is also intraspecific variation with different populations of various species ranging from wholly sedentary to extensively migratory. The full range is shown by the fox sparrow (*Passerella iliaca*) on the west coast of North America (Swarth, 1920; Lincoln, 1952). An Alaskan population winters in extreme southern California while a population in northern British Columbia winters in Oregon. A sedentary population occurs in northern Washington, coastal southern British Columbia, and Vancouver Island. This "leap-frogging" pattern is also known in other species (Salomonsen, 1955; Cox, 1968). In juncos (*Junco hyemalis*) females migrate on the average farther south than males (Ketterson and Nolan, 1976). Age correlated movements were found in blue grouse (*Dendragapus obscurus*) by Zwickel et al. (1977) who noted that colonization of new areas was mostly by yearling birds. Similar dispersal by yearlings has also been noted in other species (Salomonsen, 1955). The timing of movements also varies depending on the harshness of winter, migration being delayed or absent in mild years or if food remains abundant, e.g., in redpolls (Evans, 1969). Radar studies have shown that on a virtually daily basis throughout the winter, birds may be leaving or arriving in the British Isles (Lack, 1968a). A particularly interesting case is described by Harris (1970) who cross-fostered young of the normally sedentary (in Britain) herring gull (*Larus argentatus*) with the migratory lesser black-backed gull (*Larus fuscus*). A significant portion of these *L. argentatus* migrated, although not as far as their foster parents suggesting both gene and environmental influences. In the reverse experiment cross-fostered *L. fuscus* still migrated indicating that genetic influences on migration were dominant.

Even though patterns are diverse and often flexible, migratory routes can be extremely conservative often to the apparent detriment of the species concerned. Restriction of one part of the range, e.g., to islands, may thus result in the population becoming severely limited and the species a relict (Amadon, 1953). This may be the case with the Kirtland's warbler (*Den-*

droica kirtlandii) in North America; it breeds only in jack pines of a few counties in Michigan even though apparently suitable habitat is widespread. The restriction may come in the winter range which encompasses only a few islands in the Bahamas. The wheatear (*Oenanthe oenanthe*) has extended its range from the Palearctic to Greenland in the West and western Alaska in the East, yet at both extremes the birds migrate back to Africa across the open Atlantic, on the one hand, and most of Asia, on the other, even though apparently suitable overwintering areas occur much closer (Moreau, 1972). Moreau (1972) also gives other similar examples. Populations breeding in the same area may also divide and migrate to separate wintering areas. Dunlins (*Chalidris alpina*) in Alaska, for example, are separable into populations wintering on the Pacific coasts of Asia and North America (MacLean and Holmes, 1971). The evolution of such routes has been the subject of much speculation (Salomonsen, 1955; Lack, 1968a) but little resolution.

B. The Evolution of Bird Migration

1. Historical Theories

Early theories of the origins of bird migration concentrated on historical factors with the most popular being Pleistocene glaciation (reviewed by Landsborough Thompson, 1926; Cox 1968). Glacial theories can be roughly divided into two, one assuming that migrants originated in a southern ancestral "home," the other in a northern one. In either case migration was conceived as having developed over time to follow the routes of the receding glaciers. These theories were criticized early on by Schenk (1924, cited in Landsborough Thompson), Grinnell (1931), and Mayr and Meise (1930) all of whom pointed out that many current migration routes bear no resemblance to patterns of glacial advance and retreat. Glacial theories have been firmly laid to rest by Cox (1968) and Lack (1968a).

A more ambitious attempt to correlate migration with historical events was Wolfson's (1948) hypothesis that migration routes followed from continental drift. The model was based on two major premises: that birds frequently migrated farther than necessary to reach suitable wintering areas and that most migration was northward. Since continental drift was also presumed to be mostly to the north this would account for routes while the spread of continents would lead to apparently anomalous distances traveled if birds continued to fly to ancestral wintering grounds in the face of geographical exigencies. The hypothesis was criticized at the time by Amadon (1948) who pointed out that if continental movements occurred, the current position of the continents was reached long before the evolution of present day bird families, thereby obviating this mechanism. Amadon's criticism has been amply confirmed by recent advances in plate tectonics.

Present day migration patterns have thus evolved since the end of the last glaciation (Moreau, 1972; see above), and rather than to past events it is to present selective forces that we must look to understand their origin (Cox, 1968; Lack, 1968a). The basic questions are: (1) What makes the fitness of a bird higher if it migrates? and (2) Why do some birds migrate while others do not? (Fretwell, 1978).

2. Energetic Considerations

Since many birds leave temperate areas in winter, it is natural to consider energy balance as a selective factor in migration. Although migration is energetically costly, the expenditure is wholly or partly compensated by avoiding the temperature stresses of northern winters (Kendeigh et al., 1977). Studies of four species, the junco, *Junco hyemalis*, the white-throated sparrow, *Zonotrichia albicollis* (Siebert, 1949), the tree sparrow, *Spizella arborea* (West, 1960), and the dickcissel, *Spiza americana* (Zimmerman, 1965), all indicated an energetic advantage to southward migration in the autumn. Zimmerman (1965) further concluded that the dickcissel improved its energy balance by northward migration, but such an advantage could not be demonstrated for any of the other species. In general the costs of migration are less for nonpasserines because these species are more sensitive to cold and actually save energy by migrating, and small birds gain more by migrating than do large birds (Kendeigh et al., 1977). Cox (1961) took another tack and investigated tropical finches, concluding that they would gain little improvement in energy balance by flying north to areas of longer photoperiods. It is evident, therefore, that once certain species have adopted northern breeding areas, it behooves them to leave for energetic reasons before the onset of severe winter cold. But these considerations do not indicate how migrations to these sites evolve in the first place (Cox, 1968). Further it should be noted that small birds such as parids, which overwinter, are quite capable of evolving the appropriate thermoregulatory mechanisms (Yarbrough, 1971). It is also apparent that energetics alone cannot account for migration because different foods are important in different parts of the life cycle (Evans, 1969).

3. Competition Models

Most current models for the evolution of bird migration revolve around the notion of ecological competition (Salomonsen, 1955; Cox, 1968; Lack 1968a, 1976; Moreau, 1972; Fretwell, 1972; Karr, 1976; Rabenold, 1979, 1980). This was articulated in a general way by Lack (1968a) who, although admitting little direct evidence, argued that the number of migrating species was a density-dependent function of food availability on the wintering grounds. He cited evidence that migrant species and subspecies segregated from each other in winter (in the latter case accounting for "leap-frogging") and also segregated from closely related or ecologically similar resident species and concluded that this was the best evidence for winter competition for food. Similar arguments were also outlined by Salomonsen (1955). This view was further elaborated in a study of North American warblers (Parulidae) wintering in Jamaica (Lack, 1976) where it was shown that the migrants occupied different habitats from the residents and segregated from each other by habitat and foraging behavior. The extensive survey of the African–Palearctic migration system by Moreau (1972) revealed similar cases. Chats and wheatears (*Oenanthe* spp.) and shrikes (*Lanius*), for example, occurred in habitats devoid of resident species and segregated from each other while various sylviids overlapped broadly with residents but foraged in different microhabitats. The general and unsatisfactory lack of direct evidence regarding winter food limitation is discussed in detail by Fretwell (1978).

The question of competition was addressed for North American-Neotropical migrants by Cox (1968) who assumed the foraging niches were functions of bill morphology. If interspecific competition was irrelevant to the evolution of migration, then extent of bill differentiation should be the same for migrant and resident birds. If, on the other hand, species less able to separate ecologically by bill differentiation were able to do so by migration as an extension of niche space, then migrants should show significantly less variation in bill morphology. Three between species measures of variation showed a significant relationship when plotted against proportions of migrants in groups of nonraptorial land birds represented in North America. These were (1) coefficient of culmen variation, (2) mean culmen length, and (3) culmen variation coefficient ratio for North American and Costa Rican group members. In all three cases variation in bill length declined as the proportion of migrants increased, and Cox argued that those unable to sort out ecologically by bill length did so by migrating. This occurred in small-bodied birds and in passerines, and Cox suggested that limits in the former case were imposed by small bill size (leaving little room for variation) and in the latter by competition from phylogenetically older nonpasserines already occupying tropical niches. Migration is thus seen as a solution for species which must avoid competition by going somewhere else rather than foraging differently. Limiting resources would occur in wintering areas, as postulated by Lack (1968a), since it is in those areas where, for half the year, residents and migrants overlap.

The question of niche relationships in migrants was examined for six species of shorebirds by Baker and Baker (1973). They observed feeding behavior and habitat use on the breeding grounds in the Canadian Arctic and on the wintering grounds. Foraging methods were defined on the basis of type of bill usage and pattern of locomotion. Food density was found to be higher on the breeding grounds, suggesting an advantage to northward migration, and here foraging was more selective, but behavioral and habitat diversities were greater implying a broader niche. On the wintering grounds with low food density there was little habitat and resource overlap (narrow niches) with each species living where optimally adapted and foraging efficiency was greatest. These seasonal differences in resource partitioning imply limitation of shorebird populations through competition in wintering habitats. They also suggest that selection favors migration to reduce competition at the critical time of breeding. Shorebirds are in fact conspicuously absent as breeding birds in apparently suitable tropical habitats, e.g., tidal mudflats, indicating that perhaps because of potentially intense competition breeding cannot be supported. It should be noted, however, that there are virtually no breeding birds using tropical mudflats, suggesting that there is insufficient food for breeding with or without competition.

4. Site Dominance and Nest Site Density

Based primarily on his studies of field sparrows (*Spizella pusilla*) and dickcissels (*Spiza americana*), Fretwell (1972, 1978) has proposed a modified version of a competition based origin of migration. Fretwell stresses the difference between migrant and sedentary populations and species and asks what ecological factors contribute to each. He notes, first, that for several species the proportion of nests destroyed by predation increases as density of nests increases within a given habitat (between habitat variance also contributes to variance in predation, e.g., Robertson, 1972). In both the dickcissel and field sparrow, nesting densities reach levels where significant losses occur if all birds remain to breed in the winter habitat. In the case of the dickcissel this occurs because various tropical finches nest in the same grasslands and in the case of the field sparrow because of nesting conspecifics. The entire dickcissel and a significant portion of the field sparrow populations thus move northward to breed.

The advantage to the sedentary birds remaining is that they gain from habitat experience. The tendency to philopatry in territorial behavior implies a high survival value to familiarity with a specific area, and residents would gain relative to migrants in this respect (cf. also von Haartman, 1968; Gauthreaux, 1978). The migrants gain breeding areas with reduced competition because site dominant residents are absent during the breeding season; among migrants those individuals which are behavioral dominants gain an advantage (Gauthreaux, 1978). The return movement south results from insufficient food (or possibly other factors such as roosting sites) to support a population the size of the breeding one. Fretwell advances the hypothesis as a working model in need of testing especially with respect to its underlying assumptions. Among its predictions is that residents will display a tendency to be sexually dimorphic and/or brightly plumaged year round while migrants will do so only in the breeding season. Evidence from some species of North American warblers, orioles, and sparrows supports the prediction (Hamilton, 1961; Rohwer, 1975), although it should be noted that some resident species, especially those residing in highly seasonal tropical habitats, also show contrasting breeding and nonbreeding plumages, e.g., African ploceids.

5. Migration and Habitats

In one of the first major attempts to examine the relation between bird migration and habitat, MacArthur (1959) concluded that North American Neotropical migrants were most prevalent in deciduous forests, less so in coniferous forests, and least so in grasslands. The dominant feeding mode in forests is insectivory, especially by gleaners and flycatchers, and this suggested that foraging method played an important role in the evolution of the New World migration system. The seedeaters of grasslands were either resident or, with few exceptions, e.g., the dickcissel, migrated within the temperate zone. Willson (1976) reanalyzed MacArthur's data and came to the same conclusion with respect to forest versus grassland, but noted no difference in proportions of migrants between coniferous and deciduous forests. Most migrants in conifers were parulids, however, and total breeding numbers were different between these two habitats.

Other recent work has analyzed the niches occupied by migrants on wintering and breeding grounds and in transit. Shorebirds were discussed above (Baker and Baker, 1973). The other group receiving significant attention with respect to niche diversity are the North American parulid warblers (MacArthur, 1959; Morse, 1968, 1971; Parnell, 1969; Power, 1971; Lack, 1976; Rabenold, 1980). Warblers do segregate by foraging method and habitat, and this segregation seems to be greater on breeding or wintering grounds than in transit. The differences may be due, however, simply to more species being present in any one area during transit (Power, 1971). The general characteristics of species remain similar throughout the annual cycles. Species that show a broad niche during migration also do so during breeding and vice versa indicating that while in transit species are similar in habitat diversity to the breeding season. The same seems to apply on the wintering grounds where there is also sorting out by habitat, presumably to reduce competition both with residents and with other migrants (Lack,

1976). Some species do, however, show differences in foraging behavior between breeding and wintering areas. The black-throated green warbler (*Dendroica virens*), for example, displays much broader foraging behavior in its wintering habitat (Panama), perhaps due to increased competition for scarcer resources (Rabenold, 1980).

The degree to which passage migrants use a habitat, however, may be a function of their behavioral states, i.e., whether they are in *Zugdisposition* or *Zugstimmung* (Rappole and Warner, 1976). Rappole and Warner observed that most individuals of species that arrive and depart in "waves" are active, gregarious, feed very little, and depart quickly. Some individuals nevertheless are in *Zugdisposition* and these remain for longer periods, are highly habitat selective, and may even hold territories for a few days. Therefore, not all species seen in an area use its food resources, and even of those species that do, only a few individuals may be involved. Ecological suitability of a stopover site is paramount, and observations that less ecological segregation occurs in transit (Parnell, 1969; Morse, 1971; Power, 1971) may be misleading. Actual use of resources during passage may require as intense a habitat selection as occurs on wintering or breeding grounds.

The problem of the impact of whole migrant faunas on the residents in the tropics has been discussed by Moreau (1972) and Karr (1976, 1978). Moreau summarizes an extensive literature on African–Palearctic migrants indicating that migrants are confined virtually exclusively to savannah and scrub habitats during their overwintering in the tropics and are absent from forests. This was confirmed by Karr who also compares data from the Neotropics and the Indomalaysian region. In the Neotropics migrants occur mostly in secondary and relatively high altitude forests and almost never in mature forests. Karr (1976) suggests on the basis of this and related evidence that riverine habitats may have been important in the evolution of New World migrants while no such forest edge faunas seem to have developed in Africa. In Malaysia and India migrants are much more abundant in mature forests, but may also occur in second growth areas and gardens (Ward, 1968; Medway, 1972). It is also worth noting that overall density of birds is less in Indomalaysian forests, and with respect to India, Karr suggests the long severe dry season may have created habitats invasible by migrants.

Migrants in all three continents display two other characteristics. They are, first, likely to be transient in any one area with high densities occurring in fall and spring during peak movement and lower densities the rest of the year. This is especially noticeable in Central America and Malaysia. Second, while in the tropics migrants tend to harvest superabundant but sporadically available resources such as fruit and nectar (Leck, 1972a,b), insects stirred up by army ant swarms (Willis, 1966) and grass fires, and emerging termites (Moreau, 1972). Thrushes, for example, may be itinerant throughout their winter ranges exploiting fruit resources unable to support breeding in any species (Karr, 1978). It is also a frequent observation that migrants occur in habitats where there are apparently no resident species (Moreau, 1972; Lack, 1976) or that they are much less abundant than residents where there is co-occurrence (Tramer, 1974; Karr, 1976).

Nomadism is characteristic of Australian desert birds (Serventy, 1971). Here the patterns of rainfall are highly erratic, and many birds have developed migratory and breeding patterns keyed to the periods when rainfall does occur. Birds such as the budgerigar (*Melopsittacus undulatus*), the zebra finch (*Poephila guttata*), and the grey teal (*Anas gibberifrons*) may be absent from areas for long periods only to reappear in numbers to breed when it rains. Such breeding can occur at any time of the year. In deserts where the seasons are more predictable, such as North America and Africa, there are few nomads, although the notable *Quelea quelea* of East Africa (Ward, 1971) has been mentioned, and in North America and Europe various northern birds such as the snowy owl (*Nyctea scandiaca*) and the crossbills (*Loxia*) are well known for their erratic wanderings when food is short.

Finally, the migrations of oceanic birds deserve mention. These species require restricted breeding sites on coastlines and islands, and their breeding biology is strongly influenced as a result (for reviews, see Lack, 1968b; and Immelmann, 1971). Migrations seem to be strongly influenced by the relations between breeding sites and food supplies. Thus many northern seabirds are seasonal migrants with patterns analogous to their terrestrial counterparts (for examples, see Dorst, 1962; Ashmole, 1971). In the tropics patterns may be less clear and breeding may take place throughout the year with little migration. The factors selecting for migration in seabirds have received less attention than those influencing land birds, and little more than some obvious generalizations are now possible.

C. CONCLUSIONS: EVOLUTION AND COEVOLUTION OF MIGRATION STRATEGIES

The variety of migratory patterns displayed by birds makes a satisfactory unitary model of the evolution of migration extremely unlikely. Nevertheless are any generalizations possible? There seem to be two major ones. The first is obvious and is simply that an abundance of resources is available only during the temperate summer and only to those species able to reach the temperate zone to exploit it. The second is that the birds evolving a strategy (migration) of exploiting these resources are apt to be species whose winter habitats are also characterized by the transient nature of the resources available. In the Neotropics these are mostly secondary and high altitude forests, in Africa they are savannah and scrubland, in India forests influenced by a severe dry season, and in Southeast Asia forests of low productivity growing on nutrient poor soils (Janzen, 1974).

In the sense that they are keyed to the exploitation of superabundant but likely sporadic resources (Karr, 1976), migrant birds are "fugitive" species. Since even fugitive species must at some time settle to breed, it behooves them to do so when and where food will be available throughout the entire breeding season. Compared to the habitats occupied in the tropics, food in temperate habitats is apt to be more predictable in both space and time because these are usually in "climax," i.e., relatively stable, biomes such as deciduous and coniferous forests and prairie grasslands. Also the concentration of resource productivity in the brief summer season apparently permits "stacking" of species, such as the North American parulid warblers, with similar generalized ecology (Rabenold, 1978). I emphasize that these are broad trends as local variation does occur, and there are migratory species adapted to "pioneer" temperate habitats and others such as the Kentucky warbler (*Oporornis formosus*) whose winter habitat is mature tropical forest (Karr, 1976). The reason that some habitats in the tropics are "empty" of breeding birds may be that their productivity fluctuates too erratically in both space and time (Taylor and Taylor, 1977) to support breeding populations. Some preliminary comparisons of productivity in temperate and tropical forests are made by Karr (1975) who, along with Rabenold (1979), has expressed similar notions on the relation between productivity and migration. It goes without saying that these assertions need to be tested by assessing fluctuations in temperate and tropical areas in far more detail and with greater replication than is now available. A modern ecological tragedy is that this may no longer be possible in the tropics as man-altered habitats may be unrepresentative.

Individual migration routes, distances, and life histories are the result of the cost–benefit ratios for the particular species or population. Northern seed-eaters need move only to more southerly areas in winter where seeds are still exposed, whereas insectivorous species must migrate to the tropics and subtropics. Similarly breeding, molt, and migration must be appropriately timed. In most species the molt follows the breeding season, whereas migration occurs between molt and reproduction. There is thus temporal segregation of energetically competing physiologic functions (Immelmann, 1971). Morphology may also be altered. The wings of migrants are usually longer and narrower than those of their nonmigrant relatives (Dorst, 1962; Moreau, 1972). Other characters such as breeding plumages may also be influenced (Hamilton, 1961). There is no good evidence that life history statistics such as clutch size vary with migration independently of other factors like latitude (Lack, 1968b; Cody, 1971), but the subject could stand reexamination with migration in mind.

Finally, it should be emphasized that many factors interact to produce migration strategies (Rappole and Warner, 1976; Karr, 1978). Species vary and so too do populations and even individuals within species. The determination of the degree of flexibility of any individual of any species is specifically calibrated by natural selection.

IX. Migrations of Mammals

Mammals are unique among migrants in that extensive movements are undertaken by large numbers of species by all three means of vertebrate locomotion: swimming, walking or running, and flying. (Some birds may migrate by walking or swimming, e.g., ostriches, penguins, but few species are involved.) This of course results from the tremendous adaptive radiation of mammals into all major habitats. Therefore the adaptive advantages of migration in a group of such diverse morphology and ecology are of great interest.

A. MARINE MAMMALS

The general pattern of migration is the same for all the large whales although there may be individual variations on the general theme (Slijper, 1962). The blue (*Balaenoptera musculus*), fin (*B. physalus*), and humpback (*Megoptera novae-angliae*) whales of the Southern Hemisphere spend the summer from roughly December through March in the Antarctic, then migrate north to spend the winter in tropical and subtropical waters. Breeding takes place in the winter with humpback calves born in September (Dawbin, 1966). The breeding cycle of the humpback is fairly well known because it breeds inshore, while the offshore breeding grounds of blue and fin whales are still largely unknown (Mackintosh, 1966). Another inshore breeder is the California gray whale (*Eschrichtius gibbosus*) whose seasonal migrations between the subarctic, where it spends the summer, and subtropical breeding grounds on the coast of Baja California have become tourist attractions (Rice, 1965; Orr, 1970).

Following a winter spent in the tropics and subtropics largely without feeding (Slijper, 1962; Dawbin, 1966), the whales migrate back to the higher latitudes in the spring. In the humpback, the newly pregnant females are the first to leave the breeding area, spend the longest time in the feeding area, and are the last to return the next season to the breeding area where their calves are born. Lactating females with their new born calves are the last to leave for the feeding area, but the first to return to it after weaning their calves. In the feeding areas, whales tend to concentrate where there is the greatest abundance of krill; the combination of geographically distinct feeding and breeding areas means that these species, especially the humpback, occur as discrete populations. Variation between these populations in timing and patterns of migration remains to be investigated in detail.

Selection pressures promoting migration seem to occur with respect to both feeding and reproduction (Kinne, 1975). The harshness of the Arctic and Antarctic winters with reduction in available food makes winter at high latitudes extremely unfavorable for the large whales (Slijper, 1962). The food problem is not solved by migrations to lower latitudes because of the low productivity of tropical waters, but there is a gain in heat balance in warmer areas. In fact there is virtually no feeding during the winter breeding season which is therefore energetically costly. Oil obtained from northward migrating pregnant females of the Southern Hemisphere humpback, for example, is nearly twice that from southward moving females of equivalent size (Dawbin, 1966). Calves probably also benefit from birth in warmer waters because they suffer greater heat loss than the adults as a result of smaller size and hence greater surface to volume ratio.

Selection for migration thus seems to center on the balance between food availability and heat loss. If one is to starve anyway, it is preferable to do so in the thermally favorable tropics rather than in polar seas where energy costs are apt to be particularly high for lactating but still starving females; reduction in heat loss must in this model more than compensate for the energetic cost of migration itself. Suckling calves also can devote more energy to growth rather than maintaining body temperature against an extreme thermal gradient. The calves of those species, like the gray and humpback whales, which breed in coastal bays and lagoons, may also be protected from predation. This line of reasoning contains appealing logic, but should not be accepted uncritically. Many small whales and dolphins spend the entire year in high latitudes and have developed efficient heat conservation mechanisms (e.g., Kanwisher and Sundnes, 1966). They are piscivorous, and this may be the deciding factor. At any rate, we need measurements of the appropriate energetics both within and among species and populations, and these will unfortunately be extraordinarily difficult and costly to obtain, the more so given the current depletion of the world's stocks of whales.

Seal, sea lion, and walrus migrations appear to be prompted by the necessity for protected terrestrial rookery sites such as islands. In the northern fur seal (*Callorhinus ursinus*) breeding takes place in the summer on islands of the Bering Sea and North Pacific (e.g., Pribilofs, Commanders).

In the Eastern Pacific the population moves southward offshore with the males wintering mostly in the Gulf of Alaska but females and juveniles continuing southward for 3000 miles (Orr, 1970). Off the West Coast of North America the California sea lion (*Zalophus californianus*) and Steller's sea lion (*Eumetopias jubata*) breed on small islands with young born in early summer. Following breeding, juveniles and adults disperse with males at least often moving northward for considerable distances (Orr, 1970; Orr and Poulter, 1965). The elephant seal (*Mirounga angustirostris*) breeds from December to February off the coasts of Southern and Baja California with the adults leaving after completion of breeding in March and the pups in May. After the adults leave, immature animals arrive and stay for about 3 months. In summer adult and subadult males arrive to molt; these are again replaced by immatures until December when the breeding population returns (Radford *et al.*, 1965; Orr, 1970). In the Atlantic, breeding and seasonal movements of the gray seal (*Halicheorus grypus*) differ between the two sides of the ocean (Cameron, 1970), but the general pattern of breeding followed by dispersal and reassembly to breed occurs in both populations (Table VII).

In those pinnipeds for which information is available, it is the pre-reproductives that seem to travel farthest during migrations, and these are the individuals which found new colonies. A case in point is the founding of an elephant seal colony on the Farallon Islands, California, studied by LeBoeuf *et al.* (1974). Here the initial individual was a subadult male followed 3 weeks later by two cows one of which gave birth to a pup. In subsequent years subadult males fought and then bred with females at ages when they would have been excluded from breeding in established colonies. Other cases of greater dispersal and colony founding by subadults are reviewed by Baker (1978).

The selective forces influencing pinniped migration and breeding cycles are not well understood; information is sparse and often anecdotal. The mid-June dispersal of the Western Atlantic gray seal seems to occur at the time of the spring run of mackerel (*Scomber scombrus*) (Cameron, 1970).

TABLE VII

Seasonal Cycles of Eastern and Western Atlantic Populations of the Gray Seal, *Halichoerus grypus*[a]

Population	Breeding	First dispersal[b]	Molt	Second dispersal	Reassembly
Eastern Atlantic	Oct.–Nov.	Nov.–Dec.	Jan.–Apr.	May	Late Aug.–Oct.
Western Atlantic	Jan.–Feb.	Late Feb.?	March?–June?	Mid-June	Dec.

[a] Data from Cameron (1970).

[b] At least by yearlings, pregnant cows, and bulls.

In the Farallon elephant seals, population fluctuations are correlated with fluctuations in salinity, water temperature, and upwellings which indicate changes in phytoplankton and possibly elephant seal prey (LeBoeuf *et al.*, 1974). The opportunity to examine island sites as predator-free breeding areas by comparison with possible mainland sites may be forever lost, but future studies might well be directed at comparisons between different populations and species. The variation in the migratory and breeding cycle of the Atlantic gray seal is a case in point. Also sedentary species and populations should be studied from an evolutionary viewpoint. The harbor and ringed seals (*Phoca vitullina* and *P. hispida*), for example, are seden-tary while the related harp seal (*P. groenlandica*) is migratory (Orr, 1970). Good comparative studies are clearly needed.

B. Large Terrestrial Mammals

Large ungulates may migrate long distances. The two best studied ex-amples are the wildebeest or gnu (*Connochaetes taurinus*) and other species in the Serengeti–Mara region of East Africa (Grzimek and Grzimek, 1960; Talbot and Talbot, 1963; Pennycuick, 1975; Inglis, 1976) and various northern cervids especially the barren grounds caribou (*Rangifer tarandus*) and elk (*Cervus canadensis*) (Murie, 1935; Lent, 1966; Orr, 1970). The Serengeti wildebeest spend the wet season months of December–April in the grassy plains of the southeastern portion of the region. During this time the short grass, which is the preferred food, is green, and growing and most calving takes place. At the beginning of the dry season around May–June, the animals form large herds and move into wetter scrubland to the northwest near Lake Victoria. This movement may occur almost single file, often extending miles, and appears to be specialized migratory behavior. Around the end of July or beginning of August toward the end of the dry season, the herds move northeastward to spend the next 2 or 3 months in the northern part of the region followed by a return to the southern plains in November–December. These movements are summarized in Fig. 5. There is considerable year to year variation in the migrations with time of occupancy in any area strongly correlated with the rains that year (Pen-nycuick, 1975).

There are also two sorts of sedentary populations of wildebeest. In the nearby Ngorongoro Crater the seasons are not nearly so extreme as on the Serengeti Plains. Here the population is largely sedentary with some move-ment into or out of the crater in dry years. But on the plains, there are also sedentary animals; these are old bulls which maintain territories the year round and do not migrate with the herds (Talbot and Talbot, 1963). Year round territorial bulls also occur in the Ngorongoro, as would be expected,

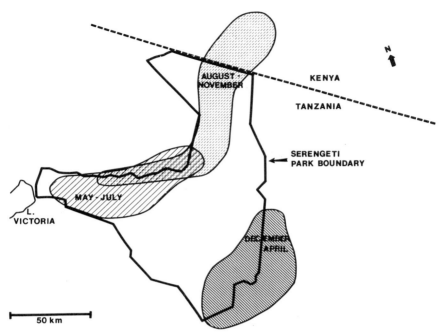

Fig. 5. Seasonal movements of the wildebeest in the Serengeti area of Tanzania for the period 1960–1973 compiled from data in Pennycuick (1975). The wet season breeding takes place in the Southeast followed in late April by migrating to wet areas near Lake Victoria to the Northwest. Later movement is to the North before remigration to the breeding habitat. Year to year variation was considerable, depending on the rains, and long term trends such as increasing use of the region north of the Kenya line (in Masai–Mara) were also apparent.

and in other areas such as Amboseli and the Athi Plains in Kenya where most animals move away in the dry season (Dingle, personal observation). Territories are also formed by bulls migrating with the herds during temporary stopovers, but quickly break up again as the herd moves on.

Sedentary populations form only where resources are sufficient. In the Ngorongoro seasonality is less extreme, and grass is usually available throughout the year. In the Serengeti sedentary territorial bulls choose sites with some vegetation available, although insufficient to support a large number of animals, and with access to a permanent water source. Permanent territories probably confer a reproductive advantage to these old bulls since they are ready to mate when females return from migration. The relative mating success of sedentary bulls versus the success plus migration costs (high mortality) of migrating bulls, however, remains to be assessed quantitatively. But it is apparent from the sedentary nature of populations in seasonally stable areas and the variation in both the timing and spacing of migration as a function of the rains that the movements are resource based (Talbot and Talbot, 1963; Pennycuick, 1975). The breeding system of the wildebeest is described by Estes (1969, 1976).

The movements of northern cervids are also resource based (Lent, 1966; Orr, 1970). The caribou spend summers in more northern areas where they feed on the leaves of low trees (e.g., willows) and forbs and then migrate to the south in winter, feeding primarily on grass and lichens. Such movements may cover as much as 300 miles. Other large cervids such as moose, elk, and deer may show considerable altitude migrations moving to high country in the summer and into adjacent lowlands for the winter (Edwards and Ritcey, 1956). Like the wildebeest these species may show specialized behavior by gathering in herds and migrating single file (compare figure on p. 228 in Orr, 1970 with Fig. 4 in Grzimek and Grzimek, 1960).

Large carnivores such as the lion and the wolf may follow their resources and migrate with the herds of ungulates. In coyotes, foxes, and wolves, young animals may disperse presumably to enhance their chances of reproduction (Bekoff, 1977). The less social the canid the more likely the chances of juvenile dispersal. Determinants of dispersal are apparently interactions between establishment of dominant–subordinate relationships and the occurrence of social play. Dominance establishment before play, e.g., in coyotes, results in rank-related differences among littermates; for example, the alpha individual has fewer social interactions whether or not related to aggression. In social species such as wolves the appearance of play before rank-related agonistic interactions may lead to coordinated pack formation with rare juvenile dispersal. The range of behaviors within litters suggests a polygenic basis. Social behavior is also influenced by other selective forces such as optimal body size, prey availability, habitat, etc. indicating the coevolutionary complexity of dispersal in the overall life history strategy. In the wildebeest synergistic selection pressures against concealment of young and in favor of rapid development of the ability to follow the mother (in the first hour), large aggregations, and synchronous breeding also suggest coevolution of migration and life histories (Estes, 1976).

C. BATS

Since they are one of only two major groups of flying vertebrates, one might expect a priori that bats would display extensive migratory capacity. Such is in fact not the case, and several reasons have been advanced as to why. First, unlike the massive pectoralis in birds, bat wing musculature is complex, involving 10 or more muscles (Vaughan, 1970a). This is apparently an adaptation to the relatively slow but highly maneuverable flight of bats necessary to catch insects on the wing (Vaughan, 1970a,b; Yalden and Morris, 1975). Indeed the wings themselves may be used to help trap insects (Griffin et al., 1960). Vaughan (1970a) also speculates that the complex muscle arrangement helped to preserve a flat sternum and narrow chest necessary for shelter seeking in crevices during early chiropteran evolution. Second, the wings are thin airfoils of high camber, effective for producing high lift at low speeds but producing excessive drag at the high speeds necessary for efficient migration (Vaughan, 1970b). Third, because of their relatively inefficient long distance flight system, bats may be unable to store enough fat for migration (Yalden and Morris, 1975). This could lead to hibernation which is less energetically costly, and this in fact occurs (Henshaw, 1970; Davis, 1970). Whatever the ultimate reason, the lack of powerful migratory flight is probably an important factor in limiting bats largely to the tropics while many birds migrate into the temperate zone. As an example of the relative difference there are some 245 breeding sites on the tropical island of Trinidad (ffrench, 1973) and about twice that number in North America north of Mexico. In contrast over 60 species of bats occur in Trinidad (Yalden and Morris, 1975) while only 39 occur in North America north of Mexico and several of these only along the Mexican border (Burt and Grossenheider, 1976).

Yet migrations in bats do occur (Griffin, 1970). Frequently these are in association with caves or other shelters used as hibernation sites. In New England *Myotis lucifugus* in Southern Vermont generally migrates southeast to spend the summer at distances up to 200 miles away before migrating back to the hibernation site in the autumn (Davis and Hitchcock, 1965). Similar migrations from hibernation sites in Kentucky northward as far as Michigan are known in *M. sodalis* (Barbour and Davis, 1969). In Europe *Nyctalus noctula* banded in Holland have been recovered as far away as Bordeaux some 900 km distant (Sluiter and van Heerdt, 1966). Interestingly, females of this species return each year to the area where they were born and breed there, while males leave their home area when mature and disperse over a wide area. This difference probably results from male territoriality and polygyny. In the tropics migrations may occur in nectar feeding bats in response to flowering season (Fenton and Kunz, 1977).

The most impressive migrations occur in the swiftest flying bats such as the hoary bat *Lasiurus cinereus* and the free-tailed bat *Tadarida brasiliensis* (Griffin, 1970; Yalden and Morris, 1975). The former species is found as far north as Alaska in summer but only south of 37°N in winter, a limit presumably set by the availability of flying insects. Interesting differences in migration exist between populations of *T. brasiliensis* in the American Southwest (Cockrum, 1969). A population occurring in southern Oregon and the northern two-thirds of California is largely resident. A second population in southeastern Nevada, western Arizona, and southeastern California in summer disappears in winter, but details of its migration are not known. A third population in southeastern Utah, southwestern Colorado, eastern Arizona, and western New Mexico seems to have a well-established flyway through the Mexican states of Sonora and Sinaloa west of the Sierra Madre. Finally, a population in most of Texas, western Oklahoma, and eastern New Mexico probably migrates in winter to northeastern portions of Mexico. These populations would be well worth further study especially with respect to climatic and other variables influencing physiological homeostasis, reproduction, and migration. Currently we have little knowledge of the selective forces acting on migration in any bat species.

D. SMALL TERRESTRIAL MAMMALS

> Our work on Microtus can be summarized by the admonition: study dispersal
>
> Krebs et al., 1973, p. 40.

Much of what we know of rodent migration derives from studies of population cycles. Rodent outbreaks have been known since Old Testament times (Krebs et al., 1973) and were brought to the attention of ecologists perhaps more than anyone else by Charles Elton (1942). Efforts to explain the causes of cycles and outbreaks have inevitably focused on the role of dispersal and how selection might act to produce dispersers. (Note: "dispersal" is used most commonly in the small mammal literature, but "migration" is often used with the same meaning. Both are consonant with the definition of "migration" in this chapter, and I use them interchangeably.) Recent research on the driving forces behind rodent population cycling has concentrated on the intrinsic factors of behavior and genetics, rather than extrinsic factors such as weather and predation. The theoretical base comes largely from Howard (1960), Lidicker, (1962, 1975), and Chitty (1967).

Howard (1960) proposed a distinction between "innate" or genetic dispersal tendency and "environmental" dispersal; the former resulted in dispersal of some individuals regardless of current environmental stresses while the latter was a function of stresses such as overcrowding. Howard's argument was rooted in group selection and was criticized by Murray (1967) who showed that migration could be produced by Darwinian selection. Lidicker (1962) postulated a system similar to Howard's and suggested that (individual) selection could favor dispersal before overcrowding because (1) it would lead to more mating contacts, (2) it would lead to increased recombination and hence more genetically diverse young (cf. Williams,

1975), and (3) perhaps most importantly, individuals could avoid the perils of population crashes by leaving before they occurred. In an updated version of the hypothesis, Lidicker (1975) used the terms "saturation" and "pre-saturation" dispersal, respectively, which were similar although not identical to Howard's "environmental" and "innate." Saturation dispersal resulted from spillover of individuals is a population near carrying capacity while presaturation dispersal could result in populations being regulated below carrying capacity. Lidicker (1975) also added increased efficiency of resource use by dispersers as an additional selective agent, and summarized evidence from a number of rodent species indicating that presaturation migration occurred. Chitty (1967) proposed that the genetic composition with respect to dispersers and sedentary individuals changes profoundly during population cycling and that the driving force behind the demographic machinery is spacing (or agonistic) behavior. Chitty's hypothesis has been the central theme in the work of Krebs and his students (Krebs et al., 1973).

A major prediction of presaturation dispersal and of Chitty's hypothesis is that migration from a cycling population will occur during the increase phase. (Here it is necessary to distinguish between seasonal fluctuations in numbers and longer-term cyclical fluctuations which are addressed here.) Various voles (Microtus) show dramatic fluctuations at intervals of 2–5 years, and it is in voles that Krebs and co-workers have studied migration (Myers and Krebs, 1971; Krebs et al., 1973, 1976; Hilborn and Krebs, 1976). The basic data were the rates of invasion of trapped out areas from neighboring populations, and the results were similar for three species of Microtus, M. pennsylvanicus and M. ochrogaster in Indiana and M. townsendii in British Columbia. In all three, dispersal rates were highest in populations in the increasing phase of the population growth cycle as opposed to the peak or declining phases. The greater the rate of increase, the more dispersers. In addition M. californicus also shows presaturation migration (Lidicker and Anderson, 1962; Krebs, 1966; Lidicker, 1973). Clear evidence of presaturation dispersal has also been found in other species such as Mus musculus (DeLong, 1967; Anderson, 1970) and Clethrionomys glareolus (Kozakiewicz, 1976).

Population fluctuations and migrations are most notorious in the Norwegian lemming, Lemmus lemmus. Studies have shown, however, two sorts of migrations in lemmings, a seasonal movement from winter drier habitats to summer wetter ones and emigrations at high population densities (Kalela, 1961; Kalela et al., 1971; Curry-Lindahl, 1962; Clough, 1965, 1968). When appropriate topographical features concentrate high density emigrants, the famous mass migrations occur (Curry-Lindahl, 1962). Whether or not emigration is saturation or presaturation is difficult to determine from the available observations, although the latter seems likely. Curry-Lindahl (1962) reports, for example, that emigration occurs before food shortage, and Clough (1968) implies the same. More research to test the Chitty–Lidicker hypotheses is clearly needed. Lemmings are highly aggressive (Clough, 1965) which is what Chitty predicted would be the driving force of fluctuations.

Of other rodents which migrate, the most frequently studied are mice of the genus *Peromyscus*. Stickel (1968) summarizes much of the older literature describing dispersal in various species. In general *Peromyscus* does not undergo the extensive population cycling of *Microtus* and mechanisms for stabilizing density seem better developed (Crowell, 1973). As a result migration is less conspicuous than in *Microtus*. Crowell (1973), for example, observed that *P. maniculatus* was a less successful colonizer of Maine islands than *M. pennsylvanicus* with *Clethrionomys gapperi* less successful still. *Peromyscus leucopus* on islands in Lake Opinicon, Ontario dispersed from island to island and from mainland to islands by swimming for 100 feet or more (Sheppe, 1965a,b), but water temperatures here were 21–27°C as opposed to 16°C or less in Maine. *Microtus* in Maine could swim across the water barriers. *Peromyscus polionotus* in Florida, a mouse of pioneer and secondary habitats, dispersed for distances of several hundred meters (Smith, 1968). Seasonal migrations also occur. In British Columbia subordinate males of *P. maniculatus* disperse in the spring (Fairbairn, 1977) and juveniles after breeding (Sullivan, 1977), and *P. boylei* in the Sierra of California changes habitats in spring and fall (Storer *et al.*, 1944).

Some data are also available from other species. On the coast of Texas the cotton rat *Sigmodon hispidus* and the harvest mouse *Reithrodontomys fulvescens* show density-dependent dispersal which is synchronized with seasonally varying environmental conditions (Joule and Cameron, 1975). Because of this density dependence, the dispersal is likely of the saturation rather than the presaturation type. The desert pocket mouse *Perognathus formosus* in the Mojave Desert of Nevada dispersed for an average distance of approximately 1400 feet, but relation to density was not determined (French *et al.*, 1968). A spectacular, apparently presaturation, migration is performed by *Rattus villosissimus* in the deserts of central Australia where individuals may move hundreds of miles (Newsome, 1975). In contrast, distances of 300 meters are significant barriers to dispersal in pikas, *Ochotona princeps*, of the Sierra Nevada of California (Smith, 1974).

Because rodents are cryptic, there have been few attempts to observe the behavior of migrants while on the move. Clough (1965, 1968) does describe the behavior of emigrating lemmings during an outbreak in central Norway and characterizes migrants as ignoring impinging stimuli, especially other lemmings, with which they would normally interact. This undistract-ableness was one of the criteria used to identify insect migrants by Kennedy (1961a) and implies that lemming migration in the outbreak phase indeed represents specialized behavior. Myers and Krebs (1971) assessed the behavior of dispersing males of *Microtus pennsylvanicus* and *M. ochrogaster* experimentally by examining agonistic interactions and exploratory activity in a laboratory "open field" (an open black wooden box) and a complex vertically tiered maze. Dispersing *M. pennsylvanicus* males at peak densities were more aggressive and in both species dispersers were less active and explored less. Residents apparently spend more time exploring to familiarize themselves with their territories.

As indicated in the discussion of insect migration, migrants are also colonists and good colonizers are characterized by being "r-strategists" with high reproductive value. In *Microtus* as in insects migrants tend to be young prereproductive animals with the highest reproductive potential as the result of both age and genetic influences (Krebs *et al.*, 1973). In general, it seems to be characteristic of small mammals that dispersers are juveniles and young adults with high reproductive value (Stickel, 1968; Lidicker, 1975). Exceptions do occur, however. In *Clethrionomys glareolus* in Poland migrants are predominantly older individuals, but those showing high reproductive activity (Kozakiewicz, 1976). In this species juveniles may remain behind and occupy vacated territories, selection favoring adults putting young in more than one place. A similar system seems to operate in prairie dogs (*Cynomys ludovicianus*) (King, 1955). In *Sigmodon hispidus* both age and sex of dispersers mirrors the age structure and sex ratio of the source population (Joule and Cameron, 1975), but in *Reithrodontomys fulvescens* males show the same age structure as the source but female migrants are skewed toward younger individuals as expected of colonists. Sex differences also occur seasonally. The winter migration away from wetter areas by *C. glareolus* is primarily by older males (Kozakiewicz, 1976), while in *Peromyscus maniculatus* in British Columbia, spring movements are primarily by smaller subordinate males (Fairbairn, 1977).

A key question of course is, are there genetic differences between migrants and residents? Such differences are implied in the Chitty, Howell, and Lidicker models. Most evidence is indirect but suggestive. For example, the heterozygote Tf^c/Tf^E at the transferrin locus is overrepresented in dispersing female voles during the population increase phase (Myers and Krebs, 1971; Kerbs *et al.*, 1973); these females also have a slight reproductive advantage over other Tf genotypes. Another approach was tried by Hilborn (1975) who examined dispersal tendency among presumptive siblings in four species of *Microtus*. Hilborn assumed that trapped individuals of similar location and size were likely members of sibships and found that these "siblings" tended to disperse together. Hilborn recognized the inherent problems in this approach, but if groupings in fact were sibs, the evidence suggests a major heritable component of dispersal behavior.

Rasmuson *et al.* (1977) have made a direct assessment of the heritability of behavior. They examined a northern cycling and a southern noncycling population of *Microtus agrestis* in Sweden and measured activity in the laboratory by determining the number of passages between sections of a laboratory enclosure. Animals were measured both singly and in groups. All measurements showed higher activity in voles from the northern cycling population and as would be expected for dispersers, were higher for group behavior and for males in northern but not southern animals. The heritability of the behavior was demonstrated by the similar behavior of wild trapped animals and their laboratory bred offspring. High proportions of additive genetic variance are suggested by the fact that hybrids between the two populations showed intermediate behavior. Intraspecific genetically controlled behavioral differences thus seem important in cycling and noncycling vole populations.

For assessing the influence of differing selection pressures on small mammal migration, islands are potentially important natural laboratories (Lidicker, 1975). In a study of *Peromyscus maniculatus* on islands off British Columbia, Redfield (1976) noted little dispersal between islands. This population was examined in greater detail by Sullivan (1977). He found that island mice dispersed at a lower rate than those on the adjacent mainland, but juveniles were recruited into the source population whereas they were not on the mainland, and that island adult males showed little aggression toward juveniles. Island mice were also larger, but on the smallest island grew faster and reproduced at higher rates, contrary to predictions of r–K theory, apparently in response to higher mortality in this particular population.

Island and mainland *Microtus* were studied in Massachusetts by Tamarin (1977a) where the larger *M. breweri* occurs on small (2.6 km^2) Muskegat island and *M. pennsylvanicus* on the adjacent mainland. On the island, *M. breweri* dispersers were a random sample of the population and moved only short distances although their home ranges were larger than for mainland *M. pennsylvanicus*. The latter, on the other hand, moved longer distances and showed a deficiency of male dispersers in the summer and an excess in the winter. Tamarin suggests that dispersal in island *M. breweri* represents simple spillover from excess reproduction while in mainland *M. pennsylvanicus* it is an evolved (presaturation) adaptation leading to population cycles. *M. breweri* showed evidence of greater K-selection in a number of traits including larger size at sexual maturity and smaller litters than *M. pennsylvanicus* (Tamarin, 1977b). In both species, however, dispersers were in reproductive condition (colonizers) to a greater extent than residents. Species on small islands in Great Britain such as *Apodemus sylvaticus*, *Clethrionomys glareolus*, and *Mus musculus* show characteristics of island populations such as large body size, high densities, and short breeding seasons (Berry, 1964; Berry and Jakobson, 1974; Jewell, 1966), but have not been compared to mainland populations with respect to dispersal.

Altitudinal variation in dispersal and social behavior has been studied in marmots (*Marmota*) by Barash (1973 a,b, 1974). In all species dispersal occurs in yearlings and appears to be initiated by agonistic behavior of the adults (cf. also Armitage and Downhower, 1974). The woodchuck (*M. monax*) occurs at low altitude where there is a long growing season; it is a highly aggressive territorial species with no social behavior beyond the family group. Yearlings disperse during the first year and may reach sexual maturity as 1-year-olds. The yellow-bellied marmot (*M. flaviventris*) occurs at intermediate altitudes with shorter growing seasons and forms colonies although individual home ranges are maintained. Aggressive behavior is less than in the woodchuck, dispersal is by yearlings, and sexual maturity occurs at 2 years. Finally, the high altitude Olympic marmot (*M. olympus*) experiences an extremely short growing season, is highly social with little aggression, and dispersal occurs at 2 years with sexual maturity at 3 years. *Marmota flaviventris* at higher altitudes converges toward *M. olympus* in both social and dispersal behavior. The social system of *M. olympus* is flexible with respect to dispersal. At high colony density juveniles experience much social interaction with other colony members (especially greetings), gravitate to the edge of the colony, and eventually disperse; at low density there is less social interaction and less dispersal with the result that colony size increases. The *M. olympus* system is thus keyed to local population size. In some areas *M. flaviventris* dispersers may aggregate in less favorable habitats as satellite populations (Armitage and Downhower, 1974). These populations are less stable than colonies and dispersal occurs in young animals less likely to reach sexual maturity than yearlings. Dispersal characteristics of marmots are, therefore, subject to short-term environmental variation as well as apparently responding to selection, although evidence concerning their heritability is lacking.

From all the above studies it is apparent that a number of environmental factors influence migration in small mammals. Of these social interactions are probably the most important, but as yet there is no conclusive evidence for Chitty's notion that spacing or agonistic behavior drives the demographic machinery. The evidence available is often conflicting, and data are needed specifically on interactions between and on the ontogeny of individual behavioral phenotypes (Bekoff, 1977). It may not in fact be agonistic behavior per se that influences dispersal, but the number of social contacts during development deriving from differing levels of agonism. In Bekoff's coyotes both highly dominant and lowly subordinate individuals had few social contacts. Differences in numbers of greetings in high and low density Olympic marmots, leading apparently to different dispersal rates, are a second example. Bekoff suggests examining social interactions of individual littermates so that the relation between the behavioral precursors of dispersal and the characteristics of dispersers with respect to variables such as age, sex, and social standing can be properly assessed. It is also important to realize that social behavior is itself responsive to environmental variation (e.g., less aggressive island mice) and is, therefore, a proximate rather than ultimate driving mechanism.

The second part of Chitty's hypothesis predicts genetic differences between sedentary and dispersing individuals. Certain genotypes such as Tf^c/Tf^E are overrepresented in dispersing female *Microtus* and appear to have higher reproductive capacity (increased fitness?) suggesting genetic influences (Myers and Krebs, 1971; Krebs *et al.*, 1973). Island–mainland differences (Sullivan, 1977; Tamarin, 1977) are also suggestive. Direct evidence of the heritability of dispersal traits, at least as measured in the laboratory, is provided by Rasmuson *et al.* (1977) for *M. agrestis.* But the heritable nature of dispersal has yet to be proved for most species because it is necessary both to demonstrate genetic differences between dispersers and nondispersers *and* to demonstrate that these differences are causally related to dispersal tendency in individuals differing only in genotype. As Bekoff indicates, this requires knowledge of behavioral ontogeny, and since the evidence strongly implies that dispersal is a polygenic trait (Rasmuson, *et al.*, 1977; cf. also insect migration), the application of quantitative genetic techniques (Dingle, 1974; Dingle *et al.*, 1977). Such techniques demand metric characters and control of environmental variance (Falconer, 1960).

Thus the development of suitable indices of dispersal, such as tethered flight in insects, which could be measured under controlled laboratory conditions should receive high priority. The work of Rasmuson *et al.* (1977) represents an excellent start and versions of the maze used by Myers and Krebs (1971) or of standard wheel running methods might also be tried. Arguments that such laboratory measurements do not truly reflect behavior in the field miss the point which is to provide an experimental correlate so that appropriate genetic techniques such as selection and parent–offspring regression can be applied, and the genetic variances determined. Only in this way can the adaptive nature of dispersal be unambiguously determined since adaptation, as Darwin noted, requires heritable traits. The fact that many small mammals can be laboratory reared opens up a wealth of possibilities for genetic and evolutionary studies.

Finally, it is worth mention that little is known of migration mechanisms in tropical small mammals. Population processes in some are similar to temperate species (Fleming, 1971), but dispersal has yet to be carefully studied. An initial prediction might be that dispersal will be greater in populations and species in seasonal (wet and dry) than in aseasonal habitats.

X. Discussion: The Evolution of Migration

In attempting to arrive at general principles concerning the evolution of migration, one at first is likely to feel overwhelmed at the bewildering diversity of patterns and processess. Is there indeed any common ground for understanding the movements of a vertically migrating copepod, a Pacific salmon, a juvenile vole, or a trans-Saharan flying bird? Can we put order into our thinking about the selective mechanisms involved? Or is the migration of each species unique requiring a unique explanation? Such questions have been examined both theoretically and with models derived from empirical comparisons.

A. THEORETICAL MODELS

Ultimately selection acting on migration is a function of the relative survival and subsequent reproductive success of migrant and nonmigrant individuals. In the simplest sense this can be expressed as the relative replacement rates (R_0 of the population ecologist) of the two types. Replacement rate is a function of survivorship, l_x, and the birth rate, m_x, and is expressed as $R_0 = \Sigma l_x m_x$ over the reproductive life of the individual. Clearly if R_0 (migrant) $> R_0$ (nonmigrant) within a population, migrants will be favored, and the key to understanding the adaptive value of migration is understanding the selective forces acting on l_x and m_x. As I have repeatedly emphasized, this makes migration an integral part of life history strategies. Migration will be favored if survival and reproduction are greater in a new habitat in spite of the risks of migrating; this frequently occurs because the risks of remaining sedentary increase as a consequence of habitat changes (Southwood, 1977; Baker, 1978). The relative stability and suitability of the habitats are functions of generation time (H/τ of Southwood, 1977). What then constitutes an optimal migration strategy?

The problem has been considered with respect to optimization theory by Cohen (1967) and Parker and Stuart (1976). Cohen examines migration as a problem of optimal choice between alternatives with randomly varying outcomes. Migration can occur as a "pure strategy," with all individuals migrating, or as a mixed strategy with only a portion doing so. Conversely remaining sedentary is also a possible pure strategy. A migratory pure strategy is favored when the variance in the viability coefficient of non-migrants increases, i.e., risks of remaining sedentary become greater, and vice versa. If viability coefficients for migrants and nonmigrants are independent, then a mixed strategy is more likely, for example, where the variability in winter survival between years is high. When some environmental cue allows prediction of unfavorability, the population will be expected to respond to that cue by emigrating. Seasonal migration in response to photoperiod is an example. As a final conclusion, Cohen's model indicates that migration can only be optimal when survival and reproduction at any one place do not remain constant; if they do organisms gain no advantage by changing habitats. However, while agreeing that migration is clearly advantageous where habitats are transient and patchy, Hamilton and May (1977) demonstrate that constancy is not necessarily a constraint and that dispersal in stable habitats can also be selectively advantageous.

Parker and Stuart (1976) devise models that consider emigration thresholds from resource patches encountered by given search strategies. They use the concept of evolutionarily stable strategies (ESS's), introduced by Maynard Smith (1974), and consider optimal investment durations, gain accumulated, and search costs between patches. Various pure and mixed strategies are possible depending on factors such as competition or resource sharing and ability to assess resources (as in Cohen's models). As an example, for patches with decreasing returns and increasing investment, the ESS is usually for all to stay until a critical threshold is reached, whereupon individuals should leave at a rate which balances the value of staying and of leaving. The model is applied to male dung flies, where the resource is availability of females, but has yet to be considered for more explicitly migratory situations (although cf. Baker, 1978). The colonization dispersals of rodents, for example, may be a possible case for wider applicability.

In a comprehensive attempt to analyze dispersal, Roff (1975) used computer simulation to determine the influence of heterogeneous environments on populations in which dispersal is genetically determined. In both simple and quantitative genetic models changes in the probability of dispersal produced changes in population size, spatial distribution of genotype frequencies, and proportion of dispersers per habitat. Roff found, however, that increasing environmental stability did not always decrease dispersal rate, nor was increased dispersal necessarily the result of an increase in dispersal genotypes. The influence of environmental stability was dependent on the nature of the genetic mechanism influencing dispersal (polygenic or "simple"), the type of dispersal strategy (density dependent or independent), and the form of the environmental changes occurring. The models also demonstrated a major role for sexual reproduction in determining genotype frequencies. In heterogeneous environments dispersers are generally at an advantage when environmental variance is high (cf. also Cohen, 1967), but with long-term stability there is a continuous loss of dispersal genotypes which are not replaced. The presence of dispersers permits colonization of new areas and persistence of the population as a whole (cf. also Taylor and Taylor, 1977). Somewhat similar conclusions with respect to populations were reached by Gadgil (1971). The various dispersal strategies of the gerrids studied by Vepsäläinen (1978) are good examples of outcomes predicted by Roff's models.

B. Some Empirical Comparisons

The most obvious conclusion from comparing diverse organisms is that migration strategies are extraordinarily flexible. Movement is a fundamental biological response to adversity (Taylor and Taylor, 1977), but the timing, duration, and other variables are subject to a variety of intrinsic and extrinsic constraints. To take an obvious but important example, long distance annual movements are most likely in large, long-lived, mobile animals such as whales and birds. Similar patterns also evolve in similar habitats; desert locusts, desert birds, and desert mammals all display migrations to ensure breeding in areas of rainfall. Other common patterns are the north–south migrations of birds, butterflies, bats, and cervids; annual altitudinal migrations of birds and mammals; and the inshore–offshore breeding migrations of whales, fish, and crustaceans. Nevertheless selective regimes do differ, and it is hardly surprising to observe differences in pattern between ocean and desert. Proximate and ultimate selective forces must also be kept distinct. Thus in Cox's (1968) competition model for the evolution of bird migration is it the case that competition leads to migration or to the occupation of marginal habitats by some species which then adopt a migration strategy? There is evidence that many migratory birds may be fugitive species in this sense (Karr, 1976), and some migratory insects may be also (Southwood, 1962). No single factor theory of migration is thus ever likely to be adequate.

Some generalizations nevertheless seem possible. Cohen (1967) and Ziegler (1976) have suggested that if cues for habitat variation are absent or unreliable, genetic determinism should be favored. Greater environmental determinism should evolve in the presence of reliable cues such as photoperiod which, if they occur at the approriate time in the life cycle, can lead to behaviorally or morphologically specialized morphs (e.g., brachyptery–macroptery in insects). When cues can occur at any time during the life cycle, a more generalized emigration response may evolve with movement a function of current conditions as in the *Tribolium* studied by Ziegler (1976) or in various temperate zone birds in winter (Lack, 1968). With respect to optimality theory, the cost/benefit ratio for migration should vary predictably from season to season, but relatively unpredictably from year to year. The former selects for physiological responses, e.g., to photoperiod, and the latter for behavioral flexibility as indeed has been observed (Lack, 1968).

In an extension of some ideas of Southwood (1977) and Solbreck (1978), I have attempted in Fig. 6 to summarize the role of migration in life histories. Included are diapause and hibernation as additional responses to environmental adversity. In general organisms can choose to breed "now" or "later" either "here" or in "near" and /or "far" environments. Immediate breeding in the current habitat requires no migration and often leads to strategies such as winglessness, asexual reproduction, and territorial behavior (used here broadly to include agonistic dominance). Variations in space and time parameters lead to other available strategies. Note that these are not mutually exclusive; for example, delayed maturity (and large body size) and round trip migration can occur in the same species (e.g., whales, sea birds). Note further that the same species, populations, or even individuals can adopt different tactics in different places at different times. These can be sequential (e.g., round trip bird migrants which are also territorial; insects which lose their wings following migratory flight) or more or less haphazard depending on circumstances. The model is clearly not all encompassing, but may serve as a useful generalization. The reader will have further conceptual modifications.

Various migration strategies may also serve as examples for the model of evolution described by Slobodkin and Rappaport (1974). They conceive of evolution as an "existential game" in which the strategy is to stay in the game (since there can be no external "payoffs") by minimizing the "stakes" of an environmental perturbation. Migration would seem to be the quintessential response in such a game, since successful evolution requires maintaining flexibility of response in the most parsimonius way. Thus organisms faced with short-term fluctuations should develop rapid response systems dependent primarily on immediate behavioral adjustments. Longer-term predictable fluctuations such as seasons will result in the evolution of physiologic responses to cues such as photoperiod and a deeper "commitment" on the part of the organism. Finally, essentially permanent alterations will result in genetic changes leading up to mor-

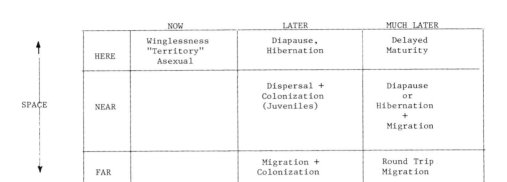

Fig. 6. Generalized "reproductive matrix" for incorporation of migration and other "escape" behaviors (hibernation, diapause) into life histories. Organisms can breed "here and now" or can postpone to avoid deteriorating habitats in space, time, or both. Categories are not mutually exclusive for species as different populations at different times and places adopt various strategies, e.g., winter diapause winged and summer wingless gerrids. Also the same populations may adopt different strategies at different times, e.g., migrant birds territorial in the breeding season. (See text for further discussion.)

phological variation such as wing polymorphism or brachyptery. Suitable adjustments in environmental periodicity should produce genetic variation influencing adaptive responses ranging through behavioral, physiological, and finally morphological mechanisms. Clearly there are numerous examples of migratory responses which have evolved at all these levels (Table II).

C. Prospects

I have tried throughout this chapter to point out areas where I felt additional research effort should be concentrated to obtain the data to fill lacunae in our knowledge. Three areas which seem to be of special importance are listed below.

1. We need more tests of theory and especially ways to test theory experimentally. For the latter, there must be suitable indices of migratory capability. For insects tethered flight is such an index if not always a satisfactory one. In spite of its limitations, however, its use has resulted in significant advances in our studies of insect migration. Suitable indices ought to be possible for mammals and fish, and a start has indeed been made (Myers and Krebs, 1971; Rasmuson *et al.*, 1977). It is also possible to fly birds (Tucker, 1972) and bats (Pennycuick, 1971) in the laboratory which

suggests indices for these animals may be possible if tedious to obtain. Investigators have already taken advantage of *Zugunruhe* in birds for studies of physiology and orientation, and it may also be possible to do so in consideration of ecological and evolutionary constraints.

2. We need to examine the interaction between genetic and environmental influences in more detail. Since migration is likely to be polygenic in most cases, we need a metric character to index it, another reason for appropriate indices. Both empirical data (Dingle *et al.*, 1977; Istock, 1978) and theoretical models (Roff, 1975) indicate the importance of understanding the underlying genetic architecture of migration. That architecture can be understood only through cognizance of the appropriate environmental context for the expression of genetic variance.

3. We need to know more about the role of migration in life history strategies. The concept of migration as an integral part of life histories is relatively new, but its importance cannot be overemphasized. New impetus is given to its development by the recent elaboration of the role of migration in population dynamics in both space and time (Krebs *et al.*, 1973; Southwood, 1977; Taylor and Taylor, 1977). Here especially we seem on the threshold of exciting discoveries and conceptual advances, and the need for data and theory is clear. Our understanding will be much more profound if we can also include the role of genetic influences.

It would be foolish to underestimate the difficulties in undertaking such studies as I have adumbrated, and I do not do so. I make no apologies, however, for the complexities of nature, since with those complexities come exciting challenges to the abilities, energies, and intuition of ecologists and evolutionary biologists.

XI. Addendum

A. Vertical Migration

Both the predation and metabolic models of vertical migration have received additional support. O'Brien (1979) has reviewed studies on the abilities of various fish species to detect and capture zooplankton. The distance at which plankters can be detected by these fish decreases markedly with decreased light intensity. In a Kansas lake several species of zooplankton migrated during daylight hours to a depth where the prevailing light levels were such that the common fish predator, the white crappie, had difficulty locating planktonic prey (Wright *et al.*, 1979). A simulation model further showed that all but the smallest zooplankters reduce their vulnerability to predation by migrating downward to lower light intensities during the day. The smallest species, or those capable of avoiding predators by evasive tactics, often display little or no vertical migration.

Giguère and Dill (1979) provide further support for a metabolic model (or energy-efficiency model) in their study of a planktonic predator, the larva of the phantom midge *Chaoborus trivittatus*. They developed a model based on a thorough study of respiration rates (Swift, 1976), assimilation efficiencies, and costs of swimming and prey capture. Their analysis was based on temperatures of 16°C at the surface, where feeding on the chief prey item, the copepod *Diaptomus kenai*, took place, and 5°C in deep water strata. They found that *Chaoborus* larvae should derive maximum benefit from vertical migration when prey density was low, but should otherwise feed near the surface. Seasonal changes in migration patterns, which could be accounted for by the model, were found in a small fishless lake in British Columbia. In the absence of fish, a predation hypothesis evidently cannot account for vertical migration patterns in *C. trivittatus*.

An interesting case of tidal, rather than diurnal, vertical migration is described by Cronin and Forward (1979). Field-caught larvae of the estuarine crab *Rhithropanopeus harrisii* display a tidal rhythm of vertical migration, while those reared in the laboratory do not. The migration is evidently cued by changing conditions in the estuary during the tidal cycle. Estuaries in which *R. harrisii* occurs have stratified currents with a high salinity current of net landward flow underlying upper seaward-flowing water. The larvae move upward at flood tide and are carried up the estuary. As the ebb begins, they move to the lower current minimizing their seaward movement. Cronin and Forward postulate that by partitioning time between layers the larvae minimize net horizontal movement, assuring that they remain in favorable habitats near the parent population while still undertaking some dispersal.

B. DEMERSAL CRUSTACEANS

Migratory and nomadic movements of palinurid lobsters have been extensively summarized by Herrnkind (1979). These movements are of several sorts, including those of juveniles from nursery areas dominated by adults, of nomadic adults, of females to areas dominated by males, of reproductive females to areas where eggs are released, and of adults to deeper water in the autumn. The relevant selective forces vary with the situation. Juveniles may need to forage more widely and find larger shelters as they become larger, and thus move out of *Thalassia* beds to seek reef habitats. The movements of ovigerous females to shelf edges often results in the release of larvae into currents flowing in the appropriate direction for future settlement of the young. Finally, the mass autumnal migrations of species such as *Panulirus argus* in the Caribbean are triggered by storms causing significant decreases in water temperatures (Kanciruk and Herrnkind, 1978). These movements to warmer water are evidently adaptive because the lobsters become moribund, cannot feed effectively, and are unable to complete molting when temperatures drop below 12°–15°C, as they sometimes do during winter in inshore areas (Herrnkind and Kanciruk, 1978). An additional advantage to all spring lobster movements may be dispersal with the colonization by individuals of new and more favorable habitats.

Migratory movements of the blue crab, *Callinectes sapidus*, on the west coast of Florida have been described by Oesterling and Evink (1977). Males tend to remain in their "home estuaries," but females display a rather dramatic along-shore postmating migration northward along the coast. The greatest number terminate their movement to spawn in the Apalachicola Bay region at the northern end of the Gulf of Mexico. This area is apparently particularly favorable for breeding because of nutrient outflow from the Apalachicola River. Furthermore zoeas are apparently flushed from the bay into the Eastern Gulf Currents which usually deposit them back at the Florida coast when they are ready to settle out as postlarval crabs. The whole reproduction–migration system therefore seems appropriately adapted to conditions both on the coast and in the Gulf and is analogous to the migration of many pelagic fishes (see Fig. 4).

C. INSECTS

Specialized migratory behavior has been nicely demonstrated for the rutherglen bug, *Nysius vinitor* (Hemiptera; Lygaeidae) (Kehat and Wyndham, 1973a,b), and for the saltmarsh aphid, *Pemphigus treherni* (Foster and Treherne, 1978; Foster, 1978). As food quality deteriorates, increasing numbers of *N. vinitor* take off on prolonged flights, but immature females are the predominant migrants of this species. Flight by young adults is, of course, characteristic of insect migrants (see Fig. 2). In addition to displaying greater flight activity, sexually immature females of *N. vinitor* were positively phototactic, a response which would be advantageous during long-distance dispersal. *Pemphigus treherni* utilizes the saltmarsh sea aster as its summer host where it lives on the root systems. It is particularly interesting because the first instar is the dispersal stage. These instars move up the rootstalk to the soil surface where they are picked up by the tide and distributed over the marsh. The photopositive response of these dispersers is reversed by 30 min of floating on sea water; the resulting photonegativity results in their moving downward to the roots of new hosts. This reversal of response to light is analogous to changes in wavelength sensitivity observed in settling black bean aphids (Kennedy *et al.*, 1961). As in other aphids, movement to the overwintering host (*Populus*) takes place by flight.

Another interesting saltmarsh migrant is the planthopper, *Prokelisia marginata* (Homoptera; Delphacidae), studied by Denno and Grissel (1979). Populations consist of both brachypters and macropters and show geographic variation in the relative proportions of each form. The proportion of macropters is significantly correlated with the proportion of unstable patches of the chief host plant, the grass *Spartina alterniflora*. Wing form is determined developmentally by factors such as crowding, but different populations display varying degrees of sensitivity to these inputs. Selection has evidently altered the thresholds at which the developmental switch occurs.

Differences in migratory flight among species have been demonstrated for *Oxycarenus* spp., bugs that feed on malvaceous hosts (Adu-Mensah and Kumar, 1977). As is the case with the malvaceous feeding *Dysdercus* spp., it is the species confined to trees with large but only seasonally available fruit crops which undertakes the most long-distance migration. Species feeding on plants whose fruits are more continuously available, albeit less abundant, fly only for short periods.

Two recent studies on aphids and crickets, respectively, demonstrate that differences in actual or potential migratory tendency within species are at least partially influenced by gene differences. Working with the pea aphid, *Acyrthosiphon pisum*, Lamb and MacKay (1979) demonstrate differential production of alates in clones from various alfalfa fields in southwestern Ontario. Spatial and temporal patterns found in nine natural populations reflected differences in the frequencies of many genetically distinct clones as determined by laboratory tests of alate production. Selection arising from differential adaptiveness of immigrant and resident clones could account for the results. An important environmental component of alate production was maternal age with older females producing fewer winged offspring (MacKay and Lamb, 1979). Harrison (1979) demonstrated by means of crosses in the laboratory that differences between long- and short-winged crickets, *Gryllus pennsylvanicus*, are the result of gene influences at many loci. The maintenance of flight polymorphism is probably a result of a balance between superior survivorship and fecundity of the short-winged form and the enhanced dispersal capability of their long-winged counterparts.

Finally, the intimate relation between migration and life history has been explored by Cheke (1978) in the desert locust, *Schistocerca gregaria*. The gregarious migratory form has a lower fecundity than the solitary, but can still attain much faster rates of population increase because of synchronous breeding and more rapid maturation. Along with greater flight capability, these are clearly adaptations for the nomadic strategy of the gregarious phase. Cheke suggests that the advantages of being solitary may occur when predation is low, creating less selective pressure for rapid development, or when the population experiences a long drought. Whatever the aspects of selection, the importance of behavior to life histories is once more evident.

D. Fish

Migration strategies in fish have been reviewed by Northcote (1978) who reaches conclusions similar to mine as discussed above in this review. Specifically Northcote suggests that fish migration has evolved to optimize feeding, to avoid unfavorable conditions, to optimize reproductive success, and to a lesser extent to promote colonization. The migrations of many species in fact contain elements of some or all of these strategies. Migrations tend to occur especially in environments or combinations of environments subject to marked temporal fluctuations; in this, of course, fish resemble several other groups in their migratory strategies. As an example Northcote notes the relatively higher proportions of migrants among certain temperate zone groups, allowing for the exploitation of habitats only seasonally available.

In an interesting review Holm and Naevdal (1978) summarize a number of studies of quantitative genetic variation in fish life history statistics, e.g., growth, size, and reproductive characteristics. Cases of dominance, epistasis, and additive variance have all been found. Estimates of additive genetic variance (heritability) are often sufficiently high to suggest that these traits would be quite responsive to selection. Since many of the characters studied also interact with migration in shaping life history strategies, there seems to be considerable potential for important and significant evolutionary studies. At present, however, with the exception of a few studies such as that of Schaffer and Elson (1975), there seems to be little melding of genetic life history analysis with studies of migration (or other behavior) so that comprehensive assessment of the evolution of fish life histories remains very much in its infancy.

E. Amphibians

The migrations of the newt, *Notophthalmus viridescens*, in mountain ponds in Virginia have been studied in detail by Gill (1978a,b). Adults of this species show classic amphibian migrations, moving out of the ponds to

88   Hugh Dingle

terrestrial hibernation sites in the fall with return movements to the aquatic habitats in the spring to breed. These adults show extreme site philopatry and strong homing behavior. Selection evidently favors homing because faithfully returning adults have the best chance for maximum reproductive success; wandering adults are likely to arrive at ponds with lower reproductive potential. The ponds, however, have a limited duration. As a result, juveniles, in contrast to adults, are selected for dispersal to other ponds, as is in fact observed. The resultant constellation of locally breeding populations constitutes a "metapopulation" internally linked by the constant gene flow produced by the dispersing young. The dynamics of such metapopulations have important implications for both migration theory and population genetics, especially with respect to the differential production of dispersing and colonizing juveniles in the various breeding subpopulations.

Most efts of a Massachusetts population of *N. viridescens* migrate to ponds in the fall where they overwinter before breeding the following spring (Healy, 1975). Individuals must reach a minimum size before migrating, and rainfall, in addition to providing conditions appropriate to migration itself, promotes feeding and growth in the terrestrial stage. Years of high rainfall therefore yield more autumnal migrants. In dry years females are underrepresented among migrating efts because they must reach a larger minimum size. Larval densities influence variation in duration of migration and mean size of emigrating newly metamorphosed juveniles. Selective forces favoring fall migration of efts and determining optimum size of migrants have yet to be carefully or experimentally defined.

F. Birds

The interactions of Palearctic migrants and local breeding birds has been examined by Sinclair (1978) in the Serengeti region of East Africa (Tanzania). The arrival of migrants coincides with that of weather fronts bringing rainstorms and a local flush of insects, themselves arriving in part with the fronts. Movements of migrants in the Serengeti depends on the distribution of local rainstorms; the birds can arrive in the vicinity of storms simply by flying downwind in much the same way as many migrant insects (e.g., locusts, *Schistocerca gregaria*). Competition between migrants and residents is minimized because habitat overlap occurs only when food is superabundant as a consequence of the rains. The "fugitive" nature of migrants in their overwintering range is thus once again demonstrated.

Factors which maintain migration patterns in shorebirds have been studied by Schneider and Harrington (1980). They found that four species of migratory shorebird significantly reduced infaunal food supplies on sand and mud flats in the vicinity of Plymouth, Massachusetts, during southward passage. Depletion resulted because renewal of these resources does not occur over the late summer time interval involved. Selection pressures would then be for early departure from breeding grounds to arrive at stopover sites before depletion of food resources. Renewable resources on tropical mud flats in the Bay of Panama were not depleted by overwintering shorebirds (Schneider, 1979). Timing of migration in different ages and sexes of rufous hummingbirds is also influenced by local patterns of food availability (Gass, 1979).

The problem of nomadism in birds is considered by Anderson (1979) in a theoretical paper. He finds that adult nomadism is favored by cyclic rather than random fluctuations in local food abundance. Other factors that should favor nomadic movements are longer intervals between successive good years (especially if they exceed individual lifespans), larger clutch sizes, and higher relative juvenile survival, although general levels of survival are not critical. There is some empirical evidence to support the predictions of Anderson's model, but critical studies of marked individuals are especially needed.

Patterns of migratory restlessness of *Zugunruhe* are examined by Gwinner and Czeschlik (1978) who note that in the autumn this usually reflects the temporal course of actual migration in nature. In caged birds in spring, *Zugunruhe* often persists throughout the normal breeding season. The most reasonable hypothesis seems to be that this results from the inability to carry out normal reproductive activity under confined conditions. Caution is thus needed if *Zugunruhe* is to be used to index migration. Circadian patterns in *Zugunruhe* may nevertheless reflect daily patterns of migratory activity during appropriate seasons (Stolt, 1978).

G. Mammals

Laws (1977) has reviewed the ecology of large marine mammals in the Antarctic. With respect to whale migration he confirms that the largest species such as the blue whale tend to arrive first in Antarctic waters with subsequent waves reflecting the extent to which various species penetrate southward to colder waters. The latter in turn is correlated with body size and the energy needs of different age, size, and reproductive classes. Intensive feeding while in the Antarctic can lead to 50% weight gain to sustain these animals during their nearly complete fast during the remaining 8 months of the year. Brodie (1975) suggests that the large body size of some Antarctic whales can be specifically attributed to selection for body proportions to reduce metabolic rate and an optimal surface area for lipid (blubber) deposition. These body proportions allow migration into areas where prey is extremely dense but only seasonally and briefly available.

Studies of *Peromyscus* (Fairbairn, 1978) and bog lemmings, *Synaptomys cooperi* (Gaines *et al.*, 1979), confirm that emigrating individuals of small rodents are generally subadults. In *P. maniculatus* studied by Fairbairn, these dispersers tend to be subordinate males in the breeding season, while in the nonbreeding season both sexes are involved without regard to social pressure or genotype. Males were generally overrepresented in *S. cooperi* dispersers, but the emigrating subadult females were more likely to be in breeding condition than the more stationary members of their age group. In isolated *Microtus* populations in habitats surrounded by unfavorable areas, Abramsky and Tracey (1979) noted annual but not multiyear cycles. They observed emigration, but virtually no immigration, and postulated that local population size was regulated by the former but that the latter contributed to the generation of the frequently observed 3 to 4 year cycles.

Finally, the Chitty hypothesis for population regulation by polymorphic behavior has been thoroughly reviewed by Krebs (1978). He carefully compares it to two other major hypotheses for population regulation, involving stress and spacing behavior, respectively. Especially valuable is the evaluation of 13 predictions of the Chitty model and the difficulty of testing these with respect to Chitty's mechanisms to the exclusion of all others. Krebs notes that while the Chitty hypothesis still remains to be adequately tested, behavior and genetics are now both considered important to the understanding of population processes.

H. General Dynamics

Taylor and Taylor (1978, 1979) have further developed their notions concerning the crucial role of movement, both dispersal and congregation, in the dynamics of populations. They consider adaptive, behavior-controlled movement an alternative to the more usually considered competition in population regulation. Much of the empirical base of their studies comes from insects in which they demonstrate constant shifting of populations in space and time in ways not predicted by conventional notions of population biology. They conclude that resources are never limiting because they are never entirely depleted. The important characteristics of individuals are their abilities to locate resources and avoid competition and in these the most important attribute is behavior-controlled movement. The role of behavior in life histories is therefore paramount in this model, especially migratory behavior in the sense that I have discussed it here.

Acknowledgments

I owe thanks to many people who have shared their expertise concerning various groups with me and have guided me to references or sent me reprints or manuscripts, in particular, J. T. Enright, Jeri Dingle, Elliott Norse, Lois Saucke, Jim Karr, Leslie Johnson, Steve Fretwell, Ellen Ketterson, William Leggett, Susan Goodner, Carol Sue Williamson, Sid Gauthreaux, Kerry N. Rabenold, and Ola Fincke. Portions of the manuscript were read by Leslie Johnson, Jim Karr, and Dick Bovbjerg. To all I am most grateful. My own research on insect migration has been generously supported for 14 years by the U.S. National Science Foundation.

References

Abramsky, Z., and Tracy, C. R. (1979). *Ecology* **60**, 349–361.

Adu-Mensah, K., and Kumar, R. (1977). *Biol. J. Linn. Soc.* **9**, 349–377.

Albrecht, F. O. (1967). "Polymorphisme Phasaire et Biologie des Acridiens Migrateurs." Masson, Paris.

Aleksiuk, M. (1976). *Copeia* pp. 170–178.

Allen, J. A. (1966). *Oceanogr. Mar. Biol.* **4**, 247–266.

Allen, K. R. (1969). *In* "Symposium on Salmon and Trout in Streams" (T. G. Northcote, ed.), pp. 3–18. University of British Columbia.

Amadon, D. (1948). *Science* **108**, 705–707.

Amadon, D. (1953). *Auk* **70**, 461–469.

Anderson, M. (1980). *J. Anim. Ecol.* **49**, 175–184.

Anderson, P. K. (1970). *Symp. Zool. Soc. London* **26**, 299–325.

Anonymous (1969). "Trout of California." California Department of Fish and Game, Sacramento.

Armitage, K. B., and Downhower, J. F. (1974). *Ecology* **55**, 1233–1245.

Ashmole, N. P. (1971). *In* "Avian Biology" (D. S. Farner and J. R. King, eds.), Vol. 1, pp. 223–286. Academic Press, New York.

Bainbridge, R. (1953). *J. Mar. Biol. Assoc. U. K.* **32**, 385–447.

Bainbridge, R. (1961). *In* "The Physiology of Crustacea" (T. H. Waterman, ed.), Vol. 2, pp. 431–463. Academic Press, New York.

Baker, J. R. (1938). *In* "Evolution: Essays on Aspects of Evolutionary Biology" (G. R. de Beer, ed.), pp. 161–177. Oxford Univ. Press, London and New York.

Baker, M. C., and Baker, A. E. M. (1973). *Ecol. Monogr.* **43**, 193–212.

Baker, R. R. (1968). *Philos. Trans. R. Soc. London* **253**, 309–341.

Baker, R. R. (1978). "The Evolutionary Ecology of Animal Migration." Holmes & Meier, New York.

Barash, D. P. (1973a). *Anim. Behav. Monogr.* **6**, 171–245.

Barash, D. P. (1973b). *Anim. Behav.* **21**, 579–584.

Barash, D. P. (1974). *Science* **185**, 415–420.

Bekoff, M. (1977). *Am. Nat.* **111**, 715–732.

Bell, M. A. (1976). *Syst. Zool.* **25**, 211–227.

Bennett, R. B., and Borden, J. H. (1971). *Ann. Entomol. Soc. Am.* **64**, 1273–1286.

Berry, R. J. (1964). *Evolution* **18**, 468–483.

Berry, R. J., and Jakobson, M. E. (1974). *J. Zool.* **173**, 341–354.

Berthold, P. (1975). In "Avian Biology" (D. S. Farner and J. R. King, eds.), Vol. 5, pp. 77–128. Academic Press, New York.
Bhakthan, N. M. G., Nair, K. K., and Borden, J. H. (1971). Can. J. Zool. 49, 85–89.
Blackman, R. L. (1971). Bull. Entomol. Res. 60, 533–546.
Blakley, N. R. (1977). Ph.D. Thesis, University of Iowa, Iowa City.
Bliss, D. E. (1968). Am. Zool. 8, 355–392.
Bovbjerg, R. V., and Bovbjerg, A. M. (1964). Proc. Iowa Acad. Sci. 71, 511–518.
Bradshaw, S. D. (1971). J. Zool. 165, 1–25.
Brinton, E. (1967). Limnol. Oceanogr. 12, 451–483.
Brodie, P. F. (1975). Ecology 56, 152–161.
Brown, E. S. (1951). Proc. Zool. Soc. London 121, 539–545.
Brown, E. S. (1965). J. Anim. Ecol. 34, 93–107.
Brown, W. S., and Parker, W. S. (1976). Copeia, pp. 225–242.
Burt, W. H., and Grossenheider, R. P. (1976). "A Field Guide to the Mammals," 3rd ed. Houghton, Boston, Massachusetts.
Caldwell, R. L. (1969). Ph.D. Thesis, University of Iowa, Iowa City.
Caldwell, R. L. (1974). In "The Experimental Analysis of Insect Behaviour" (L. Barton Browne, ed.), pp. 304–316. Springer-Verlag, Berlin and New York.
Caldwell, R. L., and Hegmann, J. P. (1969). Nature (London) 223, 91–92.
Caldwell, R. L., and Rankin, M. A. (1974). J. Comp. Physiol. 88, 383–394.
Cameron, A. W. (1970). J. Zool. 161, 15–23.
Campbell, I. M. (1962). Can. J. Genet. Cytol. 4, 272–288.
Campbell, I. M. (1966). In "Breeding Pest-Resistant Trees" (H. Gerhold, ed.), pp. 129–135. Pergamon, Oxford.
Carlquist, S. (1974). "Island Biology." Columbia Univ. Press, New York.
Carr, A. (1967). In "Animal Orientation and Navigation" (R. M. Storm, ed.), pp. 35–55. Oregon State Univ. Press, Corvallis.
Carr, A. (1975). Copeia, pp. 547–555.
Carr, A., and Coleman, P. J. (1974). Nature (London) 249, 128–130.
Carscadden, J. E., and Leggett, W. C. (1975a). J. Fish. Res. Board Can. 32, 653–660.
Carscadden, J. E., and Leggett, W. C. (1975b). J. Fish. Biol. 7, 595–609.
Carter, A. (1976). Can. J. Zool. 54, 1375–1382.
Cheke, R. A. (1978). Oecologia 35, 161–171.
Chitty, D. (1967). Proc. Ecol. Soc. Aust. 2, 313–331.
Chiykowski, L. N., and Chapman, R. K. (1965). Wis. Agric. Exp. Stn. Res. Bull. 261.
Clarke, G. L. (1934). Biol. Bull. (Wood Hole, Mass.) 67, 432–455.
Clausen, C. P. (1976). Annu. Rev. Entomol. 21, 343–368.
Clough, G. C. (1965). Am. Sci. 53, 199–212.
Clough, G. C. (1968). Pap. Norw. State Game Res. Inst., Ser. 2 28, 1–50.
Cockrum, E. L. (1969). Misc. Publ. Mus. Nat. Hist., Univ. Kans. 51, 303–336.
Cody, M. L. (1971). In "Avian Biology" (D. S. Farner and J. R. King, eds.), Vol. 1, pp. 461–512. Academic Press, New York.
Cohen, D. (1967). Am. Nat. 101, 5–17.
Cole, L. C. (1954). Q. Rev. Biol. 29, 103–137.
Corbet, P. S. (1961). Proc. Zool. Soc. London 136, 1–101.
Cox, G. W. (1961). Ecology 42, 253–266.
Cox, G. W. (1968). Evolution 22, 180–192.
Cronin, T. W., and Forward, R. B., Jr. (1979). Science 205, 1020–1022.
Crowell, K. L. (1973). Am. Nat. 107, 535–558.
Curry-Lindahl, K. (1962). J. Mammal. 43, 171–184.

Cushing, D. H. (1951). *Biol. Rev. Cambridge Philos. Soc.* **26**, 158–192.
Cushing, D. H. (1955). *J. Anim. Ecol.* **24**, 137–166.
Danthanarayana, W. (1970). *Entomol. Exp. Appl.* **13**, 236–246.
Danthanarayana, W. (1976). *Oecologia* **26**, 121–132.
Darlington, P. J., Jr. (1943). *Ecol. Monogr.* **13**, 37–61.
Darlington, P. J., Jr. (1971). *In* "Adaptive Aspects of Insular Evolution" (W. L. Stern, ed.), pp. 7–15. Washington State Univ. Press, Pullman.
David, P. M. (1961). *Syst. Zool.* **10**, 10–16.
Davis, N. T. (1975). *Ann. Entomol. Soc. Am.* **68**, 710–714.
Davis, W. H. (1970). *In* "Biology of Bats" (W. A. Wimsatt, ed.), Vol. 1, pp. 266–300. Academic Press, New York.
Davis, W. H., and Hitchcock, H. B. (1965). *J. Mammal.* **46**, 296–313.
Dawbin, W. H. (1966). *In* "Whales, Dolphins, and Porpoises" (K. S. Norris, ed.), pp. 145–170. Univ. of California Press, Berkeley and Los Angeles.
de Kort, C. A. D. (1969). *Med. Lab. Entomol., Wageningen* **159**, 1–63.
DeLong, K. T. (1967). *Ecology* **48**, 611–634.
Denno, R. F. (1976). *Ecol. Entomol.* **1**, 257–266.
Denno, R. F., and Grissell, E. E. (1979). *Ecology* **60**, 221–236.
Derr, J. A. (1977) Ph.D. Thesis, Washington University, St. Louis, Missouri.
DeWilde, P. A. W. J. (1973). *Stud. Fauna Curacao* **44**, 1–138.
Dingle, H. (1965). *J. Exp. Biol.* **42**, 269–283.
Dingle, H. (1968). *Am. Nat.* **102**, 149–163.
Dingle, H. (1969). *Crustaceana* **16**, 108–110.
Dingle, H. (1972). *Science* **175**, 1327–1335.
Dingle, H. (1974). *In* "The Experimental Analysis of Insect Behaviour" (L. Barton Browne, ed.), pp. 329–342. Springer-Verlag, Berlin and New York.
Dingle, H. (1978). *In* "Evolution of Insect Migration and Diapause" (H. Dingle, ed.), pp. 254–276. Springer-Verlag, Berlin and New York.
Dingle, H., and Arora, G. (1973). *Oecologia* **12**, 119–140.
Dingle, H., and Caldwell, R. L. (1972). *Biol. Bull. (Woods Hole, Mass.)* **142**, 417–426.
Dingle, H., Brown, C. K., and Hegmann, J. P. (1977). *Am. Nat.* **111**, 1047–1059.
Dixon, A. F. G. (1971). *Sci. Prog. (Oxford)* **59**, 41–53.
Dixon, A. F. G. (1973). "Biology of Aphids." Arnold, London.
Dixon, A. F. G. (1976). *J. Anim. Ecol.* **45**, 817–830.
Dixon, A. F. G., and Wratten, S. D. (1971). *Bull. Entomol. Res.* **61**, 97–112.
Dodson, J. J., and Leggett, W. C. (1973). *J. Fish. Res. Board Can.* **30**, 1847–1860.
Dodson, J. J., and Leggett, W. C. (1974). *J. Fish. Res. Board Can.* **31**, 1607–1619.
Dole, J. W. (1965a). *Ecology* **46**, 236–255.
Dole, J. W. (1965b). *Am. Midl. Nat.* **74**, 464–478.
Dole, J. W. (1967). *Am. Midl. Nat.* **78**, 167–181.
Dole, J. W. (1968). *Ecology* **49**, 386–399.
Dole, J. W. (1971). *Copeia*, pp. 221–228.
Dorst, J. (1962). "The Migrations of Birds." Houghton, Boston, Massachusetts.
Duffey, E. (1956). *J. Anim. Ecol.* **25**, 85–111.
Edwards, F. J. (1969). *J. Insect Physiol.* **15**, 2013–2020.
Edwards, R. J. (1977). *Copeia*, pp. 770–771.
Edwards, R. Y., and Ritcey, R. W. (1956). *J. Mammal.* **37**, 486–494.
Ellis, P. E., Carlisle, D. B., and Osborne, D. J. (1965). *Science* **149**, 546–547.
Elson, P. F. (1970). *ICES/ICNAF Salmon Doc.* **70/9**, 1–9.
Elton, C. (1927). "Animal Ecology." Sidgwick and Jackson, London.

Elton, C. (1942). "Voles, Mice and Lemmings. Problems of Population Dynamics." Oxford Univ. Press, London and New York.

Enright, J. T. (1969). *Ecology* **50**, 1070–1075.

Enright, J. T. (1977a). *Limnol. Oceanogr.* **22**, 118–125.

Enright, J. T. (1977b). *Limnol. Oceanogr.* **22**, 856–872.

Enright, J. T., and Honegger, H.-W. (1977). *Limnol. Oceanogr.* **22**, 873–886.

Estes, R. D. (1969). *Z. Tierpsychol.* **26**, 284–370.

Estes, R. D. (1976). *East. Afr. Wildl. J.* **14**, 135–152.

Evans, P. R. (1969). *Condor* **71**, 316–330.

Fairbairn, D. J. (1977). *Can. J. Zool.* **55**, 84–92.

Fairbairn, D. J. (1978). *Behav. Ecol. Sociobiol.* **3**, 265–282.

Falconer, D. S. (1960). "Introduction to Quantitative Genetics." Oliver & Boyd, Edinburgh and London.

Fenton, M. B., and Kunz, T. H. (1977). *Spec. Publ., Mus. Tex. Tech. Univ.* **13**, 351–364.

Fisher, R. A. (1958). "The Genetical Theory of Natural Selection." Dover, New York.

Fleming, T. H. (1971). *Misc. Publ., Mus. Zool., Univ. Mich.* **143**, 1–77.

Foster, W. A. (1978). *J. Anim. Ecol.* **47**, 653–659.

Foster, W. A., and Treherne, J. E. (1978). *J. Anim. Ecol.* **47**, 205–217.

French, N. R., Tagami, T. Y., and Hayden, P. (1968). *J. Mammal.* **49**, 272–280.

ffrench, R. (1973). "A Guide to the Birds of Trinidad and Tobago." Livingston Publishing Company, Wynnewood, Pennsylvania.

Fretwell, S. D. (1972). "Populations in a Seasonal Environment." Princeton Univ. Press, Princeton New Jersey.

Fretwell, S. (1978). *In* "Migratory Birds in the Neotropics" (A. Keast and E. S. Morton, eds.), pp. 517–527. Smithsonian Press, Washington, D. C.

Fryer, G. (1965). *Proc. Zool. Soc. London* **144**, 301–322.

Fuseini, B. A., and Kumar, R. (1975). *Biol. J. Linn. Soc.* **7**, 113–146.

Gadgil, M. (1971). *Ecology* **52**, 253–261.

Gaines, M. S., Baker, C. L., and Vivas, A. M. (1979). *Oecologia* **40**, 91–101.

Gauthreaux, S. A., Jr. (1978). *In* "Perspectives in Ethology" (P. P. G. Bateson and P. H. Klopfer, eds.), Vol. 3, pp. 17–54. Plenum, New York.

Gifford, C. A. (1962). *Biol. Bull. (Woods Hole, Mass.)* **123**, 207–223.

Giguère, L. A., and Dill, L. M. (1980). *In* "The Evolution and Ecology of Zooplankton Communities" (W. C. Kerfoot, ed.). University Press of New England, Hanover.

Gill, D. E. (1978a). *Ecol.Monogr.* **48**, 145–166.

Gill, D. E. (1978b). *Evolution* **32**, 839–849.

Gressit, J. L. (1970). *Pac. Insects Monogr.* **23**, 295–374.

Griffin, D. R. (1970). *In* "Biology of Bats" (W. A. Wimsatt, ed.), Vol. 1, pp. 233–265. Academic Press, New York.

Griffin, D. R., Webster, F. A., and Michael, C. R. (1960). *Anim. Behav.* **8**, 141–154.

Grinnell, J. (1931). *Auk* **48**, 22–32.

Groebbels, F. (1928). *Verh. Ornithol. Ges. Bayern* **18**, 44–74.

Grzimek, M., and Grzimek, B. (1980). *Z. Säugetierk.* **25**, 1–63.

Gwinner, E., and Czeschlik, E. (1978). *Oikos* **30**, 364–372.

Hackman, W. (1966). *Not. Entomol.* **46**, 1–16.

Hagen, D. W. (1967). *J. Fish. Res. Board Can.* **24**, 1637–1692.

Hagen, D. W. (1973). *Heredity* **30**, 303–312.

Hagen, D. W., and Gilbertson, L. G. (1973). *Heredity* **30**, 273–287.

Hagen, D. W., and McPhail, J. D. (1970) *J. Fish. Res. Board Can.* **27**, 147–155.

Hagen, K. S. (1962). *Annu. Rev. Entomol.* **7**, 289–326.

Hairston, N. G., Jr. (1976). *Proc. Natl. Acad. Sci. U.S.A.* **73**, 971–974.

Hamilton, T. H. (1961). *Am. Nat.* **95**, 121–123.

Hamilton, W. D., and May, R. M. (1977). *Nature (London)* **269**, 578–581.

Hamilton, W. J., III, and Watt, K. E. F. (1970). *Annu. Rev. Ecol. Syst.* **1**, 263–286.

Harden Jones, F. R. (1968). "Fish Migration." St. Martin's Press, New York.

Hardisty, M. W., and Potter, I. C., eds. (1971). "The Biology of Lampreys," Vol. 1. Academic Press, New York.

Hardy, A. C. (1958). "The Open Sea, Its Natural History: The World of Plankton." Houghton, Boston, Massachusetts.

Hardy, A. C., and Bainbridge, R. (1954). *J. Mar. Biol. Assoc. U. K.* **33**, 409–448.

Hardy, A. C., and Gunther, E. R. (1935). *'Discovery' Rep.* **11**, 1–456.

Harris, J. E. (1953). *Q. J. Microsc. Sci.* [N. S.] **94**, 537–550.

Harris, J. E., and Wolfe, U. K. (1955). *Proc. R. Soc. London, Ser. B* **144**, 329–354.

Harris, M. P. (1970). *Ibis* **112**, 488–498.

Harrison, R. G. (1979). *Oecologia* **40**, 125–132.

Hasler, A. D. (1960). *Science* **132**, 785–792.

Hay, D. E., and McPhail, J. D. (1975). *Can. J. Zool.* **53**, 441–450.

Healy, W. R. (1975). *Ecology* **56**, 673–680.

Heape, W. (1931). "Emigration, Migration, and Nomadism." Heffer, Cambridge, England.

Henshaw, R. E. (1970). *In* "About Bats" (B. H. Slaughter and D. W. Walton, eds.), pp. 188–232. Southern Methodist Univ. Press, Dallas, Texas.

Hernkind, W. F. (1980). *In* "Biology and Management of Lobsters" (B. F. Phillips and S. Cobb, eds.), pp. 349–407. Academic Press, New York.

Herrnkind, W. F., and Cummings, W. C. (1964). *Bull. Mar. Sci. Gulf Caribb.* **14**, 123–125.

Herrnkind, W. F., and Kanciruk, P. (1978). *In* "Animal Migration, Navigation, and Homing" (K. Schmidt-Koenig and W. T. Keeton, eds.), pp. 430–439. Springer-Verlag, Berlin and New York.

Herrnkind, W. F., and McLean, R. (1971). *Ann. N. Y. Acad. Sci.* **188**, 359–377.

Hilborn, R. (1975). *Ecology* **56**, 1221–1225.

Hilborn, R., and Krebs, C. J. (1976). *Can. J. Zool.* **54**, 1507–1508.

Hill, B. J. (1975). *Mar. Biol.* **32**, 119–126.

Hirth, H. F., Pendleton, R. C., King, A.C., Downard, T. R. (1969). *Ecology* **50**, 332–339.

Hoar, W. S. (1958). *J. Fish. Res. Board Can.* **15**, 391–428.

Holm, M., and Naevdal, G. (1978). *In* "Marine Organisms. Genetics, Ecology, and Evolution" (B. Battaglia and J. A. Beardmore, eds.), pp. 676–698. Plenum, New York.

Honêk, A. (1976a). *Zool. Jahrb., Abt. Syst. Oekol. Geogr. Tiere* **103**, 1–22.

Honêk, A. (1976b). *Zool. Jahrb., Abt. Syst. Oekol. Geogr. Tiere* **103**, 547–570.

Horn, H. S., and MacArthur, R. H. (1972). *Ecology* **53**, 749–752.

Howard, W. E. (1960). *Am. Midl. Nat.* **63**, 152–161.

Hoy, M.A. (1978). *In* "Evolution of Insect Migration and Diapause" (H. Dingle, ed.), pp. 101–126. Springer-Verlag, Berlin and New York.

Hurlbert, S. H. (1969). *Ecol. Monogr.* **39**, 465–488.

Immelmann, K. (1971). *In* "Avian Biology" (D. S. Farner and J. R. King, eds.), Vol. 1, pp. 341–389. Academic Press, New York.

Inglis, J. M. (1976). *East Afr. Wildl. J.* **14**, 17–34.

Istock, C. A. (1978). *In* "Evolution of Insect Migration and Diapause" (H. Dingle, ed.), pp. 171–190. Springer-Verlag, Berlin and New York.

Jackson, D. J. (1928). *Trans. R. Soc. Edinburgh* **55**, 665–735.

Jackson, D. J. (1956). *J. Linn. Soc. London, Zool.* **43**, 18–42.

Jackson, P. B. N. (1961). *Proc. Zool. Soc. London* **136**, 603–622.

Janzen, D. H. (1972). *Ecology* **53**, 350–361.
Janzen, D. H. (1974). *Biotropica* **6**, 69–103.
Järvinen, O., and Vepsäläinen, K. (1976). *Hereditas* **84**, 61–68.
Jewell, P. A. (1966). *Symp. Zool. Soc. London* **15**, 89–116.
Johnson, C. G. (1969). "Migration and Dispersal of Insects by Flight." Methuen, London.
Johnson, C. G. (1976). *R. Entomol. Soc. London* **7**, 217–234.
Joule, J., and Cameron, G. N. (1975). *J. Mammal.* **56**, 378–396.
Kalela, O. (1961). *Ann. Acad. Sci. Fenn., Ser. A4* **55**, 1–72.
Kalela, O., Kilpeläienen, L., Koponen, T., and Tast, J. (1971). *Ann. Acad. Sci. Fenn., Ser. A4* **178**, 1–22.
Kanciruk, P., and Herrnkind, W. (1978). *Bull. Mar. Sci.* **28**, 601–623.
Kanwisher, J., and Sundnes, G. (1966). *In* "Whales, Dolphins, and Porpoises" (K. S. Norris, ed.), pp. 397–409. Univ. of California Press, Berkeley and Los Angeles.
Karr, J. R. (1975). *In* "Tropical Ecological Systems: Trends in Terrestrial and Aquatic Research" (F. B. Golley and E. Medina, eds.), pp. 161–176. Springer-Verlag, Berlin and New York.
Karr, J. R. (1976). *Wilson Bull.* **88**, 433–458.
Karr, J. R. (1978). *In* "Migratory Birds in the Neotropics" (A. Keast and E. S. Morton, eds.), pp. 529–543. Smithsonian Press, Washington, D. C.
Karr, J. R., and Gorman, O. T. (1975). "Non-point Source Pollution: Pollution Control in Great Lakes." USEPA-905/9-75-007. pp. 120–150.
Kehat, M. and Wyndham, M. (1973a). *Aust. J. Zool.* **21**, 413–426.
Kehat, M. and Wyndham, M. (1973b). *Aust. J. Zool.* **21**, 427–434.
Kendeigh, S. C., Dol'nik, V. R., and Gavrilov, V. M. (1977). *In* "Granivorous Birds in Ecosystems" (J. Pinowski and S. C. Kendeigh, eds.), IBP Vol. 12, pp. 127–204. Cambridge Univ. Press, London and New York.
Kennedy, J. S. (1958). *Proc. Int. Congr. Entomol., 10th, 1956* Vol. 2, pp. 397–404.
Kennedy, J. S. (1961a) *Nature (London)* **198**, 785–791.
Kennedy, J. S. (1961b). *Symp. R. Entomol. Soc. London* **1**, 80–90.
Kennedy, J. S. (1965). *J. Exp. Biol.* **43**, 489–509.
Kennedy, J. S. (1966). *J. Exp. Biol.* **45**, 215–228.
Kennedy, J. S. (1975). *In* "Insects, Science and Society" (D. Pimentel, ed.), pp. 103–119. Academic Press, New York.
Kennedy, J. S., and Booth, C. O. (1963a). *J. Exp. Biol.* **40**, 67–85.
Kennedy, J. S., and Booth, C. O. (1963b). *J. Exp. Biol.* **40**, 351–369.
Kennedy, J. S., and Booth, C. O. (1964). *J. Exp. Biol.* **41**, 805–824.
Kennedy, J. S., Booth, C. O., and Kershaw, W. J. S. (1961). *Ann. Appl. Biol.* **49**, 1–21.
Kennedy, J. S., and Ludlow, A. R. (1974) *J. Exp. Biol.* **61**, 173–196.
Kerfoot, W. B. (1970). *Am. Nat.* **104**, 529–546.
Kerfoot, W. B. (1972). *Am. Nat.* **106**, 548–553.
Ketterson, E. D., and Nolan, V., Jr. (1976). *Ecology* **57**, 679–693.
Kimura, T., and Masaki, S. (1977). *Kontyu* **45**, 97–106.
King, J. A. (1955). *Contrib. Lab. Vert. Biol., Univ. Mich.* **67**, 1–123.
Kinne, O. (1975). *In* "Marine Ecology" (O. Kinne, ed.), Vol. II, Part 2, pp. 829–844. Wiley, New York.
Kisimoto, R. (1956). *Nature (London)* **178**, 641–642.
Kisimoto, R. (1971). *In* "Rice Insects," pp. 201–216. Ministry of Agriculture and Forestry, Tokyo.
Kisimoto, R. (1972). *In* "Viruses and Invertebrates" (A. J. Gibbs, ed.), pp. 138–156. North-Holland Publ., Amsterdam.

Kozakiewicz, M. (1976). *Acta Theriol.* **21**, 321–338.

Krebs, C. J. (1966). *Ecol. Monogr.* **36**, 239–273.

Krebs, C. J. (1978). *Can. J. Zool.* **56**, 2463–2480.

Krebs, C. J., Gaines, M. S., Keller, B. L., Myers, J. H., and Tamarin, R. H. (1973). *Science* **179**, 35–41.

Krebs, C. J., Wingate, I., LeDuc, J., Redfield, J. A., Taitt, M., and Hilborn, R. (1976). *Can. J. Zool.* **54**, 79–95.

Kutsch, W. (1973). *J. Insect Physiol.* **19**, 763–772.

Lack, D. (1968a). *Oikos* **19**, 1–9.

Lack, D. (1968b). "Ecological Adaptations for Breeding in Birds." Chapman & Hall, London.

Lack, D. (1976). "Island Biology Illustrated by the Land Birds of Jamaica." Univ. of California Press, Berkeley and Los Angeles.

Lamb, R. J., and MacKay, P. A. (1979). *Oecologia* **39**, 289–299.

Landahl, J. T., and Root, R. B. (1969). *Ecology* **50**, 734–737.

Landsborough Thomson, A. (1926). "Problems of Bird-Migration." Witherby, London.

Laws, R. M. (1977). *Philos. Trans. R. Soc. London, Ser. B* **279**, 81–96.

LeBoeuf, B. J., Ainley, D. G., and Lewis, T. J. (1974). *J. Mammal.* **55**, 370–385.

Leck, C. F. (1972a). *Auk* **89**, 842–850.

Leck, C. F. (1972b). *Condor* **74**, 54–60.

Lees, A. D. (1966). *Adv. Insect Physiol.* **3**, 207–277.

Leggett, W. C. (1972). *Trans. Am. Fish. Soc.* **101**, 549–552.

Leggett, W. C. (1976). *In* "The Connecticut River Ecological Study," Monogr. No. 1, pp. 169–225. Am. Fish. Soc.

Leggett, W. C. (1977a). *J. Fish. Res. Board Can.* **34**, 1422–1426.

Leggett, W. C. (1977b). *Annu. Rev. Ecol. Syst.* **8**, 285–308.

Leggett, W. C., and Whitney, R. R. (1972). *Fish. Bull.* **70**, 659–670.

Lent, P. C. (1966). *Z. Tierpsychol.* **23**, 701–756.

Levins, R. (1968). "Evolution in Changing Environments." Princeton Univ. Press, Princeton, New Jersey.

Lewontin, R. C. (1965). *In* "The Genetics of Colonizing Species" (H. G. Baker and G. L. Stebbins, eds.), pp. 77–94. Academic Press, New York.

Lewontin, R. C. (1974). "The Genetic Basis of Evolutionary Change." Columbia Univ. Press, New York.

Lidicker, W. Z., Jr. (1962). *Am. Nat.* **96**, 29–33.

Lidicker, W. Z., Jr. (1973). *Ecol. Monogr.* **43**, 271–302.

Lidicker, W. Z., Jr. (1975). *In* "Small Mammals: Their Productivity and Population Dynamics" (F. B. Golley, K. Petrusewicz, and L. Ryszkowski, eds.), pp. 103–128. Cambridge Univ. Press, London and New York.

Lidicker, W. Z., Jr., and Anderson, P. K. (1962). *J. Anim. Ecol.* **31**, 503–517.

Lincoln, F. C. (1952). "Migration of Birds." Doubleday, Garden City, New York.

Lindroth, C. H. (1945). *Medd. Göeteborgs Mus. Zool. And.* **109**; 707.

Lindroth, C. H. (194). *Medd. Goeteborgs Mus. Zool. And.* **110**, 277.

Lindroth, C. H. (1949). *Medd. Goeteborgs Mus. Zool. And.* **122**, 1–911.

Lowe-McConnell, R. H. (1975). "Fish Communities in Tropical Freshwaters." Longmans, Green, New York.

McAllister, C. D. (1969). *J. Fish. Res. Board Can.* **26**, 199–220.

MacArthur, R. H. (1959). *Auk* **76**, 318–325.

MacKay, P. A., and Lamb, R. J. (1979). *Oecologia* **39**, 301–308.

Mackintosh, N. A. (1966). *In* "Whales, Dolphins, and Porpoises" (K. S. Norris, ed.), pp. 125–144. Univ. of California Press, Berkeley and Los Angeles.

McLaren, I. (1963). *J. Fish. Res. Board Can.* **20**, 685–727.

McLaren, I. (1974). *Am. Nat.* **108**, 91–102.

MacLean, S. F., Jr. and Holmes, R. T. (1971) *Auk* **88**, 893–901.

Mauchline, J., and Fisher, L. R. (1969). *Adv. Mar. Biol.* **7**, 1–454.

Maynard Smith, J. (1974). *J. Theor. Biol.* **47**, 209–222.

Mayr, E., and Meise, W. (1930). *Vogelzug* **1**, 149–172.

Medway, L. (1972). *Bull. Br. Mus. (Nat. Hist.) Entomol.* **23**, 105–154.

Menge, B. A. (1972). *Ecology* **53**, 635–644.

Menge, B. A. (1974). *Ecology* **55**, 84–93.

Menge, B.A. (1975). *Mar. Biol.* **31**, 87–100.

Michael, E. L. (1911). *Univ. Calif. Berkeley, Publ. Zool.* **8**, 21–186.

Mikkola, K. (1971). *Acta Entomol. Fenn.* **28**, 107–120.

Miller, C. B., Pearcy, W. G., and Schonzeit, M. H. (1972). *Am. Nat.* **106**, 545–547.

Mills, D. (1971). "Salmon and Trout: The Resource, Its Ecology, Conservation, and Management." St. Martin's Press, New York.

Moodie, G. E. E. (1972). *Heredity* **28**, 155–167.

Moreau, R. E. (1972). "The Palearctic-African Bird Migration Systems." Academic Press, New York.

Morse, D. H. (1968). *Ecology* **49**, 779–784.

Morse, D. H. (1971). *Annu. Rev. Ecol. Syst.* **2**, 177–200.

Morton, E. S. (1977). *Auk* **94**, 97–106.

Moyle, P. B. (1976). "Inland Fishes of California." Univ. of California Press, Berkeley.

Mundie, J. H. (1969). *In* "Symposium on Salmon and Trout in Streams" (T. G. Northcote, ed.), pp. 135–152. University of British Columbia.

Murie, O. J. (1935). *USDA, North Am. Fauna* **54**, 1–93.

Murray, B. G., Jr. (1967). *Ecology* **48**, 975–978.

Myers, G. S. (1949). *Copeia* pp. 89–97.

Myers, J. H., and Krebs, C. J. (1971). *Ecol. Monogr.* **41**, 53–78.

Nelson, K. (1964). *Univ. Calif., Berkeley, Publ. Zool.* **75**, 59–152.

Newsome, A. E., and Corbett, L. K. (1975). *Monogr. Biol.* **28**, 117–153.

Nicholls, A. G. (1933). *J. Mar. Biol. Assoc. U. K.* **19**, 139–164.

Nichols, J. D., Conley, W., Batt, B., and Tipton, A. R. (1976). *Am. Nat.* **110**, 995–1005.

Nikolsky, G. V. (1963). "The Ecology of Fishes." Academic Press, New York.

Norse, E. A. (1975). Ph.D. Dissertation, University of Southern California, Los Angeles.

Norse, E. A. (1977). *Bull. Mar. Sci.* **27**, 440–447.

Norse, E. A., and Fox-Norse, V. (1977). *Mar. Biol.* **40**, 374–376.

Northcote, T. G. (1969). *In* "Symposium on Salmon and Trout in Streams" (T. G. Northcote, ed.), pp. 183–203. University of British Columbia.

Northcote, T. G., (1978). *In* "Ecology of Freshwater Fish Production" (S. D. Gerking, ed.), pp. 326–359. Halstead Press, New York.

O'Brien, W. J. (1979). *Am. Sci.* **67**, 572–581.

Odhiambo, T. R. , and Arora, G. K. (1973). *Entomol. Exp. Appl.* **16**, 455–470.

Oesterling, M. L. and Evink, G. L. (1977). *In* "Proceedings of the Conference on the Apalachicola Drainage System" (R. J. Livingston and E. A. Joyce, Jr., eds.), Fla. Mar. Res. Publ. No. 26.

Oldham, R. S. (1966). *Can. J. Zool.* **44**, 63–100.

Omori, M. (1974). *Adv. Mar. Biol.* **12**, 233–324.

Orr, R. T. (1970). "Animals in Migration." Macmillan, New York.

Orr, R. T., and Poulter, T. C. (1965). *Proc. Calif. Acad. Sci.* **32**, 377–404.

Packer, W. C. (1963). *Copeia* pp. 378–382.

Pajunen, V. I. (1970). *Ann. Zool. Fenn.* **7**, 270–272.

Parker, G. A., and Stuart, R. A. (1976). *Am. Nat.* **110,** 1055-1076.

Parnell, J. F. (1969). *Auk* **86,** 505-521.

Pennycuick, C. J. (1971). *J. Exp. Biol.* **55,** 833-845.

Pennycuick, L. (1975). *East Afr. Wildl. J.* **13,** 65-87.

Pienkowski, R. L., and Medler, J. T. (1964). *Ann. Entomol. Soc. Am.* **57,** 588-591.

Power, D. M. (1971). *Ecology* **52,** 434-443.

Pritchard, P. C. H. (1973). *Anim. Behav.* **21,** 18-27.

Pullen, E. J., and Trent, W. L. (1969). *FAO Fish. Rep.* **3,** No. 57, 1001-1014.

Rabenold, K. N. (1979). *Am. Nat.* **114,** 275-286.

Rabenold, K. N. (1980). *In* "Migratory Birds in the Neotropics" (A. Keast and E. S. Morton, eds.), pp. 297-307. Smithsonian Press, Washington, D. C. (in press).

Radford, K. W., Orr, R. T., and Hubbs, C. L. (1965). *Proc. Calif. Acad. Sci.* **31,** 601-612.

Rainey, R. C. (1976). *Symp. R. Entomol. Soc. London* **7,** 75-112.

Rainey, R. C. (1978). *In* "The Evolution of Insect Migration and Diapause" (H. Dingle, ed.), pp. 33-48. Springer-Verlag, Berlin and New York.

Ralph, C. P. (1976). *Oecologia* **26,** 157-175.

Rand, A. S. (1968). *Copeia* pp. 552-561.

Rankin, M. A. (1974). *In* "The Experimental Analysis of Insect Behaviour" (L. Barton Browne, ed.), pp. 317-328. Springer-Verlag, Berlin and New York.

Rankin, M. A. (1978). *In* "Evolution of Insect Migration and Diapause" (H. Dingle, ed.), pp. 5-32. Springer-Verlag, Berlin and New York.

Rankin, M. A., and Riddiford, L. M. (1977). *Gen. Comp. Endocrinol.* **33,** 309-321.

Rappole, J. H., and Warner, D. W. (1976). *Oecologia* **26,** 193-212.

Rasmuson, B., Rasmuson, M., and Nygren, J. (1977). *Hereditas* **87,** 33-42.

Raymont, J. E. G. (1963). "Plankton and Productivity in the Oceans." Macmillan, New York.

Redfield, J. A. (1976). *Can. J. Zool.* **54,** 463-474.

Reese, E. S. (1968). *Science* **161,** 385-386.

Reid, R. W. (1958). *Can. Entomol.* **90,** 464-468.

Rice, D. W. (1965). *J. Mammal.* **46,** 504-505.

Richard, J. D., and Hughes, D. A. (1972). *Mar. Biol.* **16,** 297-309.

Riegert, P. W. (1962). *Nature (London)* **194,** 1298-1299.

Robertson, R. J. (1972). *Can. J. Zool.* **50,** 247-263.

Roff, D. A. (1975). *Oecologia* **19,** 217-237.

Roff, D. A. (1977). *J. Anim. Ecol.* **46,** 443-456.

Rohwer, S. A. (1975). *Evolution* **29,** 593-610.

Rose, D. J. W. (1972). *J. Anim. Ecol.* **41,** 589-609.

Rose, M. (1925). *Arch. Zool. Exp. Gen.* **64,** 387-649.

Rowley, W. A., and Graham, C. L. (1968). *J. Insect Physiol.* **14,** 719-728.

Ruello, N. V. (1975). *Aust. J. Mar. Freshwater Res.* **26,** 343-354.

Russell, F. S. (1927). *Biol. Rev. Cambridge Philos. Soc.* **2,** 213-262.

Rygg, T. D. (1966). *Entomol. Exp. Appl.* **9,** 74-84.

Salomonsen, F. (1955). *Dan. Biol. Medd.* **22,** 1-62.

Schaffer, W. M. (1974a). *Ecology* **55,** 291-303.

Schaffer, W. M. (1974b). *Am. Nat.* **108,** 783-790.

Schaffer, W. M., and Elson, P. F. (1975). *Ecology* **56,** 577-590.

Scheltema, R. S. (1971). *Eur. Mar. Biol. Symp. [Proc.], 4th, 1969* pp. 7-28.

Schneider, D. (1980). *Oecologia* (in press).

Schneider, D., and Harrington, B. A. (1980). *Auk* (in press).

Scholz, A. T., Horrall, R. M., Cooper, J. C., and Hasler, A. D. (1976). *Science* **192,** 1247-1249.

Scudder, G. G. E. (1964). *Am. Zool.* **4,** 331.

Serventy, D. L. (1971). In "Avian Biology" (D. S. Farner and J. R. King, eds.), Vol. 1, pp. 287-339. Academic Press, New York.

Shapiro, A. M. (1976). Evol. Biol. 9, 259-333.

Sheppe, W. (1965a) Evolution 19, 480-495.

Sheppe, W. (1965b). J. Mammal. 46, 336-337.

Shoubridge, E. A. (1977). M.Sc. Thesis, McGill University, Montreal.

Siebert, H. C. (1949). Auk 66, 128-153.

Sinclair, A. R. E. (1978). Ibis. 120, 480-497.

Slijper, E. J. (1962). "Whales." Basic Books, New York.

Slobodkin, L. B., and Rappaport, A. (1974). Q. Rev. Biol. 49, 181-200.

Sluiter, J. W., and van Heerdt, P. F. (1966). Arch. Neerl. Zool. 16, 423-439.

Smith, A. T. (1974). Ecology 55, 1112-1119.

Smith, B. R. (1971). In "Biology of Lampreys" (M. W. Hardisty and I. C. Potter, eds.), Vol. 1, pp. 207-247. Academic Press, New York.

Smith, M. H. (1968). Bull. Ga. Acad. Sci. 26, 45-51.

Solbreck, C. (1972). Entomol. Scand. 3, 267-274.

Solbreck, C. (1974). Oecologia 17, 265-275.

Solbreck, C. (1978). In "Evolution of Insect Migration and Diapause" (H. Dingle, ed.), pp. 195-217. Springer-Verlag, Berlin and New York.

Southwood, T. R. E. (1962). Biol. Rev. Cambridge Philos. Soc. 37, 171-214.

Southwood, T. R. E. (1977). J. Anim. Ecol. 46, 337-365.

Stearns, S. C. (1976). Rev. Biol. 51, 3-47.

Stegwee, D., Kimmel, E. C., deBoer, J. A., and Henstra, S. (1963). J. Cell Biol. 19, 519-527.

Stickel, L. F. (1968). In "Biology of Peromyscus (Rodentia)" (J. A. King, ed.), Spec. Publ. No. 2, pp. 373-411. Am. Soc. Mammal.

Stolt, B. -O. (1978) Oikos 30, 413-418.

Storer, T. I., Evans, F. C., and Palmer, F. G. (1944). Ecol. Monogr. 14, 165-192.

Strathmann, R. (1974). Am. Nat. 108, 29-44.

Sullivan, T. P. (1977) Ecology 58, 964-978.

Sutcliffe, W. H. (1953). J. Mar. Res. 12, 173-183.

Swarth, H. S. (1920). Univ. Calif., Berkeley, Publ. Zool. 21, 75-224.

Swift, M. C. (1976). Ecology 57, 900-914.

Taisson, N. E. (1974). Bol. Cent. Invest. Biol., Univ. Zulia 10, 9-50.

Talbot, L. M., and Talbot, M. H. (1963) Wildl. Monogr. 12, 1-88.

Tamarin, R. H. (1977a). Ecology 58, 1044-1054.

Tamarin, R. H. (1977b). J. Mammal. 58, 536-548.

Taylor, L. R. (1975). J. Anim. Ecol. 44, 135-163.

Taylor, L. R., and Taylor, R. A. J. (1977). Nature (London) 265, 415-421.

Taylor, L. R., and Taylor, R. A. J. (1978). In "Population Control by Social Behaviour" (F. J. Ebling and D. M. Stoddart, eds.), pp. 181-212. Institute of Biology, London.

Taylor, R. A. J., and Taylor L. R. (1979). In "Population Dynamics" (R. M. Anderson, B. D. Turner, and L. R. Taylor, eds.), pp. 1-27. Blackwell, Oxford.

Thorson, G. (1950). Biol. Rev. Cambridge Philos. Soc. 25, 1-45.

Thorson, G. (1964). Ophelia 1, 167-208.

Tramer, E. J. (1974). Condor 76, 460-464.

Tucker, V. A. (1972). Am. J. Physiol. 222, 237-245.

Twitty, V. C. (1966). "Of Scientists and Salamanders." Freeman, San Francisco, California.

Urquhart, F. A. (1960). "The Monarch Butterfly." Univ. of Toronto Press, Toronto.

Uvarov, B. (1966). "Grasshoppers and Locusts," Vol. 1. Cambridge Univ. Press, London and New York.

Uvarov, B. (1977). "Grasshoppers and Locusts," Vol. 2. Cambridge Univ. Press, London and New York.

Vance, R. (1973). *Am. Nat.* **107**, 339–352.

Vannini, M. (1976). *Monit. Zool. Ital., Suppl.* **7**, 145–185.

Vaughan, T. A. (1970a). *In* "About Bats" (B. H. Slaughter and D. W. Walton, eds.), pp. 127–143. Southern Methodist Univ. Press, Dallas, Texas.

Vaughan, T. A. (1970b). *In* "Biology of Bats" (W. A. Wimsatt, ed.), Vol. 1, pp. 195–216. Academic Press, New York.

Vepsäläinen, K. (1973). *Ann. Zool. Fenn.* **10**, 419–444.

Vepsäläinen, K. (1974a). *Ann. Acad. Sci. Fenn., Ser. A4* **202**, 1–18.

Vepsäläinen, K. (1974b). *Acta Zool. Fenn.* **141**, 1–73.

Vepsäläinen, K. (1978). *In* "Evolution of Insect Migration and Diapause" (H. Dingle, ed.), pp. 218–253. Springer-Verlag, Berlin and New York.

Vlymen, W. J. (1970). *Limnol. Oceanogr.* **15**, 348–356.

von Haartman, L. (1968). *Ornis Fenn.* **45**, 1–7.

Ward, P. (1968). *Ibis* **110**, 239–255.

Ward, P. (1971). *Ibis* **113**, 275–297.

Waterman, T. H., Nunnemacher, R. F., Chace, F. A., Jr., and Clarke, G. L. (1939). *Biol. Bull. (Woods Hole, Mass.)* **76**, 256–279.

Wellington, W. G. (1960). *Can. J. Zool.* **38**, 289–314.

Wellington, W. G. (1964). *Can. Entomol.* **96**, 436–451.

West, G. C. (1960). *Auk* **77**, 306–329.

Weygoldt, P. (1969). "The Biology of Pseudoscorpions." Harvard Univ. Press, Cambridge, Massachusetts.

Whitehead, P. J. P. (1959). *Rev. Zool. Bot. Afr.* **59**, 329–363.

Whitford, W. G., and Vinegar, A. (1966). *Copeia* pp. 515–519.

Williams, C. B. (1958). "Insect Migration." Collins, London.

Williams, G. C. (1975). "Sex and Evolution." Princeton Univ. Press, Princeton, New Jersey.

Williams, G. C., Koehn, R. K., and Mitton, J. B. (1973) *Evolution* **27**, 192–204.

Willis, E. O. (1966). *Living Bird* **5**, 187–231.

Willson, M. F. (1976). *Wilson Bull.* **88**, 582–587.

Wolfson, A. (1948). *Science* **108**, 23–30.

Wootton, R. J. (1976). "The Biology of the Sticklebacks." Academic Press, New York.

Wratten, S. D. (1977). *Ann. Appl. Biol.* **85**, 319–331.

Wright, D., O'Brien, W. J., and Vinyard, G. L. (1979). *In* "The Evolution and Ecology of Zooplankton Communities" (W. C. Kerfoot, ed.). Univ. Press of New England, Hanover, New Hampshire.

Wynne Edwards, V. C. (1962). "Animal Dispersion in Relation to Social Behaviour." Oliver & Boyd, Edinburgh and London.

Yalden, D. W., and Morris, P. A. (1975). "The Lives of Bats." Demeter Press, New York.

Yarbrough, C. G. (1971). *Comp. Biochem. Physiol. A* **39**, 235–266.

Young, E. C. (1965). *J. Anim. Ecol.* **34**, 353–390.

Zaret, T. M., and Suffern, J. S. (1976). *Limnol. Oceanogr.* **21**, 804–813.

Ziegler, J. R. (1976). *Evolution* **30**, 579–592.

Zimmerman, J. L. (1965). *Physiol. Zool.* **38**, 370–389.

Zwickel, F. C., Redfield, J. A., and Kristensen, J. (1977). *Can. J. Zool.* **55**, 1948–1957.

2 *The Influences of Long-Term and Short-Term Climatic Changes on the Dispersal and Migration of Organisms*

Sidney A. Gauthreaux, Jr.

Animal Migration, Orientation, and Navigation

I. Introduction

The environment is constantly changing in time and space, and in order to discuss meaningfully the influence of these changes on organisms, one must do so with reference to particular temporal and spatial scales. The scales are organized hierarchically (Fig. 1). At the highest level in the hierarchy the temporal scale may have a periodicity or wavelength of several hundred million years and may encompass spatially the entire planet. At much lower level in the hierarchy the temporal scale may have a periodicity or wavelength of a few minutes and may encompass spatially only a few centimeters. Because various critical components of the environment of an organism are increasing and decreasing in favorability at different rates, the suitability of a particular environment changes. When this happens a population of organisms with limited tolerances to environmental change faces three alternatives: migration, evolutionary adaptation, or extinction *in situ* (Grant, 1977, p. 271). The first alternative, migration, is by far the most likely to occur if the organisms have the ability to move. The persistence of biological populations is dependent largely on the ability of individual organisms to move so that they can reproduce in the most appropriate place at the most appropriate time. When habitats elsewhere have a higher mean favorableness than the habitat already occupied, selection will favor those traits that serve to maximize the expectancy of arriving and surviving to breed in the new habitat, e.g., the ability to disperse or migrate, the ability to survive the dangers of dispersal and migration, and the ability to find or arrive in the new habitats (Southwood, 1977). Every organism then must retain the ability for it or its descendants to change location eventually or face extinction. Movement therefore is a fundamental biological response to adversity, and all populations are spatially fluid in some measure (Taylor and Taylor, 1977).

The types of environmental changes that make dispersal and migration necessary are diverse and range from physical factors such as temperature, moisture, and photoperiod to biological factors such as food, mates, and predators. The types of environmental changes and their timing, magnitude, and spatial properties have shaped in large part the types of migratory behavior found in nature. My task in this chapter is to examine the environmental factors that have contributed to the shaping of the temporal and spatial patterns of dispersal and migration. In doing so I have organized the topics according to temporal and spatial scales beginning with a treatment of biogeographical migrations in response to long-term environmental changes and ending with an emphasis on migration patterns in response to short-term environmental changes.

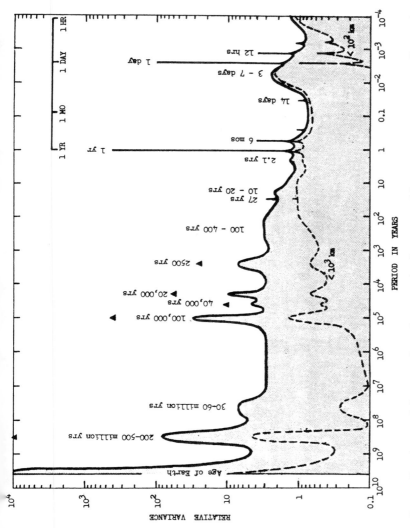

Fig. 1. Distribution of climatic cycles and variation for the time period from 10^{10} years to 10^{-4} years. Sharp spikes indicate astronomically dictated periodic variation in climate while broader peaks indicate quasiperiodic or aperiodic variations. The solid triangles relate the amplitudes of the different types of variations to one another. The shaded area represents total climatic variability on all spatial scales of variability, the dashed lines delimit the contributions to total climatic variability from different spatial scales. (From Mitchell, 1976; courtesy of the Quaternary Research Center, University of Washington.)

II. Environmental Determinants of Distribution and Abundance

The interrelationships of geophysical and geochemical processes and the flora and fauna are complex and can best be examined in terms of a hierarchy of processes operating at different temporal and spatial scales (see Haskell, 1972, pp. 52–62; Simon, 1973). The distribution and the abundance of organisms (and their patterns of dispersal and migration) are dependent on the environment they occupy. Abiotic and biotic processes are important in this regard, but it is the environment as the organism perceives it through its sensory capacities (von Uexküll, 1909) that is ultimately important in the organization of the elements of the biota. For terrestrial and aquatic autotrophs, geophysical and geochemical processes having different temporal and spatial scales constitute in large part the "habitats" that these organisms track. Heterotrophs, on the other hand, are dependent in large measure on the distribution and abundance of autotrophs, or in the case of parasites and predators, the distribution and the abundance of hosts and prey, respectively. Because of the geophysical and geochemical foundations of biotic organization, meteorological processes and climatic factors exert a major influence on the distribution and the abundance of the biota.

A. CLIMATOLOGY, METEOROLOGY, AND THE BIOTA

The climate of the earth is ultimately dependent on the amount of solar radiation impinging on the planet, the planet's shape and rotation, and the characteristics of the planet's orbit, particularly the angle of the ecliptic (Imbrie and Imbrie, 1980). Within this framework, climatic zonation is dependent on (1) the radiation received in relation to latitude, altitude, and albedo, (2) atmospheric water, (3) atmospheric pressure and wind belts, and (4) the spatial distribution of land and water. Latitudinal heat exchange and moisture flux drive the dynamic general circulation of the atmosphere. Because climatic components vary in space and time, it is not surprising that heat exchange and moisture flux can vary with a wide range of wavelengths and amplitudes (Herman and Goldberg, 1978).

The influence of physical and dynamic meteorology on the biota must be examined in terms of the appropriate meteorological scales. The atmosphere is constantly in motion and diverse atmospheric processes are constantly changing in time and space. The changes constitute the weather conditions present at a given locality at a certain time. The physical condition of the atmosphere is usually expressed in terms of meteorological elements including temperature, atmospheric pressure, humidity, wind velocity and direction, condensation products and precipitation, solar radiation, and visibility. Weather is characterized by changes in these meteorological elements within certain limits, depending on location and season. The climate of an area is comprised of the average values of the meteorological elements and certain statistics about their distribution.

Weather systems can best be discussed in terms of a spectrum of motions containing long and short wavelength phenomena. Large-scale disturbances occur in the general atmospheric circulation of the planet and are mapped on synoptic weather charts as long waves in the westerlies (cyclonic depressions or troughs and anticyclonic ridges of the middle and high latitudes) and as short waves in the easterlies (tropical cyclones). These synoptic phenomena may encompass a spatial scale of several thousand to a few hundred kilometers, and they greatly influence the weather of the troposphere (Riley and Spolton, 1974). At an intermediate scale, showers, thunderstorms, tornadoes, land and sea breezes, and mountain and valley winds are the important systems that influence the weather. Small-scale turbulence (eddies moving upward and downward) exchanges the momentum of the air between atmospheric levels and represents the dominant small wavelength system.

The role of meteorology in the quest for facts about climatic change is to identify the true nature of the weather events and the atmospheric circulation processes at work, and in doing so account for the distribution of weather phenomena at any time and place—an approach based on the principle of uniformitarianism (Lamb, 1977, p. 22). The circulation of the atmosphere has distinctly monthly patterns that interact with surface thermal and topographic characteristics to give each place on earth a particular annual sequence of dominant airstreams. These airstreams with their physical properties determined by their previous history, plus the dynamics of the locale and its radiation regime, largely dictate the climatic character of a location (Bryson and Hare, 1974, pp. 18–24). The regions are characterized by an annual sequence of assemblages of meteorological parameters and are delineated by the preferred location of the airstream boundaries. Bryson and Hare call these distinctive climatic assemblages *climata,* and the regions they occupy, *genetic climatic regions.* The latter correspond quite well with many of the major biotic regions. This agreement is hardly surprising, because the *biota* have developed under the influence of corresponding *climata* (Bryson, 1966).

The climate of the oceans is closely related to atmospheric motions. The major surface and near-surface circulation of the ocean in horizontal and vertical dimensions, and the effects of these patterns on the distribution of heat, nutrients, and life have been examined by Reid *et al.* (1978). The gross pattern of ocean circulation is geostrophic and is largely driven by global wind belts in relation to the positions of continental land masses.

Drift at lower latitudes is dependent on the trade winds, and drift at higher latitudes is dependent on the westerlies. The wind patterns and the continental boundaries generate a north–south flow at the eastern and western continental boundaries, and the currents form large gyral systems within the major oceans. In general each ocean contains (1) an anticyclonic gyre within the subtropical latitudes, (2) a cyclonic gyre in the middle or subarctic or subantarctic latitudes, and (3) a system of east–west flows near the equator. Each system has its own range of light, nutrients, and temperatures, and each has its own biomass and typical species assemblages. In general the anticyclonic gyres show less seasonal variation than the cyclonic gyres. When the subsurface and vertical flow systems are added to the major patterns of horizontal flow, an array of separate oceanic habitats results (Platt, 1978). Recently considerable attention has been devoted to the distribution and the abundance of the biota in relation to ocean habitats and their dependence on meteorological and climatological conditions (Cushing, 1978; Haury et al., 1978; Newell, 1979; Smith, 1978; Cushing and Dickson, 1976; Garrod and Colebrook, 1978). Although similar data on lakes are scarce, some information on phytoplanktonic communities in relation to physical variability in time and space is available (Harris, 1980; Harris and Smith, 1977).

Climatic regions are not static but undergo dynamic and unpredictable changes (Bryson, 1974; Gribbin, 1978; Kutzbach, 1976, 1978; Lamb, 1977; Mason, 1976; Mitchell, 1976; Pittock et al., 1978; Shutts and Green, 1978), and these changes have rather dramatic effects on terrestrial and aquatic ecosystems by altering responses of the constituent biotic elements. Truly equilibrium states probably do not exist, and the climatic system can be regarded as being in a continual state of transient adjustment (Mitchell, 1971).

Climatic changes show a wide range of periodicities, and an overview of climatic variability and its causal mechanisms has been provided by Mitchell (1976). He presents a variance spectrum of climatic change (Fig. 1) that spans all time scales of variability from about the age of the earth (4×10^9 years) to about 1 hour (10^{-4} years). In his overview of the spectrum Mitchell groups all potential sources of climatic variability into three categories: (1) internal stochastic mechanisms that involve interactions (i.e., feedbacks) between different parts of the climatic system, (2) external forcing mechanisms that involve forcing of the climatic system or the statistical behavior of the system by environmental events or changes that are independent of the state of the climatic system, and (3) processes that involve some form of resonance between internal modes of climatic system behavior and external forcing of a repetitive or cyclical character. The elements of the climatic system (atmosphere, oceans, land surface, ice masses or cryosphere, and the biosphere) may freely interact either physically or chemically, and a change in any one results in changes in all the others.

According to Mitchell some variability of the atmospheric state is from processes with time constants of minutes or hours, but considerably larger contributions come from various slower-acting processes internal and external to the climatic system, each having different time constants. As one moves from the short-wave end toward the long-wave end of the spectrum in Fig. 1, one finds regions of the spectrum corresponding to time constants of internal stocastic processes that add variance to the climate. In each of these regions the amplitude of the spectrum increases incrementally, thus generating a hierarchy of individual stochastic processes, each characterized by its own unique time constant, and each adding to the variability of climate on all time scales longer than that time constant. In Fig. 2 the internal stochastic processes that contribute to the total variability of the climate are indicated below the curve, and those features of the curve that are not accounted for by internal stochastic processes are identified phenomenologically above the curve.

Because climatic elements show periodicities with different time constants, habitats that are strongly influenced by these climatic elements also show similar periodicities with comparable time constants. Habitats will change with long wavelengths (e.g., successional changes), and at the same time show short wavelength periodicities (e.g., seasonal or diel changes) of favorability and unfavorability for survival and reproduction. The responses of the organisms to climatically induced changes in their habitat will be dependent on their particular adaptations (tolerances) as well as their generation times. Southwood (1977) has discussed the importance of the level of favorableness of the general physical and chemical environment and its constancy within the normal climatic cycle and points out that when values are low, the habitat has greater adversity (*sensu* Whittaker, 1975a) and lower biomass and productivity (Reichle *et al.*, 1975).

Animals can respond to changes in climate and weather directly (Lowry, 1969; Munn, 1970), or they can respond indirectly through the effects on their habitat (Crisp, 1959, 1965). In terms of the regulation of animal distribution and abundance, climatic factors are considered density independent, but changes in climate may influence the operation of density dependent factors (Krebs, 1978, pp. 283–296). Throughout the range of a species, the density independent actions of climate can be expected to produce place-to-place differences in carrying capacity, and this same phenomenon, operating in a continuous manner, should contribute toward setting the distributional boundary of a species (Enright, 1976). Climate might also alter the action of density dependent processes, for example, by

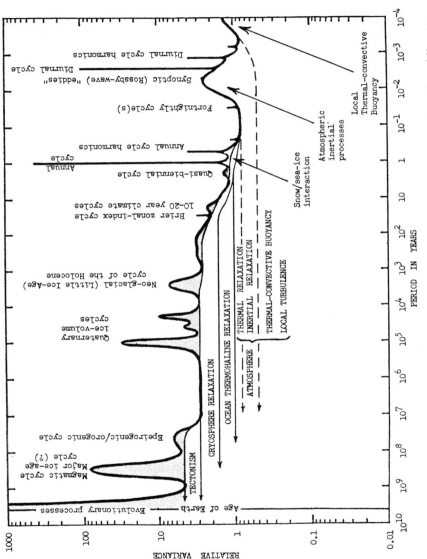

Fig. 2. The spectrum of Fig. 1 containing indications of the contributions to total variance in climate from different internal stochastic processes (indicated below curve). Contributions to the spectrum not necessarily accounted for by internal stochastic processes are indicated above the curve. Shaded portions indicate excess variance associated with each of these features. (From Mitchell, 1976; courtesy of Quaternary Research Center, University of Washington.)

affecting resource availability, but one need not presume such effects to account for the correlations between population density and climatic variables (Enright, 1976).

B. Tolerance to Environmental Changes

If animals and plants can tolerate fluctuating environments or change tolerances as the environment changes, they do not have to move in response to adversity. The tolerance of a species can be viewed from the standpoint of the population as well as the individual. Individuals in a population may have very limited ranges of tolerance that differ slightly from individual to individual. The total range of tolerance is represented by the extreme individuals in the population. In the face of environmental change certain individuals can remain at the location, but others that cannot tolerate the new conditions must move or die. Alternatively, individual organisms may have a wide range of tolerance because they can alter their physiology (e.g., capable of hibernating or diapausing) and survive a stressful periodic environmental change without moving. The duration and amplitude of environmental change in relation to the life span of an organism dictate the types of tolerance found in a species. If life spans are considerably shorter than the periodicity of environmental change, then clearly different individuals in the population (or their descendants) must adapt to the change over time or track a suitable environment. Organisms with life spans considerably longer than the periodicities of environmental change likely have relatively wide ranges of tolerance that permit physiological adjustment during the periods of adversity, or for the duration of the adversity they move to more suitable areas.

Considerable information on the tolerances of extant animals, plants, and microorganisms to climatic and other geophysical and geochemical variables has been gathered by physiological ecologists. General discussions and overviews can be found in Allee et al. (1949, pp. 73–227, particularly pp. 206–215), Budyko (1974), Gates (1979), Itoh et al. (1972), Krebs (1978, pp. 65–129), Newell (1976), Robertshaw (1974, 1977), Rosenberg (1974), Vernberg (1975a), Yousef et al. (1972), Andrewartha and Birch (1954, pp. 129–332), and Gates and Schmerl (1975). Studies emphasizing the tolerances of plants to diverse climatic variables have been reviewed by Bannister (1977), Etherington (1975), Grace (1978), Larcher (1980), Levitt (1972), Monteith (1975), Mooney (1975), Pigott (1975), and Walter (1979). The literature on the tolerance of animals to climatic factors is vast and summaries can be found in Bligh et al. (1977), Hill (1976), Irving (1972), and Maloiy (1972). The tolerances of freshwater and marine organisms to changing geophysical and geochemical conditions must be considered in terms of particular features of the aquatic environment (e.g., temperature effects on salinity and oxygen tension, nutrients, light), and reasonably thorough reviews can be found in Chavin (1973), Harris (1980), Vermeij (1978), Vernberg (1975b), Sharp and Dizon (1978), and Vernberg and Vernberg (1972). Physiological ecologists have examined the tolerances of organisms to a number of climatic and other geophysical variables (e.g., temperature, moisture, pH, soils, topography, solar radiation, salinity) and in so doing have provided evidence that can be used to generate a more precise picture of the climatic limits for existing species. The evidence suggests that complex biotic communities can be correlated with multiple geophysical factors and that community gradients (coenoclines) can be related largely to climatic gradients, in terms of major ecoclines and geographic patterns (Whittaker, 1975b, pp. 111–191).

C. Movement in Response to Environmental Change

If animals and plants cannot tolerate fluctuating environments and cannot change tolerances as their environment changes, then these organisms must move in response to changes in their physical and chemical environment or face extinction. Consequently, movement by organisms (or their offspring) is an important means of adjustment to major changes in their geophysical and biotic environment. The movements may be in response to gradual, long-duration changes and accomplished by many successive generations (migrations in response to paleoclimatological changes) or the movements may be in response to rather abrupt, short-term climatic changes and accomplished several times during the lifetime of an organism (e.g., migrations in response to seasonal climatic changes). The results of such movements may be spatially identical, but the time frames involved are dramatically different.

From the standpoint of migratory movements, the climatic and meteorological factors responsible for changing the favorability of a habitat in space and time may be in large part the same factors that shape the spatial and temporal characteristics of movements from or to that habitat. Climatic changes can control the spatial distribution and relative favorability of habitats and by doing so influence either directly or indirectly the direction and extent of the movement patterns. Likewise climatic changes can dictate the heterogeneity of habitats in time (length of favorable and unfavorable periods, and length of time a location remains suitable) and influence the phenology (timing and rate) of the movements. Thus by tracking climatic and meteorological changes through movement, organisms can maximize the expectancy of finding a new habitat or returning to a habitat after a period of unfavorability. The remainder of this chapter will be devoted to an examination of the movement patterns of animals in response to climatic and meteorological changes beginning with paleoclimatological events and biogeographical migrations and ending with short-term geophysical influences on the vertical migrations of planktonic organisms.

III. Paleoclimatological Changes and Biogeographical Migrations

A. PALEOCLIMATIC CHANGE AND BIOTIC RESPONSE

The climatic limits of extant biological species can be used to help reveal climates of the past based on the historical distribution and abundance of the same or very similar species, provided no evolutionary change in tolerance has occurred. Paleoecologists and paleoclimatologists have concentrated on many such organisms in their quest for information about the biota and climates of the past, and in doing so they have gathered considerable information on the biogeographical migrations of plants and animals. Over the last 300,000 years through several ice ages and warm interglacial times, the remains of botanical species, land vertebrates and invertebrates, and marine life are basically identifiable with species still living today somewhere in the world (Lamb, 1977, p. 174). Lamb further suggests that the assumption that the species used as indicators of past climates have not changed their tolerance is most likely to be true where the same assemblage of fauna and flora are found in the past as at present. If the tolerance of certain organisms does not change or changes relatively slowly with respect to climatic change, that is their evolutionary rate is bradytelic (Simpson, 1953), then the changes in distribution and abundance of these organisms can be of great importance in documenting the migrations of the biota in response to climatic change. Several workers have examined the climatic tolerances of species that lived in the past and are still present today (see Section II,B and Lamb, 1977, pp. 176–183). These studies suggest that many extant species have had rather slow rates of evolutionary change in tolerance and are consequently quite useful in monitoring recent climatic changes.

Species analysis of the pollens found preserved in peat and lake bed sediments (Aaby, 1976; Anderson, 1973; Bonny, 1972; Davis, 1969; Wright, 1976), and similar analyses of the remains of microscopic organisms that lived in the surface waters or in the depths of the oceans largely constituting the organic fraction of the sediments on the ocean bed (Emiliani et al., 1975; Weyl, 1978) have contributed an enormous number of time cross sections registering the effects of climatic history in most parts of the world. These cores provide documentation of the changing populations and distributions of the various species, the changes being only partly (but often largely) a direct response to climate and partly determined by such things as biological interactions and the effects of previous climatic, geophysical, and biotic conditions in modifying the soil or substrate.

Although there is a clear diagnostic advantage in dealing with microfauna and the microscopic pollen and spores of plants because they commonly produce very large populations, and variation in their abundance may be firmly established by statistically significant changes in the density of populations, the fauna in general from the smallest forms to the larger vertebrates are found to accompany the vegetation which provides their preferred habitat (Lamb, 1977, pp. 174–175). Fossil remains of mollusks (Frey, 1964, 1965; Leonard, 1953; Taylor, 1965), arthropods (Coope, 1975, 1977a,b; Frey, 1965; Ross, 1965), and vertebrates (Lundelius, 1976) offer additional evidence of climatic change. The use of yearly growth rings in the stems of trees that grow in seasonal climates of the middle and higher latitudes has provided considerable information on past weather and climatic regimes (Fritts, 1976, 1978; Hughes *et al.*, 1978; LaMarche, 1978; Moore, 1978; Dyer and Curtis, 1978), and just as this information can be used to infer climatic changes, it can also be used to infer changes in other aspects of the biota (e.g., reduced productivity). Even though considerable evidence of past climates has been derived from the changing distribution and abundance of biological species (brief summaries in Bowen, 1978; Butzer, 1976; Lamb, 1977), examination of the effects of climatic change on the biota is possible without the problem of circularity. This is so because of the abundance of corroborative climatic data derived from non-biological sources such as oxygen isotopes and paleotemperatures; changes in sea, lake, and river levels; ice sheet stratigraphy and glacier variations; and changes in geomorphology (Bowen, 1978, pp. 109–136).

B. Paleoclimatology

The chronology of climatic change during the past 100,000 years has only become clear since about 1970, and the main features of the chronology of the past one million years has become established only since 1973 (U. S. Committee for GARP, 1975). One of the most interesting features of the climatic record of the past 150,000 years is that both the present interglacial and the Eemian interglacial began with a rather abrupt termination of a very cold glacial interval. Because of the impact of these events on the climatic record, Broecker and van Donk (1970) have named these deglaciations Termination I (Holocene or present interglacial) and

Termination II (Eemian or penultimate interglacial). Excellent reviews of the Quaternary ice ages and interglacial periods can be found in Bowen (1978), Butzer (1976), Frenzel (1973), Gribbin (1978), Lamb (1977), Pearson (1978), and USC-GARP (1975).

1. The Eemian Interglacial

The Eemian interglacial reached its warmest period about 120,000 years ago, and this period lasted approximately 10,000 years. During this period the sea level was up to 18 m higher than at present, Scandinavia was an island, England was separated from the European continent for the first time, and all of northern Europe had considerably more of an oceanic climate than during postglacial times. An abrupt drop in temperature triggered rapid growth of ice on land about 110,000 years ago followed by another warm period that was somewhat less pronounced, and this warm period ended about 90,000 years ago when a sharp cold period occurred. After about 2000 years the temperatures once again warmed and a relatively warm period occupied the succeeding 10,000 to 20,000 years until approximately 75,000 years ago. The total length of the Eemian interglacial was about 50,000 years, but brief cold periods of approximately 2000 years duration interrupted the warm periods every 10,000 years. The Eemian interglacial is known as the Sangamon in North America, the Ipswichian in Great Britain, and the Riss–Würm interglacial in the Alps. The Eemian is used in northern Europe.

2. The Weichselian Glacial

The Weichselian (Wisconsin in North America, Würm in the Alps, Devensian in the British Isles, and the Gamblian and Makalian in Africa) Glacial Period in northern Europe began approximately 70,000 years ago. The ice sheets quickly formed over northern and northwestern Europe probably including parts of the British Isles and over the northern portions of North America within a time span of 1000 to 5000 years. The ice sheets were developed enough to survive warm interstadials that occurred during the early, middle, and late glacial periods although some retreat of the ice sheets during the interstadials has been documented (Dreimanis and Raukas, 1975). The major warm interstadials included the Amersfoort (65,000 BP), the Brørup (62,000 BP), and the Odderade (45,000 BP) during the Early Glacial; the Upton Warren or Moershoofd (45,000–40,000 BP), the Hengelo (40,000–35,000 BP), and the Denekamp (35,000–30,000 BP) during the Middle Glacial; and the Bølling (12,000 BP) and the Allerød (11,400 BP) during the Late Glacial. According to some authors (e.g., Lamb, 1977) the Upton Warren interstadial includes both the Hengelo and Denekamp interstadials.

The main peak of the glaciation was reached about 14,000 to 22,000 years ago, although an earlier peak (45,000–50,000 years ago) was nearly as extensive. Although the maximum extent of glaciation occurred between 22,000 to 14,000 years ago, the dates of accumulation and decline are not identical for the various global ice sheets. The ice sheets that covered parts of eastern North America (Laurentide) and parts of northern Europe (Scandinavian) reached their maximum extent between 22,000 and 18,000 years ago. The ice sheet over the Rocky Mountains (Cordilleran) reached its maximum only 14,000 years ago (USC-GARP, 1975).

At the last glacial maximum, continental ice sheets covered extensive areas of the northern hemisphere, sea level was lowered about 85 m, and sea-surface temperatures in the North Atlantic fell by as much as 10°C. Mean annual temperatures were considerably lower than at present (e.g., in northern Tennessee and the adjacent regions to the west in the United States, temperatures were probably 13°–15°C lower than today). At the peak of the glacial phases deciduous forests were largely eliminated from Europe north of the Alps (48°N) and in North America the Laurentide ice sheet extended as far south as 40°N, particularly in the midwestern portions of the United States (Bryson *et al.*, 1970).

3. The Holocene Interglacial

The last deglaciation began at different times at different locations throughout the world. Warmer oceans were largely responsible for bringing the first warming to high latitudes. The earliest postglacial warming probably took place on continental land surfaces at low latitudes and in the southern hemisphere, areas farthest from the extensive ice sheets (Lamb, 1977). In New Guinea evidence of glacial retreat and of vegetation changes establish incontrovertibly that temperatures had begun to rise about 15,000 years ago (Bowler *et al.*, 1976; Hope *et al.*, 1976). In New Zealand, Suggate (1965) presents evidence that deglaciation began as early as 14,000 years ago, and Walker (1978) suggests the beginning of deglaciation in Australia and in New Guinea occurred about 16,000 ± 2000 years ago. Walker (1978) mentions that in the Snowy Mountains of New South Wales glaciers had begun to retreat about 20,000 years ago, but the vegetational response to temperature rise seems not to have occurred until about 16,500 years ago. The Cordilleran ice sheet started a rapid recession about 14,000 years ago and was gone by 10,000 years ago. The Scandinavian ice sheet retreated at a rate of 1 km per year between 10,000 and 9,000 years ago. The ice conditions of Europe and North America had reached essentially their present extent about 8500 and 7000 years ago, respectively. The declines were characterized by substantial marginal fluctuations and the climatic instability during these fluctuations has been well documented by the records from fossil pollen, deep-sea cores, sea-level variations, and lacustrine records in western North America and Africa (USC—GARP, 1975). Several periods of widespread cooling and glacial expansion, approximately 2500 years apart, in the regions bordering the Atlantic Ocean have occurred since deglaciation began. One such expansion in Europe took place about 10,800 to 10,100 years ago (Younger Dryas) and lasted for approximately 700 years. The present interglacial (Holocene) apparently reached its peak in Europe and North America about 7500 to 8000 years ago, and temperatures were warmer than today. Since that time there have been colder intervals about every 2500 years, the last occurring about 300 years ago, that have punctuated the general cooling trend. During the last 1000 years there has been basic agreement among the climatic fluctuations in Western Europe, Southeastern Asia, and along the west coast of North America (Libby and Pandolfi, 1977, but see Ingram et al., 1978). Based on mean annual temperatures of the northern hemisphere, from about 100 to 1400 A.D. a warm epoch was present followed by a period known as the Little Ice Age from about 1430 to 1850 A.D. with cold maxima in the fifteenth and seventeenth centuries. In the middle of the seventeenth century, temperatures were sufficiently low that many glaciers in Alaska, Scandinavia, and the Alps advanced close to their maximum positions since the last major ice age thousands of years earlier. Since the middle 1800s temperatures have generally climbed, and only in the last few decades (1940s) have temperatures started to show a gradual decline in the Northern Hemisphere (Brinkmann, 1976). In New Zealand temperatures are still climbing slowly (Wilson et al., 1979). These changes are consistent with concomitant changes in the large-scale atmospheric circulation as reflected in sea-level pressure patterns (Mitchell, 1963, 1971).

C. Biogeographical Migrations

The major climatic changes that accompanied the various ice ages and interglacial periods of the last one million years, and the warmer interstadials and colder stadials within them have had a profound effect on the distribution of animals and plants throughout the world (Deevey, 1949). The biogeographic migrations of flora, fauna, and marine forms in response to paleoclimatic change during the end of the Pleistocene and the Holocene were generally dramatic. The movements involved primarily range retreats and expansions in response to colder ice ages and warmer interglacials and interstadials, respectively. The migrations consisted of major and minor shifts in the biota, and most were along a north–south axis in response to temperature changes, but some changes in the distribution and abundance of terrestrial biota were along a west–east axis in response to moisture changes.

1. Migrations of the Flora

When one considers the time spans involved in biogeographical migrations, the organisms do not have to be markedly mobile. Animals that are basically sedentary follow the slow movements of their habitat, and as Seddon (1971, Chapter 8) has emphasized, mobility does exist in the dispersal stages of a plant's life history when the reproductive body or propagule is transported by such means as wind, water, birds, fruit-eating bats, and the like. The distance and direction of transport can be haphazard and depend largely on fortuitous circumstances at the time the propagules are released, but the direction and velocity of prevailing winds or water currents can dramatically alter and bias the distance and direction to which a propagule is dispersed (Cremer, 1977; Daubenmire, 1978, pp. 5–32; Porter, 1976; Salisbury, 1976; Sorensen, 1978). Ordinarily it is a seed or detached fruit that is transported, but a detached vegetative organ of the plant can also be involved. The general topic of dispersal in higher plants has been reviewed by Van der Pijl (1972).

Because animals are dependent in large part on changes in the distribution and the abundance of plants, the biogeographical migrations of plants have been important in regulating the movement patterns of animals. For plants to migrate, that is change their range, they must produce viable propagules that are transported to areas outside that occupied by the parent population and become successfully established. According to Seddon (1971, p. 136), cases in which the distribution contracts by the elimination of portions of the population through death, and there is no occupation of new territory, should not be referred to as migration. He refers to such cases as examples of migrational retreat.

The features of plant and animal biogeographical migrations in response to long-term climatic changes can best be discussed by making reference to Good's theory of plant migration (Good, 1953). Basically Good's theory is one of an organism tracking its climatic environment (Fig. 3). If a tree species has a tolerance range between two isotherms T_{min} and T_{max} and a gradual climatic change occurs involving a general lowering of temperature (isotherms shift southward), the species will eventually occupy an area outside of its range of temperature tolerance. At the northern portions of the species' range the climate will not be favorable for production of ripe seed because of the inadequate warmth during the summer, or if the trees can reproduce, winter temperatures are too cold to permit survival of seedlings. The result is migrational retreat (Fig. 3A). At the southern edge of the species' range, southward expansion was limited originally by high summer

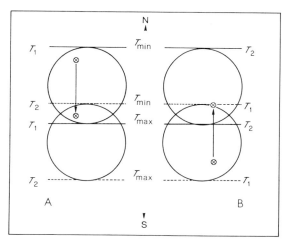

Fig. 3. Diagrammatic representation of biogeographical migrations in response to a geographical shift in a weather element (e.g., mean maximum and minimum temperatures). (A) Shift in ranges during an extended period of lower temperatures; (B) return of former range during a long-term warming trend. (Based on Good, 1953.)

temperatures before the climatic shift in isotherms, but after the shift, southward expansion or migration can be accomplished through seed dispersal (Fig. 3A). In the event that the climatic shift causes the isotherms to move northward as during an interglacial period, the changes in the range of the species will eventually follow suit (Fig. 3B). Just as climatic shifts can ultimately change the latitudinal distribution of a species, the shifts can also cause altitudinal changes in the distribution of a species in mountainous regions (Frenzel, 1968; Simpson, 1974; Van der Hammen, 1974; Wijmstra, 1978; Wright, 1971; Wulff, 1943). In general the rate of plant migration lags behind the rate of climatic shift. The rates of biogeographical migrations can vary widely among species depending on such factors as dispersal range, reproductive age, and a host of other geophysical and biotic factors (e.g., topography, soil nutrients, previously established species). The rates of plant migrations vary widely, and the characteristics of the disseminules are not necessarily good predictors of that rate. For example, *Corylus* has a heavy nut and *Pinus* a small-winged seed, but *Corylus* migrated at a faster rate than *Pinus* to reoccupy deglaciated terrain when the ice receded in Europe during the present interglacial (Faegri, 1963). Thus a community that originally existed under stable climatic conditions cannot migrate as a unit in response to a recent climatic change because of the differential rates of migration of its constituent species. New associations of species may result, and new communities may be formed, and it is distinctly possible that the former community will never be reconstituted in its original form.

The literature on plant biogeographical migrations is vastly superior to that available on animals because of the pollen profile record (Bowen, 1978; Lamb, 1977). Distribution maps for the different tree pollens during the late Pleistocene and the Holocene indicate changes in species compositions and often the routes that different plant species followed in response to climatic changes. General treatments of this topic can be found in Frakes (1979), Frenzel (1967, 1973), Pearson (1978), West (1977), Wijmstra (1978), and Wright (1976). More specific regional coverage can be found in Brunnschweiler (1964), Davis (1976), Dillon (1956), Nichols (1974), Whitehead (1965), Wright (1971, 1976), Bernabo and Webb (1977), and Van Devender and Spaulding (1979) for North America; in Ab'Saber (1977) and Van der Hammen (1974) for South America; in Farrand (1971), Firbas (1949), Frenzel (1968), Godwin (1956, 1975), Iversen (1973), Van der Hammen et al. (1971), Wijmstra (1969), Woillard (1974), and Zagwijn (1974, 1975) for Europe; in Coe (1967), Coetzee (1967); Kendall (1969), and Livingstone (1975) for Africa; in Frenzel (1968), Fuji (1973), and Tsukada (1966, 1967) for Asia; and Bowler et al. (1976) and Pittock et al. (1978) for Australia and New Guinea.

Evidence gathered from pollen profiles suggest that during the Eemian interglacial the Siberian forest species (e.g., *Picea obovata*) retreated rapidly eastward from Europe, and *Pinus* initially became the dominant forest type with some *Betula* present. The July temperatures rose to, or even slightly above, present values. The first return migrations of *Quercus* and *Ulmus* into the European lowlands north of the Alps then occurred, and Siberian and European *Picea* were absent. Later during the moist *Quercus–Ulmus* phase, European or Norway spruce (*Picea abies*) spread rapidly over a wide area of Europe. Following the oak–elm phase, a hazel–basswood (*Corylus–Tilia*) phase and hornbeam (*Carpinus*) phase occurred. During the second part of the oak–elm phase, basswood first began to immigrate to central Europe and replace oak and elm rapidly. As soil deterioration continued, hornbeam began its invasion. As the climate became even wetter, health (*Ericaceae*) rapidly became important and basswood and elm were eliminated. At the end of the Eemian as cooling began, extensive spruce (*Picea*) and fir (*Abies*) forests returned to Europe. Dabrowski (1971) has documented the rapid increase in species diversity of forest trees as they migrated into northern Europe during the Eemian interglacial in Poland.

Although equivalent data are lacking for North America during the Sangamon (= Eemian), pollen data from northeastern United States suggest that during the climatic optimum, woodlands of oak, beech (*Fagus*), and hickory (*Carya*) were extensive (Donner, 1964), and swamp cypress

(*Taxodium distichum*) was abundant in the vicinity of Washington D.C. (Knox, 1962). It is believed that the following events took place during the Sangamon in the northcentral United States. According to Terasmaë (1960) and Kapp and Gooding (1964) pine and spruce woods initially predominated, and as the climate became drier pine became the dominant tree. As the climate continued to warm pine was replaced by hardwoods and bushes that migrated into the area. Oak and hickory were dominant, but maple (*Acer*), beech, ash (*Fraxinus*), walnut (*Juglans*), sweet gum (*Liquidambar*), tulip tree (*Liriodendron*), hornbeam, and elm also occurred. A bit later hemlock fir (*Tsuga*) and larch (*Larix*) appeared. Finally spruce and fir once again became established as the climate became cold and moist. By the end of the Eemian, boreal forests in the Northern Hemisphere began to expand southward in response to cooling temperatures.

Following the Eemian interglacial in northern Europe, the onset of the Weichselian Glacial Period began with an abrupt cold period and boreal forest became more open, and a subarctic park landscape prevailed (Wijmstra, 1978; Woillard, 1974). *Pinus* forests returned during the Amersfoort interstadial and rather high values of *Quercus*, *Fraxinus*, and *Ulmus* pollen strongly suggest that even thermophilous trees showed a return migration by the end of the interstadial. Following the Amersfoort the climate deteriorated once again and the forests opened. Although some arctic and alpine plants were in the area, genuine tundra conditions probably did not occur during this cold phase. By the time of the Brørup interstadial, spruce forests were present with only scattered thermophilous elements (*Quercus* and *Ulmus*) and following this interstadial the flora became basically open arctic tundra. The last interstadial of the Early Glacial Period, the Odderade, was characterized by a return of pine forests with some thermophilous elements. During the beginning of the Middle Glacial Period (Lower Pleniglacial) a polar desert prevailed over much of northern Europe, and the "warm" interstadials during this period (Moershoofd, Hengelo, and Denekamp) permitted only the return of vegetation types similar to those typical of arctic and polar areas. In northeastern France steppelike vegetation chacterized the Lower and Middle Glacial Periods except for the return migration of pine and birch during the warm interstadials of the Middle Glacial Period.

The climate once again turned colder during the last of the Middle Glacial Period (Upper Pleniglacial) and polar desert conditions prevailed. At the peak of the Glacial Period nearly all of Europe north of the Alps supported tundra vegetation only. In fact during long intervals of time between 60,000 and 50,000 BP and between 25,000 and 15,000 BP polar deserts with little or no vegetation covered extensive areas. In the Mediterranean region the landscape was also largely unforested with cold-steppe vegetation similar to that occurring in central Asia at present. Only during the moist, warm interstadials did forests of pine and deciduous trees expand northward to recolonize once forested areas. In Macedonia the forests had largely disappeared by 60,000 BP, and the area had a steppe landscape. Forests did not return to the area until after the end of the Last Glacial Period, about 14,000 years ago (Farrand, 1971; Wijmstra, 1969). In the Netherlands during the Late Glacial, the Bølling and Allerød interstadials were characterized by a parklike birchwood landscape and pine forests, and between these interstadials herbaceous plants were more important elements of the flora.

In North America the vegetational responses during the Early Glacial Period are poorly known, but the few studies of vegetational changes in the Middle and Late Glacial Periods indicate a surprising distribution of species (Davis, 1976). The vegetational changes during the interstadials of the Wisconsin (55,000 to 22,500 years ago) in the eastern Great Lakes region has been examined by Berti (1975). During the Port Talbot I interstadial (55,000 to 50,000 BP) *Pinus* domination gave way to *Pinus*, *Quercus*, and other nonarboreal species, suggesting warm and dry conditions (mean July temperatures between 15° and 21°C). During the Port Talbot II (48,000 to 36,000 BP) and the Plum Point (32,500 to 22,500 BP) interstadials, *Pinus* and *Picea* dominated, indicating moister, cooler conditions (mean July temperatures between 10° and 15°C). Prior to 30,000 years ago, prairie predominated at a southern Illinois site (Gruger, 1972a,b) and in the piedmont of Georgia an open xeric woodland with oak and pine was present (Watts, 1973). During a warm interstadial (28,000 to 22,000 BP) oak and hickory forests replaced the oak and pine woodlands, and cypress was present. Farther north spruce and pine extended southward to central Illinois. Mixed forests of pine, birch, alder, and spruce occurred in southern Illinois. For a site in northeastern Kansas, Gruger (1973) suggests that prior to 23,000 years ago, the vegetation was rather open with some pine, spruce, birch, and local stands of alder (*Alnus*) and willow (*Salix*). Between 23,000 BP and 15,000 BP, spruce forests prevailed, and by 11,000 BP mixed deciduous forests and prairie characterized the area.

At the culmination of glaciation about 20,000 years ago, the ice sheet was bordered by a tundra belt extending from Minnesota and Wisconsin, where it was relatively narrow, to central New York State and Pennsylvania, where it was more than 100 km in width. Tundra and permafrost extended at the higher elevations in the Appalachians southward to the Carolinas (Maxwell and Davis, 1972). During the full glacial the forests of southeastern Virginia were primarily boreal (45% spruce, 40% pine, and 3% fir) while those of southeastern North Carolina were dominated by red pine (*Pinus resinosa*) and jack pine (*Pinus banksiana*) with relatively little spruce (Whitehead, 1965). The vegetation that characterized the plains and Midwest regions has been reviewed by Davis (1976). Much of the central plains were forest covered, and spruce grew in pure stands in Kansas east to central Illinois and north to Iowa. Pine extended into the plains from the Rocky Mountains. Mixed stands of spruce, with prairie vegetation and mixed oak forests nearby, characterized the southern Illinois region. During the late Wisconsin glacial maximum (22,000 to 17,000 years ago) woodland communities of pinyon (*Pinus edulis*) and junipers (*Juniperus*) occurred at middle elevations of 1525 to 550 m in areas now occupied by desert scrub communities in most of the present Chihuahuan, Sonoran, and Mohave deserts in the southwestern United States (Van Devender and Spaulding, 1979).

During the Late Glacial Period (15,000 to 10,000 years ago) in the Chesapeake Bay region boreal forests gradually changed to a hemlock–northern hardwoods type, and in southeastern North Carolina the pine–spruce forests were initially replaced by beech–hemlock and eventually by oak, hickory, and other deciduous forest species (Whitehead, 1965). At the beginning of the late glacial phase in the Southwest warm desert species were common in the woodlands at lower elevations, and mixed conifer and subalpine forests were present at higher elevations. As the climate changed and glacial retreat began, montane communities acquired modern aspects, and the more mesophytic species disappeared from lower woodlands about 11,000 years ago (Van Devender and Spaulding, 1979).

In Europe and North America, forests showed return migrations at about 30 km a century following the retreat of the ice sheets at the beginning of the Holocene (Moreau, 1955; Flint and Deevey, 1951). The forests returned in northwest Europe in three stages. Birch–pine (*Betula–Pinus*) forests were the first to show a return migration into the steppe and tundra characteristic of the preceding cold period. During the peak warming period oak and elm forests returned, and finally the oak–birch and oak–beech forests that are typical of present European forests developed. The actual distributions of these trees once they returned were largely governed by edaphic factors. The northward expansion of forests during the warm period was halted, and forests were destroyed in many places during the cold period—the Younger Dryas (Denton and Karlen, 1973).

In the northern portion of the Mediterranean (Macedonia) as the climate warmed at the beginning of the Holocene, *Quercus cerris*, *Q. robur*, *Pinus*, and *Juniperus* returned to the area, but during the Younger Dryas a more steppelike vegetation prevailed. Following this cold period *Quercus robur* and *Carpinus* returned and these sub-Mediterranean elements were eventually replaced by a more typical Mediterranean forest of evergreen oaks, maquis elements, and a fir (*Abies*) zone at higher elevations (Wijmstra, 1969).

In North America the return migrations of forest elements during the Holocene are rather well known (Davis, 1976; Bernabo and Webb, 1977). The number of forest species migrating northward in North America was considerably greater than that known for Europe. It is generally agreed that during the glacials many European species went extinct because of the blockage by the east–west orientation of the Alps. The north–south orientation of the mountains in North America did not prevent the southward migration of forest elements in response to the changing climate of the glacials. Bernabo and Webb (1977) have shown that after 11,000 years ago the boreal forests of spruce decreased in extent as pine moved rapidly westward and northward. Oak dominated deciduous forests moved northward from the South, and prairie flora expanded rapidly eastward between 10,000 and 9000 years ago, reaching its maximum eastward extent by 7000 BP. At this time a reversal occurred and boreal elements migrated southward once more, and the prairie flora retreated westward. Since the last 500 years man has influenced the vegetation through land use practices, and this has been documented by the greatly expanded distribution and abundance of herb pollen (Bernabo and Webb, 1977).

Davis (1976) has prepared maps based on 26 pollen sites that show the migration sequence for several forest species in the eastern United States. The maps show the different source areas and directions of movement and document the varying rates of return migration for each species. At the beginning of the Holocene, spruce moved northward following ice retreat. Tundra was present for several thousand years after glacial retreat in southern New England, and the first spruce forest to return to the area did so 11,500 years ago. One thousand years later spruce was dense and alder (*Alnus*), fir, and jack pine (*Pinus banksiana*) had returned. Jack pine from the southeastern United States followed the northward advancement of spruce. White pine (*Pinus strobus*) moved northward and westward from the mid-Atlantic coast about 1000 years after jack pine. The northward limit of white pine was at its maximum 5000 years ago; since that time the northward limit has been contracting, populations in New England and southern Canada have shown a decline, and a westward expansion has been recorded. The northward migration of oak (*Quercus*) closely followed white pine. Hemlock (*Tsuga canadensis*) movement followed white pine after a lag of approximately 500 to 1000 years. This slow-growing shade-tolerant species invaded the closed forest of pine, oak, birch, and maple. The migratory pattern of hemlock is similar to that of white pine in that it has been retreating southward for the last 5000 years and expanding west-ward. Beech (*Fagus grandifolia*) as it moved north penetrated closed forest communities, because of its adaptation to seed in the shade of a closed forest canopy. Hickory (*Carya*) slowly moved northward and eastward from southern midcontinent, and arrived in Connecticut only 5000 years ago. Chestnut (*Castanea*) moved very slowly northward and westward from the mid-Atlantic region and arrived in the Alleghenies 3000 years before it arrived in Connecticut.

In contrast to the gradual migrations of the central and eastern forests of the United States during the Holocene, the adjustment of the flora in the Southwest occurred rapidly, probably because of the dependence of the desert and woodland species on precipitation (Van Devender and Spaul-ding, 1979). Current vegetational and climatic regimes in the Southwest became established after about 8000 years BP as xeric species exploited in-creasingly drier microhabitats at low elevations and forests moved into more mesophytic and cooler microhabitats at higher elevations in the mountains.

2. Migrations of the Fauna

Information on the biogeographical migrations of faunal elements during the late Pleistocene and Holocene is not as abundant as that available for plants, but fossil remains Foraminifera, Radiolaria, Mollusca, Arthropoda, and Vertebrata document the past movements of marine and terrestrial animals in response to paleoclimatological changes. General reviews and collections of papers on this topic can be found in Bowen (1978), Colbert (1953), Crisp (1959), Frenzel (1973), Hibbard (1949), Lamb (1977), Shotton (1977), West (1977), Zeuner (1959), Dort and Jones (1970), Wright and Frey (1965), Johnson and Smith (1965), and Sparks and West (1972). Unlike the biogeographical migrations of plants, the biogeographical migrations of animals took place quite rapidly in response to climatic change because of the mobility of most animals. Consequently, during in-terstadials faunal components frequently outpaced floral elements in return migrations to areas made habitable by climatic amelioration, and during stadials faunal elements quickly retreated from areas of climatic deteriora-tion (Sissons, 1979).

a. Foraminifera and Radiolaria. As global temperatures changed during the glacials and interglacials, ecological water masses defined by a range of temperature and salinity values with precise modern analogues (Ruddiman and McIntyre, 1976) shifted positions in the oceans. Different assemblages of Foraminifera and Radiolaria tracked particular water masses, and as

these organisms died and accumulated at the bottom of the ocean, they left
in the sediment a record of their presence in the surface waters (Schopf,
1980). Five major assemblages can be recognized: (1) tropical, consisting of
Globigerinoides ruber, *Globigerinoides sacculifer*, *Globorotalia menardii*,
and *Pulleniatina obliquiloculata*; (2) subtropical, consisting of *Globorotalia
truncatulinoides*, *Globigerinoides falconensis*, *Globigerinoides inflata*, and
Globigerinoides calida; (3) subpolar, including *Globigerinoides pachyderma*, *Globigerinoides bulloides*, and *Globigerinita glutinata*; (4) polar, consisting of *Globigerinoides pachyderma*; and (5) a gyre-margin, including
Globorotalia menardii and *Pulleniatina obliquiloculata* . A certain species
(*Globorotalia menardii*) is strongly associated with warm water masses and
its absence in sediments at the time of deposition suggests cold conditions
(Corliss, 1975). The sole polar species, *Globigerinoides pachyderma*, is an
excellent indicator of cold polar surface waters (Ruddiman *et al.*, 1977).

During the Eemian interglacial (120,000 BP) subtropical species assemblages reached 50°N in the Atlantic, and at peak of the Weichselian
glacial (13,000 BP), polar forams moved southward to 40°N. By 9300 years
ago polar water was retreating northward, and at present it is restricted to
the position it occupied during the Eemian (Ruddiman and McIntyre,
1976). Additional studies of Foraminifera movements in response to
climatic change for all the oceans of the world can be found in Ericson *et al.*
(1961, 1964), Glemarec (1979), McIntyre (1967), McIntyre *et al.* (1972),
Moore (1973), Ramsay (1977), Ericson and Wollin (1970), Imbrie and Kipp
(1971), Steuerwald and Clark (1972), Thunell and Lohmann (1979), and
Williams and Johnson (1975).

The absolute ages of Quaternary radiolarian assemblages in the equitorial Pacific have been examined by Johnson and Knoll (1975), and they
suggest that the migration of newly evolved species from their place of
origin to other favorable regions is dependent on their Darwinian fitness
and on physical factors, principally the motion of ocean currents. Griggs *et
al.* (1970) have pointed out that in the Eastern Pacific Ocean and in the
western North Atlantic between 11°N and 41°N, Foraminifera are more
abundant relative to Radiolaria in cold waters today, while Radiolaria are
more numerous in warmer waters. As water temperatures change the proportions of radiolarians to foraminiferans also change as each group
migrates to track its appropriate environment. Documentation of the
movements of radiolarians in response to long-term climatic changes and
migrating biogeographic provinces within the oceans can be found in Hays
(1967), Moore (1973), Nigrini (1971), Sachs (1973a,b), Johnson and Knoll
(1974, 1975), and Keany and Kennett (1972). The temporal fluctuations in
benthic communities in relation to climatic changes has been discussed by
Glemarec (1979).

b. Mollusca. The record of the changes in the distribution and abundance of terrestrial and marine mollusks in response to long-term climatic changes is strongly dependent on an alkaline environment and the proper conditions for preservation. In acid soils mollusks deposits rapidly degenerate and disappear. According to Zeuner (1958) the rate of molluscan evolution is slower than that found in mammals, and consequently the deposits of mollusks, when present, provide a good record of the biogeographical migrations of members of this group in response to changes in their environment. Little is known about the ecology of marine mollusks in response to long-term climatic change so the record of their migrations is poor (see Norton, 1977), but terrestrial mollusks have left a considerably better record of their patterns of movement in response to environmental change (Sparks, 1961). Kerney (1977) has pointed out that because mollusks are not a rapidly evolving group, changes in their assemblages over time are attributable more to their ability to migrate to favorable environments and from unfavorable ones than to evolutionary change. The changes in the distribution and abundance of various snail assemblages are reasonably well documented, and Sparks (1961) recognizes four types of assemblages: (1) slum group, (2) catholic group, (3) ditch group, and (4) a moving water group. Certain species are particularly sensitive to climatic change and the biogeographical migrations of these species have provided a good record of climate changes in the past. Arctic–alpine species (e.g., *Columella columella*) showed southward movements during the glacial advances and more thermophilous species (e.g., *Discus rotundatus*) moved northward during interstadials and interglacial periods. Kerney (1976a,b) has discussed the changing distribution and abundance of the snail fauna of England during the last 120,000 years. During the Ipswichian (Eemian) Interglacial, a number of southern, freshwater species of snails (e.g., *Belgrandia marginata, Segmentina nitida, Potomida littoralis, Corbicula fluminalis, Pisidium moitessierianum,* and *Pisidium supinum*), moved into Britain.

During rapid climatic change assemblages mix because of the different levels of tolerance of the species comprising particular assemblages and the different rates of biogeographical migrations. In Britain it is not clear that any terrestrial mollusks survived during the period of maximum cold during the Devensian (Weichselian) between 26,000 BP and 14,000 BP but assemblages of catholic palaearctic and arctic–alpine species were present during the middle Devensian (between 42,000 BP and 27,500 BP). At the end of the Devensian the snail fauna of southeastern England is well known and consisted of a mixture of biogeographical elements with no modern analogue. Elements of climatically tolerant species of wide range, arctic–alpine species, western European species absent from Scandinavia, and species of limited range in northern Europe are represented in the fauna. During the Flandrian Interglacial (present postglacial) terrestrial snail faunas reflect the return of open forest with the appearance and expansion of more thermophilous species, and freshwater snail assemblages also document the warmer climate (9000 BP). According to Kerney (1977), the last of the remaining late-glacial open-ground species disappeared during the warmest portion of the Flandrian (about 7000 BP). The snail *Pomatias elegans* has shown a southward retreat to its present northeasterly limits in Europe in response to a drop in winter temperatures since the Flandrian optimum (Kerney, 1968). A review of the movements of land snails in central Europe during the Holocene interglacial can be found in Lozek (1972). The snail record also clearly documents the arrival of the Romans in England (ca. 14 A.D.) with the appearance of introduced Mediterranean elements (e.g., *Helix aspersa*).

The changing distribution and abundance of mollusks have been correlated with climatic changes in North America during the last 125,000 years by Frey (1964, 1965), Leonard (1953), Ortmann (1913), Taylor (1965), Wagner (1977), and Wright (1932). In general the northward movement of certain mollusks is correlated with increasing temperatures during interglacials and interstadials and southward movements occurred in response to colder climates during the glacials and stadials. This is particularly true for the shallow-water mollusks of the continental shelf in the Texas–Louisiana region (Curray, 1960; Parker, 1960). During the peak of the Wisconsin glaciation in the Gulf of California, macroinvertebrates, chiefly mollusks, probably were able to migrate as much as 1100 km south of their present position (Parker, 1964a,b).

c. Arthropoda. The paleomigrations of arthropods in response to climatic changes have received considerable attention from paeloclimatologists. The glacial advances and retreats of the late Pleistocene and Holocene have greatly affected the distribution and abundance of terrestrial crustaceans and insects (Frey, 1964, 1965), and the biogeographical migrations of ostracods, cladocerans, amphipods, decapods, and insects have tracked closely the changes in environmental conditions during the last 125,000 years.

During the maximum glaciation, the destruction of drainages in North America probably resulted in forced southward migrations of amphipods (Bousfield, 1958). Following glacial retreat some thirty species and subspecies of the Grammaridae (e.g., *Crangony*) performed return migrations into the regions previously covered by the glaciers. A southern species of amphipod (*Hyalella azteca*) also moved northward and invaded much of the area vacated by glacial retreat and is presently the most common and widely distributed freshwater amphipod in North America. Frey (1958) has studied the fossil remains of cladocerans in a northern German lake and found that during the Allerød Interstadial there was a rapid return of Cladocera, followed by a sharp decline when tundra conditions returned. Following the Weichselian Glacial, permanent temperate lakes acquired very rapidly essentially modern communities of Cladocera (Frey, 1961), suggesting a strong northward immigration in response to climatic warming. During the Loch Lomond Stadial of the British Isles, a colder period roughly equivalent to the Younger Dryas of Scandinavia, arctic cladocerans such as *Chrydorn sphaericus* and *Acroperus harpae* increased in abundance (Harmsworth, 1968). The changes in the distribution and abundance of freshwater (Staplin, 1963) and marine ostracods (Valentine, 1976; Kaesler and Mulvany, 1977) have been particularly useful in the determination of past climates, because of their slow rate of evolution and their movements in response to climatic change. Some work on the biogeographical migrations of crayfishes in relation to drainage patterns during the late Pleistocene and Holocene has been done in the southeastern United States (Hobbs, 1969), but, in general, there are very few fossil records of freshwater decapods. Frey (1965) has provided a summary of other studies of freshwater crustacean paleomigrations during the Quaternary, and Dadswell (1974) has reviewed the postglacial dispersal of selected crustaceans in eastern North America. A discussion of the changes in the ranges of marine crustaceans in response to climatic changes during the late Pleistocene and Holocene can be found in Briggs (1974), Crisp (1965), and Ekman (1953). The responses to more recent changes in climate are treated by Cushing and Dickson (1976).

The biogeographical migrations of insects, particularly the Coleoptera are relatively well known, and Coope (1977a,b) has published considerable information on this subject. In his work he has emphasized the enormous scale of the changes in geographical distribution of species in response to changes in climate. Within the last glacial–interglacial cycle, some species of beetles changed the limits of their distributions by as much as 7000 km (Ullrich and Coope, 1974). Since the Younger Dryas nearly 40% of the Coleopteran fauna in the British Isles has disappeared. Such rapid rates of change suggest that faunal assemblages are very transient with elements moving rapidly into and out of an area in response to changes in climate.

During the Ipswichian or Eemian the beetle fauna of Britain contained about 29% more thermophilous species than at present (Coope, 1975, 1977a). These species (*Oodes gracilis*, *Cybister lateralimarginalis*, *Airaphilus elongatus*, *Caccobius schreberi*, *Onthophagus opacicallis*, and *Valgus hemipterus*) presently occur in southern Europe and indicate a climate at the time of the interglacial about 3° above that found in England today. During the last glaciation (Weichselian, Devensian), coleopteran elements that are characteristic of colder climates moved into Britain (Coope, 1968). When interstadials occurred (e.g., Upton Warren), thermophilous coleopterans moved northward rapidly in response to warming temperatures and returned to areas that were treeless because of the lag time in vegetational migration (Morgan, 1973; Coope and Angus, 1975). Similarly during the Windemere Interstadial (14,000–10,600 BP) more thermophilous beetles moved in Britain well ahead of the arrival of birch woodlands (Coope, 1977b). About 10,000 years ago, the Loch Lomand Stadial in Britain strongly influenced the biogeographical movements of insects. According to Sissons (1979) northern species of beetles moved into Britain (e.g., *Nebria nivalis* which is currently found only on the highest mountains of Scotland, and *Bembidion dauricum* which is presently limited to arctic Norway and Swedish Lapland), and thermophilous species moved southward from Britain. About 10,300 BP a thermophilous fauna began to return (Ashworth, 1972) and the new fauna was firmly established by 9500 BP (Osborne, 1974). In southwestern Scotland, the beetle assemblage indicated a climate even warmer than at present (Bishop and Coope, 1977). As the climate warmed the insects moved ahead of the woodland vegetation. Walker (1957) has shown that biogeographical migrations of insects have occurred in the last 60 years in the Ontario region of Canada at Lake Simcoe. Seven orthopteran species of northern affinity have disappeared while five species of southern affinity have become established. Similarly, DeWorms (1958) has discussed the northward movement of certain species of European lepidoptera into the British Isles in relation to relatively recent climatic changes, and Mere (1961) has studied the recent colonization of England by new species of macrolepidoptera. Additional details of insect biogeographical migrations can be found in Ashworth (1980), Barr (1969), Byers (1969), Coope (1975, 1977a,b), Frey (1965), Henriksen (1933), Ross (1965, 1970), and Theobald (1952).

   d. Vertebrata. During the late Pleistocene and the Holocene, the movements of the vertebrate fauna in response to climatic changes were as dramatic as those recorded for invertebrates. Because the numbers of vertebrate fossil remains are considerably less than those for invertebrates and vegetation, the record of vertebrate biogeographical migrations during the last 125,000 years contains gaps in space and time. However, the existing record clearly indicates movements of considerable magnitude in response to changing temperature and moisture regimes during glacial advances and retreats. Information on the changes in distribution of Pleistocene vertebrates in the eastern, central, and western United States can be found in Hay (1923, 1924, 1927), respectively, and similar information on the past and present distribution of vertebrates in the southern United States is reviewed by Blair (1958). A summary of Pleistocene changes in the fauna and flora of South America can be found in Vuilleumier (1971). Stuart (1974, 1976, 1977a,b) has examined the changes in the distribution of vertebrates during glacial and interglacial environments in Britain, and Degerbøl (1964) has remarked on the late and postglacial vertebrate fauna and its ecological relations in northern Europe. The Pleistocene and Holocene zoogeography of Africa has been discussed by Monod (1964) and Moreau (1969), and Lundelius (1976) has summarized the changes in vertebrate distribution during the Pleistocene and Holocene throughout the world. Because climatological changes of different periodicities are occurring continuously, biogeographical migrations continue at present, and records of vertebrate faunal shifts during the last two centuries have been documented for many different regions of the world (e.g. May, 1979).

   i. Fishes. Although little evidence of faunal shifts exists in freshwater fishes during the major climatic changes of the late Pleistocene and Holocene (Miller, 1965), drainage and climatic changes associated with Pleistocene glaciation clearly had major effects on the distribution and evolution of the North American fish fauna (Cross, 1970; Deevey, 1949; Smith and Fisher, 1970). According to Jenkins *et al.* (1971) many fish stocks were displaced southward into refugia during the peak of the glaciation, and faunas were enriched by dispersal of species through conjoined river drainages, such as those affected by stream capture and by extended rivers on the Atlantic slope during Pleistocene periods of lowered sea level. A similar analysis for the Central Great Plains of North America can be found in Cross (1970). Cross presents evidence that several species of freshwater fishes (e.g., *Dionda nubila, Fundulus sciadicus*) probably dispersed southward during the glacial advances, and when warmer conditions returned, the species moved northward, except for some populations that found suitable habitat in the South. A number of species of fish have shown multiple crossings of the Continental Divide of western North America, and most of these crossings probably occurred in postglacial times; no crossing is thought to be earlier than the latter part of the Pleistocene (Miller, 1958). The postglacial migrations of some species of freshwater fish in eastern North America has been studied by Dadswell (1974).

   Ekman (1953) has summarized the shifts in the distribution of marine fish in response to climatic changes at the end of the Pleistocene and during the Holocene, and additional useful discussions of changes in the oceanic environment during the Quaternary can be found in Briggs (1974) and Schopf

(1980). Cushing (1975, 1976a, 1978) and Cushing and Dickson (1976) have reviewed the shifts in the distribution of marine fishes in response to climatic changes during the last two centuries. There is increasing evidence of a 100-year periodicity in the wind system of the northern hemisphere and an associated change in the Scandinavian herring fishery, and the Japanese sardine appears to show a similar cycle. Similarly, Hubbs (1948, 1960) has discussed the northward movement of tropical and subtropical fishes and invertebrates in coastal California waters during warm periods in the mid-1800s and from 1957 to 1959.

ii. Amphibians and reptiles. Zoogeographical distributions of amphibians and reptiles during the Quaternary indicate distinct north–south changes in their distributions in response to glacial and postglacial conditions (Smith, 1957). At the time of the Sangamon (Eemian) Interglacial in the Great Plains Region of the United States, the climate was moist, temperate, and equitable; and it was considerably less continental than today, with winter temperatures rarely reaching freezing and summer temperatures slightly less warmer than today (Hibbard, 1955, pp. 199–204; Kapp, 1965). These environmental conditions permitted the northward movements of a coastal–plain biota from the southeast and Gulf Coast into the southern High Plains. The southern turtle and lizard fauna that invaded the southern High Plains has been examined by Etheridge (1958, 1960), Milstead (1967), and Schultz (1969), and this fauna included *Geochelone, Gopherus, Terrapene carolina llanensis, Phrynosoma modestum,* and *Holbrookia texana.* The snake fauna of the High Plains during the Sangamon included *Agkistrodon contortrix* (Brattstrom, 1967), and Tihen (1962) has commented on the Bufonids present at this time. During the Wisconsin the herpetofauna shifted southward in response to glacial advancement (Holman, 1963, 1964; Mecham, 1959; Milstead, 1967; Tihen, 1962). The herpetofauna record also suggests the eastward movement of several species during the xerothermic phase of the postglacial 10,000 to 9000 years ago (Schmidt, 1938), and additional studies of amphibian faunal shifts during the Wisconsin glacial and the postglacial support these findings (Blair, 1958, 1965). Blair points out that presently disjunct southern populations of relatively northern species reflect southward displacement during the Wisconsin glaciation and that northern disjuncts are rare. Blair attributes the existence of various east–west pairs of anurans in the eastern United States to splitting of previous ranges in these thermophilous animals during southward shifts in climatic zones during glacial advances, and he believes the east–west disjunctions of urodeles occurred during interglacial conditions following the spread during glacial maxima. In Europe the postglacial migrations of the fire–bellied toad (*Bombina bombina*) and the yellow-bellied toad (*Bombina variegata*) have been studied by Arntzen (1978). These species have moved from refugia to the Hungarian Plains and the plain around Prague, following the Danube and Elbe Rivers.

Jopson (1971) has studied the changes in turtle distribution during the late Pleistocene and Holocene and concluded that some turtles were forced from the southern uplands by cold in the Pleistocene and have only recently returned. Because of the mobility of turtles there are few species that were isolated physiographically during the last glacial–interglacial cycle. *Clemmys* is the most northerly genus in distribution and *C. muhlenbergi* (the bog turtle) has a disjunct range in the southern uplands and is more generally distributed in the northern part of its range. The southern uplands form is a postglacial relict that has extended its range northward of the terminal moraine and has become scarce in the South as cold bog habitats have disappeared in recent times. Similarly, *Chelydra serpentine* and *Sternothaerus odoratus* are both relatively cold tolerant, and both have extended their ranges well north in recent geological time.

There is additional evidence from the herpetofauna that the maximum warmth (hypsithermal) of the present interglacial occurred about 8000 to 9000 years ago. Bleakney (1958) in studying the recolonization of glaciated country in eastern Canada by amphibians and reptiles pointed out that if recolonization was from the Nova Scotia peninsula after glacial recession, then the present climatic zone must have been 320 km farther north than at present. Additional details of amphibian and reptilian zoogeographic migrations during glacial and interglacial times can be found in Holman (1976), Karlstrom (1962), Martin (1958), Porter (1964, 1972), Savage (1960), Auffenberg and Milstead (1965), and Peabody and Savage (1958).

iii. Birds. Most of the fossil birds from the Pleistocene come from the middle and late periods (Selander, 1965), and this record documents the shifts in the ranges of several species in response to glacial and interglacial climatic changes. At the peak of the last glaciation (Wisconsin) the distributional limits of many bird species were shifted considerably farther south than at present. The influence of climatic changes during the late Pleistocene and Holocene on the avifauna of the Southern Appalachian Mountains has been discussed by Hubbard (1971). During the maximum glaciation bones of spruce grouse (*Canachites canadensis*) at Ladds, Georgia (Wetmore, 1967) and bones of spruce grouse, sharp-tailed grouse (*Pedioecetes phasianellus*), and gray jay (*Perisoreus canadensis*) at Natural Chimneys in central–western Virginia (Guilday, 1962) indicate a major southward shift in the ranges of these boreal species. Although these species are sedentary their presence in the Southern Appalachians during the height of the Wisconsin glaciation correlates well with the boreal habitat that characterized the area at that time.

Miller (1937, 1940) has reviewed the ecological associations and life-zones of Pleistocene birds in California, and he concluded that the distribution and abundance of a number of passerine species indicated a southward shift of approximately 300 km at the peak of the glaciation in North America. During the Late glacial (13,000 to 10,500 BP) an eastward expansion of some western species of birds occurred accompanying the expansion of prairie eastward. Today the sharp-tailed grouse does not occur south and east of the Great Lakes Region, but during the period following maximum glaciation, it ranged through the valleys of the northern Appalachians in Pennsylvania (Wetmore, 1959) and in Virginia (Wetmore, 1962). The magpie (*Pica pica*) is presently restricted to areas from Manitoba, Kansas, and Oklahoma westward, but during late-Wisconsin times it ranged eastward into Virginia (Wetmore, 1962).

The influence of climatic changes on the avifauna of France during the last glacial and the present interglacial has been examined by Mourer-Chauvire (1974, 1975, 1976) and Demarcq and Mourer-Chauvire (1976). In France, the vegetation during the glacial maximum was boreal and open steppe, and the avifauna was closely correlated with the vegetation. Similar information for Poland can be found in Bochenski (1974), and Stuart (1977a) has reviewed some of the changing avifauna of the British Isles, particularly during the last glacial stage (Devensian). The changing climate of the Pleistocene and its effects on the vegetation and the avifauna of South America has been examined by Haffer (1974), and Monod (1964) and Moreau (1963, 1969) have done the same for Africa. Moreau (1963) in discussing the avifauna of Africa has pointed out that during the Pleistocene the southern margin of the Sahara fluctuated 480 km south and north of its present position, and 7000 years ago during the interglacial optimum the southern margin was approximately 480 km farther north than at present. The Kalahari avifauna extended considerably farther northeast than at present only 10,000 years ago.

Serventy (1958) has discussed the shifts in the marine avifauna in response to changes in climate during the late Pleistocene and Holocene. As different water masses shifted southward and northward during the last glacial and interglacial, respectively, faunal boundaries shifted accordingly. During the last glacial maximum the tropical zone was at its narrowest in the eastern Atlantic and eastern Pacific oceans, and this enabled some cold-water species to transgress southward and northward and to establish transequitorial populations (Serventy, 1960, pp. 123–124). The present Baltic populations of three species of Alcidae [razorbill (*Alca tor-*

da), common murre (*Uria aalge*), black guillemot (*Cepphus grylle*)] probably originated from immigrations from the south about 4000 years ago
(Løppenthin, 1963).

Lamb (1977) has pointed out that by a long succession of extensions of
range, the global pattern of seasonal migrations of birds must have been
established (or reestablished) in the 15,000 to 20,000 years since the maximum extent of the last glaciation. Lamb (1977, pp. 184–186) has emphasized that birds are sensitive indicators of climatic change because of their
mobility, and changes in the ranges of numerous resident and migratory
species have been closely associated with climatic trends during the first
half of the present century. A detailed account of this subject for Europe
can be found in Kalela (1949), and Serventy (1977) has done the same for
Australia. Within the last 1000 years, a notable warm period occurred between A.D. 1000 to 1300, and a Little Ice Age was present between A.D. 1550
to 1800. During the Little Ice Age some species of birds (e.g., the bald ibis
Geronticus eremita and the rock partridge *Alectoris graeca*) showed sharp
declines in central Europe and their ranges shifted southward (Hamm,
1951). For approximately the last 200 years, the ranges of birds have been
expanding northward in response to a general warming trend (Crisp, 1959;
Harris, 1964; Salomonsen, 1948; Williamson, 1975, 1976).

The record of the biogeographical migrations of birds in response to
relatively recent climatic changes is rather well documented. Since the late
1800s the climate of the Holarctic region has become warmer, in general
(Lamb, 1977). In Finland because of the warmer springs and milder winters
which the Scandinavian peninsula has experienced since the 1880s, a
number of resident and migratory birds have expanded their ranges northward (Hustich, 1952; Kalela, 1949). Some 50 species of birds have extended
their ranges northward in northern Siberia (Uspenskii, 1969) and treeline
has advanced on to the tundra at a rate of 200 to 700 m per year. In North
America a number of "southern" species have advanced northward in
response to ameliorating climatic conditions since the late 1800s (see Boyd
and Nunneley, 1964). Several species [e.g., red-bellied woodpecker (*Cen-
turus carolinus*), tufted titmouse (*Parus bicolor*), Carolina wren
(*Thryothorus ludovicianus*), blue-gray gnatcatcher (*Polioptila caerulea*),
white-eyed vireo (*Vireo griseus*), prothonotary warbler (*Protonotaria
citrea*), and cardinal (*Richmondena cardinalis*)] has shown range expansions
northward during the last century. The expansions, particularly among
migrants, are facilitated by strong flows of maritime tropical airmasses in
spring (Gauthreaux and LeGrand, 1974) and when these conditions occur,
numerous species "overshoot" the northern limit of their breeding range.
The northward expansion of three bird species (*Streptopelia decaocta, Den-
drocopos syriacus*, and *Hippolais pallida*) into Hungary since 1925 has

been discussed by Keve (1963). The three species moved into Hungary from the south through the valleys of the Wardor and Morava rivers, and the range expansions required different periods of time. Recent warming of the Canadian Arctic east of 100° W since 1930 has produced notable changes in the distribution of several avian species, and the blue phase of *Chen caerulescens* has moved rapidly into newly available suitable areas of the post-Pleistocene emergence (Cooch, 1963).

Even within the last two decades movements of the avifauna have been well documented and although it is too soon to say that these are climatically induced, in some cases there can be little doubt that this is the case. Morel and Morel (1978) have reported that following the recent drought in West African savannas the population of *Quelea quelea* has dramatically decreased while the southern limit of *Passer tuteus* has shifted southward in central West Africa. Since 1960, even though some bird species continue to expand northward, a few species have shown notable southward expansions in the southern United States [e.g. whip-poor-will (*Caprimulgus vociferus*), barn swallow (*Hirundo rustica*), indigo bunting (*Passerina cyanea*)] and in Scotland a number of more northern species [e.g., snowy owl (*Nyctea scandiaca*), northern diver (*Gavia immer*)] have reappeared since 1960 after an absence of a half century or more and started to breed. Lamb (1977, p. 186) attributes these southward expansions of more northern species to a general cooling of the Arctic in the last two decades.

iv. Mammals. Many details of mammalian faunal changes in Europe during the Quaternary are treated in detail by Stuart (1974, 1976, 1977a,b) Kowalski (1967), Mayhew (1975), and Sutcliffe and Kowalski (1976) for the British Isles; by Frenzel (1968), Kurten (1968), and Zeuner (1959) for central, Europe; by Degerbøl (1964) and Gromov et al. (1978) for northern Europe; by Agadjanian and Agadzhanyan (1977) for the Russian Plains; and by Moreau (1969), Monod (1964), and Cooke (1972) for Africa. In North America many details of the changes in the distribution and abundance of mammals during the late Pleistocene and Holocene can be found in Kurten and Anderson (1980) for all of North America; in Guilday (1967, 1971), Guilday and Parmalee (1972), Graham (1976), Webb (1974), Handley (1971) for the eastern United States; in Hibbard et al. (1965), Hibbard (1970), Schultz and Martin (1970), Hoffmann and Jones (1970), Schultz and Hillerud (1977) for the Great Plains; and Anderson (1974) for Wyoming.

The mammalian faunal changes that occurred during the Eemian interglacial in Central Europe have been reviewed by Frenzel (1973). During this time *Hippopotamus amphibius* moved northward to Britain and France but was absent in Central Europe implying cooler conditions at the latter location (Kurten, 1968; Zeuner, 1959). The wooly mammoth (*Mammuthus primigenius*) occurred in Scandinavia, but was absent from Central Europe. Stuart (1976, 1977b) has discussed the vertebrate fauna of the British Isles in relation to vegetational changes during the Ipswichian, and additional detailed studies of the Pleistocene mammalian fauna of Britain can be found in Kowalski (1967) and Mayhew (1975). In general some species of mammals occurred well north of their present ranges [e.g., *Hippopotamus* and *Crocidura* (lesser white-toothed shrew)]. Another straight-tusked elephant (*Palaeoloxodon antiquus*) also ranged in Britain, and other mammals present in Britain during the Ipswichian included fallow deer (*Dama dama*), spotted hyaena (*Crocuta crocuta*), lion (*Panthera leo*) and rhinoceros (*Dicerorhinus hemitoechus*). These species were associated with regional mixed oak forests.

Studies of the mammalian fossil record for the Sangamon (Eemian) Interglacial in North America are not numerous, but some documentation of the northward advances of certain species characteristic of more southern latitudes has been published (Hibbard *et al.*, 1965; Hibbard, 1970; Hoffmann and Jones, 1970). Most of the records come from the Great Plains region and include: yellow bat (*Lasiurus intermedius*), capybara (*Hydrochoerus*), rice rat (*Oryzomys*), water rat (*Neofiber*), hog-nosed skunk (*Conepatus*), ocelot (*Felis pardalis*), and jaguar (*Panthera onca*). These species are in keeping with the climate characteristic of this period at this locality (see section on the herpetofauna, pp. 132–133).

In the British Isles during the latter half of the Ipswichian Interglacial, the temperate taxa were gradually replaced by taxa usually considered to be characteristic of cold stages [e.g., mammoths (*Mammuthus primigenius*) and horses (*Equus caballus*)]. Somewhat later as forests deteriorated and the landscape became increasingly open, northern voles (*Microtus oeconomus*), arctic collared lemmings (*Dicrostonyx torquatus*), Norway lemmings (*Lemmus lemmus*), wooly rhinoceroses (*Coelodonta antiquitatus*), and musk oxen (*Ovibos moschatus*) moved into the British Isles (Stuart, 1976). By the early and middle Devensian, polar bears (*Ursus maritimus*) and reindeer (*Rangifer tarandus*) occurred at a locality now near London (Stuart, 1977a). During the Devensian a number of northern species moved to lower latitudes and into Britain, including the arctic fox (*Alopex lagopus*), the wooly mammoth, the wooly rhinoceros, the reindeer, the arctic collared lemming, the tundra vole (*Microtus gregalis*), and the horse (*Equus caballus*). In the region of Morocco–Tunisia during the late Pleistocene, the bear (*Ursus arctos*) and the elk (*Cervus elephas*) occurred according to Moreau (1963). Monod (1964) and Moreau (1969) give additional details of the changes in the distribution and the abundance of mammals in Africa during the Quaternary. In Africa, unlike Europe, there were no barriers to faunal movements during periods of climatic change (Cooke, 1972). Starkel (1977) has studied the paleogeography of middle and eastern Europe during the last cold stage and made a comparison with western Europe. In general the climatic changes were more severe at the former locations than at the latter location. The migrations of the mammalian fauna in the Scandinavian steppe zone in response to paleoclimatic change and glaciation has been examined by Gromov *et al.* (1978), and Agadjanian and Agadzhanyan (1977) have reported on the changes in the small mammal fauna of the Russian Plains during the late Quaternary.

During the Wisconsin glacial period in North America, the ranges of many mammalian species shifted southward (and eastward and westward) in response to glacial advancement and colder climatic conditions (Kurten and Anderson, 1980). The shifts in the mammalian fauna of the Great Plains during the Wisconsin and Holocene have been summarized by Hoffmann and Jones (1970). In central Texas from 40,000 to 25,000 years ago, the presence of black-tailed prairie dogs (*Cynomys ludovicianus*) and prairie voles (*Microtus ochrogaster*) resulted from eastward and southward displacements, respectively (Slaughter and Ritchie, 1963), and the presence of the meadow voles (*Microtus pennsylvanicus*), and the short-tail shrews (*Blarina brevicauda*) resulted from southward and westward biogeographical migrations, respectively (Dalquest, 1962). Dalquest (1964) also reports that the bog lemming (*Synaptomys cooperi*) was present in central Texas from 30,000 to 20,000 years ago.

About 16,700 ± 560 years ago full-glacial climatic conditions occurred in central Texas, and short-tail shrews, meadow voles, bog lemmings, cotton rats (*Sigmodon hispidus*), and possibly the prairie vole, and Richardson's ground squirrel (*Citellus richardsoni*), were present (Dalquest, 1965). Some additional species [masked shrew, northern water shrew (*Sorex palustris*), and northern pocket gopher (*Thomomys talpoides*)] were also resident in the area, representing a considerable southeastward shift in relation to their present distributions. In southwestern Kansas the presence of the masked shrew (*Sorex cinereus*) and the meadow vole indicate a boreal or cold steppe condition at this location during the peak of the Wisconsin glaciation (Hibbard and Taylor, 1960). Nebraska had steppe species (*Cynomys* and the pronghorn, *Antilocapra americana*), boreal species [caribou (*Rangifer*)], and scrub-tundra species [muskox (*Ovibos moschatus*)] during the middle Wisconsin period (Banfield, 1962; Hibbard *et al.*, 1965; Schultz *et al.*, 1951).

Cool climates continued over the Southern Great Plains during the late Wisconsin period (13,000 to 10,500 years ago) as indicated by the presence of masked shrews and meadow voles. Other boreal species included the short-tail weasel (*Mustela erminea*), the Franklin ground squirrel (*Citellus franklini*), and the bog lemming (Slaughter and Hoover, 1963; Lundelius, 1967). Cool steppe conditions prevailed in southwestern Kansas during the Late Glacial and the mammalian fauna included masked and northern water shrews, meadow voles, meadow jumping mice (*Zapus hudsonius*), and bog lemmings (Schultz, 1967). In Nebraska, steppe species began replacing tundra species during the last stages of the Wisconsin. The mammalian fauna of eastern Wyoming during the late Pleistocene and Holocene has been reviewed by Anderson (1968, 1974) and Guilday et al. (1967), and Guilday and Adam (1967) have done the same for west of the Continental Divide in Idaho. Steppe conditions prevailed at the bases of mountains with pronghorn, sagebrush vole (*Lagurus curtatus*), and black-footed ferret (*Mustela nigripes*). Higher in the mountains, montane and subalpine forests occurred with pygmy shrews (*Microsorex hoyi*), red-backed voles (*Clethrionomys gapperi*), and martens (*Martes americana*), and alpine tundra was present at the highest elevations in the mountains with pikas (*Ochotona princeps*), mountain goats (*Oreamos americanus*), and least shrews (*Cryptotis parva*).

Guilday (1971) has examined the Pleistocene history of the Appalachian mammalian fauna. The yellow-cheeked vole (*Microtus xanthognathus*), presently characteristic of taiga areas, occurred as far south as Natural Chimneys, Virginia. The Hudson Bay collard lemming (*Dicrostonyx hudsonius*) was present at New Paris, Pennsylvania, during the peak of the Wisconsin glaciation. This species is now found only in tundra areas characterized by permafrost conditions and north of the continental treeline. During the late Wisconsin (13,460 ± 420 years ago) the Greenland caribou (*Rangifer tarandus*) occurred at Saltville, Smyth County, Virginia, and at Baker's Bluff Cave, Sullivan County, Tennessee. This species normally inhabits tundra, taiga, and coniferous forests today. Graham (1976) has also discussed the late Wisconsin mammalian faunas in relation to environmental gradients of the eastern United States. As is the case for other faunal groups, some western mammals moved eastward toward what is now the eastern deciduous forests during the Late Glacial (ca. 11,000 to 10,500 years ago). Guilday (1971) has reported on fossils of grizzly bear (*Ursus horribilis*), badger (*Taxidae taxus*), and thirteen-lined ground squirrel (*Citellus tridecemlineatus*) at several sites in Pennsylvania and Virginia.

About 11,300 ± 1000 years ago the local fauna in central Pennsylvania was still under the influence of a boreal climatic regime. By 9340 ± 1000 years ago the extensive mammal fauna at Hosterman's Pit in central Pennsylvania was completely modern with no hint of boreal species (Guilday, 1967, 1971). Thus, during approximately 2000 years there was a major geographical shift in ranges that accompanied the warming climatic conditions that marked the onset of the Holocene.

At the end of the late-Wisconsin period and the beginning of the Holocene (10,500 years ago), the continental ice sheets retreated rapidly in response to abrupt climatic warming (Bryson and Wendland, 1967). The mammalian fauna responded rapidly (Hoffmann and Jones, 1970), and many species began biogeographical migrations that would eventually return them to preglacial distributions. Between 10,500 and 8500 years ago, masked shrews and meadow voles left the central Texas mammalian fauna and bog lemmings were gone from the area by 7000 years ago. The climate in central Texas continued to warm and become drier from 5000 to 1000 years ago, and since about 1000 years ago a modern faunal composition has been recorded in the area (Lundelius, 1967). The Appalachian mammalian geography during the Holocene has been extensively reviewed by Handley (1971), and the probable post-Pleistocene routes of return migrations for several species have been mapped. Handley emphasizes the mobility of mammals and points out that until about 12,500 years ago or later, all mammalian ranges had to be south of the Wisconsin glaciation. The ranges of a number of mammals north of this margin since 12,500 BP, or much more recently than that indicate that many species moved northward rapidly following the gradual retreat of the glaciers and return of a more thermophilous flora. At the peak of the last interglacial the tapir migrated as far north as Indiana and Ohio and the manatee inhabited the Atlantic coast as far north as New Jersey (Hay, 1924). Between 6500 and 3500 BP the spotted skunk (*Spilogale putorius interrupa*) occurred in Illinois (Parmalee, 1968), indicating a warmer period than at present.

The geographical source areas and the movement patterns of mammals in relation to their present distribution in the Northern Great Plains of North America have been analyzed by Hoffmann and Jones (1970). By mapping a sequence of distributional events beginning about 10,000 years ago, they have attempted to explain the present distributional patterns exhibited by mammals inhabiting the Northern Great Plains. Although somewhat hypothetical, the analysis is logical and agrees with data on the current changes in the distribution and the abundance of mammals, there and elsewhere, in response to changing climatic and floral conditions. A similar analysis for boreal and desert mammals of the Southwest can be found in Findley (1969).

The present distributional patterns of mammals are changing constantly and biogeographical migrations are in evidence in many regions of the world. Hall (1958) has reported that cotton rats (*Sigmodon hispidus*) and opossums (*Didelphis virginiana*) have been expanding their ranges northward, and Vaughan (1978, p. 353) has mentioned that the raccoon (*Procyon lotor*) has expanded its range from the riparian habitats of the plains of eastern Colorado well into the foothills of the Rocky Mountains in the last 50 years. A brief overview of the movements of arctic mammals in response to climatic changes during the last century can be found in May (1979). The recent migrations of the nine-banded armadillo (*Dasypus novemcinctus*) have been documented by Schultz (1972) and Humphrey (1974). Between the late 1800s and 1950s the species moved northward, but since that time it has been retreating southward. This pattern agrees closely with the change in atmospheric circulation patterns since 1950. The change has resulted in the renewed increase of the Arctic ice and an enhanced variability in the character of the long spells of weather occurring from one year or one run of years to another (Brinkmann, 1976).

The biogeographical migrations of flora and fauna occur in response to climatic changes with varying periodicities. The longer periodicities that influence the relatively slow biogeographical migrations of organisms occur at intervals about 100,000, 40,000, and 25,000 years and correspond to all types of variation in the earth's orbit (Imbrie and Imbrie, 1980). The biogeographical movements of organisms may be influenced also by shorter periodicities of climatic change, and these may range from time scales of 2500–2000 years to 5.5 and 2.2 years (Mitchell, 1976, Figs. 1 and 2). The influence of interannual climatic fluctuations on biological systems has been considered by Clark *et al.* (1975). Even though a species may be undergoing a biogeographical migration in response to climatic changes of one or more of the above periodicities, the individuals that comprise the species may also show seasonal migrations in response to annual cycles of climatic change. Seasonal migrations may be spatially identical to biogeographical migrations but the periodicities are of course markedly different.

IV. Annual Climatic Periodicities and Seasonal Migrations

A. ANNUAL CLIMATIC CHANGE AND BIOTIC RESPONSE

Because the earth's orbit is an ellipse with the sun at one focus, the distance between the earth and the sun varies from about 146 million km at perihelion (on about 3 January) to 151 million km at aphelion (on about 4 July). Although these distances do cause some changes in the amount of solar radiation impinging on the earth, the distances contribute relatively little to the annual forcing of climate. The annual and daily forcing of climate arises primarily from the inclination of the earth's rotational axis with respect to the plane of the ecliptic and from the rotation of the earth, respectively. The north pole is tilted 23.5° toward the sun on 21 or 22 June (summer solstice) and the south pole is tilted 23.5° toward the sun on 22 or 23 December (winter solstice). When the earth reaches points halfway between these positions and its axis is perpendicular to a line drawn to the sun (no pole is closer to the sun), these points are called equinoxes (vernal on 20 or 21 March and autumnal on 22 or 23 September).

An organism responds ultimately to fluctuations in its physical and biotic environment, and in the case of the annual climatic cycle, an organism may show seasonal responses to changes in the weather elements (air temperature; air pressure; winds, direction and strength; atmospheric moisture including humidity, clouds, and fog; and precipitation) directly, or indirectly through changes in its biotic environment. Because the amount and duration of solar radiation varies markedly from winter to summer at higher latitudes in the northern hemispheres, many organisms have proximal photoperiodic mechanisms that regulate the phases of their annual cycle (e.g., Beck, 1968; Danilevskii, 1965; Saunders, 1976; Tauber and Tauber, 1976, for insects; Elliott, 1976; Farner et al., 1973; Wurtman, 1975; Zaslavsky, 1976; Farner and Follett, 1979; Farner and Lewis, 1971, for vertebrates).

The phases of the annual cycle of an organism (e.g., migration, reproduction) are closely tied to the annual variations in the environmental conditions that influence the organism's chances for survival and reproduction. The timing of these recurring biological events, the cause of their timing with regard to biotic and abiotic forces, and the interrelation among phases of the same or different species comprise the study of phenology (Lieth, 1974). Because annual climatic changes have a high predictability (the major causative factors are deterministic), there is increasing evidence that many organisms possess circannual clocks and show endogenous annual rhythms of migratory and reproductive behavior (Pengelley, 1974; Saunders, 1976).

The climate of the ocean also shows marked annual periodicities (Pocklington, 1972, 1978; Reid et al., 1978; Robinson, 1976; Veronis, 1978; Cushing and Dickson, 1976), and in the sea, the most important seasonal effect in the waters of the higher latitudes is the delay or advance in the spring outburst which is dependent on annual northward and southward shifts in the local pressure systems. On a global scale seasonal changes in the position of the subtropical anticyclone influence the system of equatorial currents, and these seasonal changes exert a profound influence on the productivity of local marine habitats and the migration patterns of the animals dependent on them.

B. Annual Climatic Changes

This section will not present a detailed discussion of the aspects of annual climatic change, because a number of excellent books on this topic have been published and are readily available (Boucher, 1976; Lamb, 1972;

Smith, 1975; Barry and Chorley, 1978; Barry and Perry, 1973). The emphasis of Section IV is on the seasonal migrations of animals, and two important points about the annual climatic cycle and meteorological events should be made before addressing the seasonal movements of selected organisms in response to seasonal climatic sequences in the annual climatic cycle. The climatic environment that an animal experiences during the year is generated by two types of mechanisms (Mitchell, 1976). As mentioned earlier the primary external forcing factor that drives the annual climatic cycle is the inclination of the earth's rotational axis in relation to the plane of the ecliptic. This relationship generates the predictable sequence of seasons and photoperiods. Superimposed on this deterministic forcing of the climatic system are internal stochastic generating mechanisms that introduce variability or variance to the combination of weather elements that generally characterize a particular season. The majority of the stochastic generating mechanisms that contribute to the variability of climate are on time scales of an hour or less and involve local atmospheric turbulence and thermal–convective buoyancy. Additional important sources of variability can be attributed to synoptic weather processes (cyclones and anticyclones) and the undulating upper-atmospheric wind streams, both having time constants of a few days. The last source of stochastic variability of the annual climatic system involves the relatively slow thermal adjustments in the atmosphere and the thermodynamic transactions between the atmosphere and the ocean surface. The time constants of these processes are on the order of weeks and months.

Section IV,C will concentrate on seasonal climatic processes that strongly influence the seasonal migrations of animals (e.g., monsoons, intertropical convergence zones) and also cover the influence of certain specific weather elements (e.g., temperature, wind, moisture) as they relate to the day-to-day timing of seasonal migrations. The seasonal changes in the ocean environment have marked effects on the movements of marine organisms, and some of these relationships are also included in the following section.

C. Seasonal Migrations

The annual timing or phenology of the seasonal migrations of animals is strongly correlated with the deterministic forcing function of the earth's orientation during its orbit around the sun. This is hardly surprising because the favorability of terrestrial habitats (e.g., productivity) is dependent largely on such factors as increased duration of solar radiation at higher latitudes, increased moisture, and warm temperatures. As in biogeographical migrations, organisms tend to track their preferred environment, and when their environment changes seasonally, their pattern of migration is also seasonal, provided they do not remain where they are and survive the rigors of the unfavorable period (e.g., diapause, hibernation, or another form of tolerance).

In discussing seasonal migrations, I will not attempt an exhaustive summary of the seasonal migrations of representatives from as many taxa as possible but will emphasize selected cases that demonstrate the interactions of climatic and meteorological factors and their influence on seasonal migratory behavior. The first case concerns the seasonal migrations of aquatic and marine organisms, and the second examines the migrations of insects and birds.

1. Aquatic and Marine Migrations

Although the durations of the total migratory journey of many aquatic and marine organisms range from a few months to several years, the migrations are best regarded as seasonal, because the reproductive period in most instances is linked to a seasonal optimum. Even though maturation may take varying lengths of time, and the organisms may perform extended migrations during this time, the period when (and the place where) the organisms can reproduce is relatively restricted by the seasonal favorability of the breeding habitat. The maturation schedule of an organism is also probably dependent in large part on seasonally dependent factors such as food availability (see Jones, 1973), but this topic is beyond the scope of this discussion.

In the Norwegian Sea the marine copepod, *Pseudocalanus minutus*, shows distinct seasonality in its migration circuit (McLaren, 1963). It spawns during the period of peak phytoplankton abundance in April and May. The older stage young then migrate downward from surface waters, and by August the entire population is below 600 m, where a strong change in temperature occurs. The stage V copepodids containing large quantities of stored lipid overwinter at temperatures 7°–8°C cooler than the surface waters, and in March and April molt into adults and migrate to surface waters to reproduce. Similar seasonal vertical migrations are well known for the high-latitude copepods, *Calanus*, in the North Atlantic (Sømme, 1933; Marshall and Orr, 1960). Reid *et al.* (1978) discuss these movements as well as those that involve seasonal vertical movements that result in displacements equatorward in summer in the upper layer, and poleward in winter in the deeper layer (see Mackintosh, 1937). The seasonal distribution of benthic invertebrates with pelagic larvae is reflected in the oceanic circulation patterns operating temporally as a dispersal mechanism (Zinsmeister and Emerson, 1979). Certain oceanic currents thus can act as one-way corridors of dispersal or migration and at the same time serve as barriers to movements in the opposite direction. An organism's ability to change vertical position in the ocean permits it to exploit countercurrent systems in the ocean and perform return migrations in certain instances.

The banana prawn, *Panaeus merguiensis*, has both a marine and estuarine phase, and adults spawn in the open sea (Munro, 1975). From November to February, in the vicinity of the Gulf of Carpentaria, Australia, the majority of adolescent prawns migrate from rivers into the shallow inshore coastal zone. From February to September a slow seaward migration takes place and spawning occurs mainly during the spring–summer period (September–Feburary). The postlarvae move into the rivers that surround the Gulf from October to May. Staples (1979) has shown that the mass movement of adolescent prawns from the nursery area rivers coincides with the northwest monsoon period (November–February) and that the main recruitment into the Gulf fishery occurs during the period from December to April. The movements of the postlarvae are strongly influenced by seasonal currents that change from a northerly to a southerly direction during October–November (Cresswell, 1971).

On the basis of tagging populations that inhabit the outer continental shelf, Cooper and Uzmann (1971) have demonstrated a seasonal migration in the northern lobster, *Homarus americanus*. In the spring individuals move distances of 200 km or more from the edge of the continental shelf to shallower waters along the southern coast of New England. In the fall the lobsters return to deeper water on the edge of the continental shelf where they overwinter. The migrations are performed year after year by the same individuals. The mass migrations of the spiny lobster, *Panulirus argus*, occur in the autumn on the shallow Bahama bank and deeper reefs surrounding Bimini. During a two- to three-week period the net movement of the lobsters is westward, and an autumnal storm-induced drop in water temperature is thought to be the triggering stimulus for the movements (Kanciruk and Herrnkind, 1978). Johnson (1976) has exmained the seasonality of the rocky intertidal amphipod, *Oligochinus lighti*, in response to microenvironmental variables, and found that changes in temperature, salinity, and moisture content in the algal clump inhabited by the amphipod were of primary importance in deterimining the distribution and abundance. The algal microclimate was found to be related in a complex fashion to seasonal changes in the coastal macroclimate mediated by the irregular semidiurnal tidal cycles of the region.

The seasonal migrations of the crayfish, *Pacifastacus leniusculus*, in Lake Tahoe, California, have been studied by Flint (1977). The peak bathymetric distribution of crayfish during the summer and autumn is in shallow, near-shore waters. In the autumn as water temperatures and daily solar radiation decrease, the adults migrate into deeper waters. Flint points out that this species shows peak activity during the warmer months and very low activity during the winter months, a pattern typical for most invertebrates living in an environment with large seasonal variations in temperature. The seasonal migrations of another crayfish species *Orconectes virilis*, in marl lakes has been examined by Momot and Gowing (1972).

The onshore–offshore seasonal migrations of *Littorina* have been examined and reviewed by Hamilton (1978). In the fall when temperatures decrease the snails migrate offshore and return in spring when the water temperatures increase in the intertidal zone. Hamilton also discusses the differences in migratory behavior between *Littorina* of the northeast coast and those inhabiting low-energy beaches along the northern coast of the Gulf of Mexico. Fresh water snails also change their distributions in response to seasonal climatic changes, and Wall (1977) has studied the seasonal movements of the pond snail, *Lymnaea catascopium*, in a northern lake. Baker (1978, pp. 830–841) gives several additional examples of the seasonal and ontogenetic return migrations of sublittoral and littoral invertebrates, and he points out that no generalizations can be made concerning the timing and direction of these movements even though temperature and light intensity effects should tend to favor a similar migration pattern for a wide range of species. Baker's comment serves to emphasize a common problem in the understanding of the variability of the migration systems that occur in nature, and this important topic will be treated in Section VI.

The seasonal migrations of fish are well known (Baker, 1978, pp. 786–823, 843–851; Cushing, 1976b; Harden Jones, 1968, 1977; Harden Jones *et al.*, 1978) and the timing of the migrations of many species is closely tied to the spawning seasons. Cushing (1969, 1971) has examined the regularity of the spawning season of such species as the salmons (*Oncorhynchus* and *Salmo*), cod (*Gadus*), herring (*Clupea*), and plaice (*Pleuronectes*) and found that derived indices for the mean dates of peak spawning has standard deviations of about 7 days. In temperate waters where production shows a strong seasonal cycle, fish spawn at a fixed season and the growth of the offspring parallels growth of their food as Jones (1973) has suggested. In areas where food is available at all times (upwelling areas, subtropical and tropical seas) spawning is less exact in timing and shows less seasonality (Cushing, 1976b). The seasonal changes of planktonic organisms and the number of generations at different latitudes has been studied by Bogorov (1960).

The spawning grounds of many species of fish are fixed, and the fish tend to return to the location of first spawning year after year. In the case of Pacific salmon (*Oncorhynchus*) the mature fish return from the sea to the locations where their parents spawned and died. Atlantic salmon (*Salmo salar*) may survive for two or more spawning seasons and return to the same spawning site.

Near-shore and off-shore currents are important determinants of migratory pathways for both larval and adult fish, and these currents are in many instances strongly influenced by seasonal climatic factors, in much the same way as are atmospheric currents. Climate also influences the general structure of planktonic communities, a major food supply for growing fish (Steele and Frost, 1977), and governs the patterns of their seasonal migrations. Thus for many species of fish, climatically driven water currents are important in all phases of their life history, because the currents direct their migrations during different phases of their life history and also control the distribution and abundance of the food present in nursery areas during development and growth.

Larval drift from spawning areas to nursery areas in relatively shallow water is largely current controlled (Cushing, 1976b) and the recruitment of immature fish to the adult stock on the feeding grounds in deeper water is also influenced by currents. The mature fish migrate from feeding localities to spawning localities and back again each year, and in several species (with the obvious exceptions of fish that swim up rivers against the current) the patterns of these movements are closely correlated with specific currents. Recent work on the spring movements of smolts of the Atlantic salmon indicate that tidal currents are important in the seaward migrations from open waters of a bay (LaBar *et al.*, 1978). The smolts moved rapidly seaward on ebb tides with some reverse migration occurring on flood tides, but some continued to move seaward although at half the speed on flood tides. Vector analysis of the directions of movement indicated a significant clustering around the direction of water currents, and water current drift drogues at 2 m and 5 m closely approximated the routes of smolts.

Harden Jones (1977, 1979), Harden Jones *et al.* (1978), and Walker *et al.* (1978) have shown that plaice and cod use a selective tidal transport mechanism when on migration on the continental shelf where the tidal streams are relatively strong. The mechanism involves changing vertical position in the waters in a semidiurnal manner so as to take advantage of a certain tidal flow. Plaice were found to come off the bottom at slackwater and move downcurrent with the tide in midwater and return to the bottom at the next slack water when tidal current flow reversed, thus avoiding any movement during the opposite tide. The semidiurnal vertical movements were highly correlated with the tidal cycle. DeVeen (1978) and Harden Jones *et al.* (1979) have shown that North Sea plaice and other flatfish species use selective tidal stream transport when migrating from feeding grounds to spawning grounds in the winter and on their return in the spring.

From the studies of the migration of the American shad, *Alosa sapidissima*, Neves and Depres (1979) conclude that circulation patterns along the Atlantic coast do not account for the seasonal distribution of chad according to survey data (Talbot and Sykes, 1958) or the coastal migration routes based on tagging studies (Leggett, 1977a). Neves and Depres believe that seasonal shifts in isotherms, as influenced by circulation patterns, exert a greater influence on the migratory route. Several studies have shown a strong relationship between oceanic temperature patterns and the movement of fish (Leggett, 1977b, pp. 293–294), but as Leggett points out the relationship may be indirect, for a number of authors have commented on the occurrence of preferred food sources at certain temperatures (e.g., Berenbeim *et al.*, 1973; Pavshtiks, 1959). Seasonal changes in the position of currents can affect the timing and directions of fish migrations and also influence the abundance of the food upon which the fish are dependent. The arrival of warm currents in spring not only causes the onset of phytoplankton and zooplankton blooms, but also transports migrant fish to these areas. Additional recent information of the seasonal movements of fish include studies by Haynes *et al.* (1978) on the white sturgeon (*Acipenser transmontanus*) in the mid-Columbia River, by Austin and Custer (1977) on the striped bass (*Morone saxatilis*) in Long Island Sound, by Guelpen and Davis (1979) on the winter founder (*Pseudopleuronectes americanus*) in Newfoundland, by Laurs and Lynn (1977) on the North Pacific albacore (*Thunnus alalunga*) in North American coastal waters, and by Johnson and Muller (1978) on juvenile pike (*Esox lucius*) migration to the northern part of the Bothnian Sea.

Ocean fronts and currents are important determinants of pathways during oceanic fish migrations and even rather complex migration circuits can be explained on the basis of changing convergences and divergences and shifts in oceanic frontal zones. Leggett (1977b) emphasized the need for more concurrent observations of fish abundance and hydrographic conditions over large areas, but the somewhat limited available data suggest that the migratory routes of fishes often follow ocean discontinuities (fronts) characterized by steep gradients of temperature, salinity, and nutrients (see Smith, 1976). Fujii (1975) has demonstrated that seasonal changes in the temperature and salinity barrier to the Bering Sea are of great significance to the homing migrations of the western Alaska sockeye salmon (*Oncorhynchus nerka*). The fish enter the Bering Sea via the Aleutian Passes only after spring tides and wind action in June permit a northbound flow of Alaskan Stream water (with a higher temperature and a lower salinity) into the Bering Sea and then eastward toward Bristol Bay. Once in Bristol Bay the salmon move into rivers to spawn. The literature on the influence of currents on the migration of fishes is extensive and further information on this topic can be found in Arnold (1974) and in Harden Jones (1968). The circulation and geochemical characteristics of oceanic waters have likewise received considerable attention, and several recent papers provide overviews (e.g., Reid et al., 1978; Smith, 1976; Veronis, 1978; Zinsmeister and Emerson, 1979).

2. Terrestrial and Atmospheric Migrations

The annual climatic cycle produces seasonal environmental changes that influence strongly the migrations of terrestrial organisms over land and through the atmosphere. The seasonal climatic changes and the concomitant changes in weather elements are important determinants of the timing and the direction of the movements. This section will emphasize the seasonal migrations of insects and birds, because much is known about their migration systems, and their movements through the atmosphere are in many respects quite analogous to those of organisms in the ocean (see Rainey, 1978).

The major features of insect migration have been summarized in several books (e.g., Baker, 1978; Dingle, 1978; Johnson, 1969; Williams, 1958; Rabb and Kennedy, 1979) and in a number of review papers (e.g., Dingle, 1972, 1979; Kennedy, 1975; Schneider, 1962; Southwood, 1962, 1975; Rabb and Stinner, 1979). The movements of insects through the atmosphere in relation to seasonal climatic and meteorological factors have received considerable attention (e.g., Johnson, 1969; McManus, 1979; Rainey, 1973, 1976, 1978, 1979; Wellington, 1979; Michel and Albrecht, 1978; Snow and Copeland, 1969). Seasonal vertical migrations are well known also for soil inhabiting invertebrates in response to colder temperatures during autumn and winter (Dowdy, 1944) or to decreasing humidity during the dry seasons (Strickland, 1947), but these movements will not be examined in this section. The annual climatic cycle dictates directly or indirectly the availability and suitability of habitats for insect survival and reproduction. Many insects escape the rigorous phases of the cycle by diapausing, but many others undergo migrations that carry them to suitable habitats. Changes in the dominance of different air masses with their characteristic weather elements (temperature, moisture, and rainfall) are associated with seasonal changes in climate, and the movements of these air masses are closely correlated with the seasonal migratory movements of insects.

The annual cycles of insect abundance and distribution in the tropics led Golding (1928) to hypothesize a link between the periodicity of insect migration and the movements of the intertropical convergence zone (ITCZ). This hypothesis was further developed by Rainey (1951), and Bowden (1973), and subsequently the migrations of a number of insects have been shown to be associated with the seasonal movements of the ITCZ, particularly in Africa (e.g., Bowden, 1973, Duviard, 1977; Rainey, 1976, 1978).

The ITCZ is an intertropical zone of low atmospheric pressure between two anticyclonic air masses at higher latitudes above and below the equator. The airflows from the anticyclonic systems converge along a low-pressure axis, and the surface separating the northeasterlies and the southeasterlies is the intertropical front (ITF). The ITCZ represents a belt 600–900 km wide while the ITF is the edge of the two air masses and has a typical frontal morphology. The annual climatic cycle influences the movements of the ITCZ and the ITF so that the southernmost position for the ITF is about 7°30'N in February and the northernmost position for the ITF is about 22°N in August. The slope of the frontal surface is greater when the ITF is at its southernmost position so that the ITCZ is narrower during the period between December and February and widest between July and September. North of the ITF a continental dry climate prevails, but immediately south of the ITF an oceanic dry climate with predominantly southwesterly winds is found. Farther behind the front a transition to rainy climatic conditions occurs, and near the southernmost portion of the ITCZ there is a return to drier conditions. As the ITCZ moves northward in the spring and summer and southward in the fall and winter it exerts a strong influence on the distribution and abundance of many insect species and the wind patterns are important in directing the migrations of insects from deteriorating environments to improving ones.

Desert locusts (*Schistocerca gregaria*) migrate from areas that are drying up and into areas of new rain (Rainey, 1951, 1979), and they use the winds that generate these conditions as a means of directing their movements in the ITCZ. The movements of Diptera, Hemiptera, Thysanoptera, and Coleoptera have been found to be correlated with movements of the ITCZ in the Sudan (e.g., Bowden, 1976; Bowden and Gibbs, 1973; Joyce, 1976; Duviard, 1977), and the outbreaks of other species (African armyworm, *Spodoptera exempta*) have been linked to movements of the ITCZ (Haggis, 1971; Rose and Law, 1976). Schaefer (1976, 1979), Lecoq (1978), and Riley and Reynolds (1979) have examined the migratory movements of grasshoppers in Africa in relation to wind currents and changes in the position of the ITF.

The migrations of *Dysdereus voelkeri* and *D. melanoderes* (Hemiptera: Pyrrhocoridae) have been related to movements of the ITCZ in West Africa by Duviard (1977). The migratory flight activity of *D. voelkeri* is associated with warm, wet, and sunny conditions that occur a short distance behind the ITF. This species can survive only a short time north of the ITF because of the temperature and humidity conditions, and it cannot survive in the southern portions of the ITCZ because heavy rains drown colonies and lethal fungal diseases curtail populations of the species. There is then a window of climatic conditions that permit migrations and reproduction and the timing of these activities is dependent on the seasonal movements of the ITCZ. Many other studies have shown that insects not only use the wind and precipitation patterns characteristic of the ITCZ, but that the short- and long-range migratory movements of insects in general are tied to the climatic and meterological changes (Rainey, 1974, 1976) associated with the planet's annual climatic cycle.

The transport of migrating insects by "preferred" wind currents has been documented in a number of studies (e.g., Basedow, 1977; Campion et al., 1977; Danthanarayana, 1976; Kisimoto, 1976; Solbreck, 1975; Meyer and Appleby, 1973; also in Johnson, 1969, for earlier references). The movements are triggered by certain meteorological conditions (Taylor, 1974) or by entrained rhythms (Saunders, 1976) and the wind currents associated with these conditions direct the subsequent movements of the migrants (see Riordan, 1979). Thus in fall in the north temperate zone, the increasing frequency of increasingly strong cold fronts with northerly winds aids insects migrating southward to overwinter, and in the spring the increasing frequency of the penetration of warm, moist air from the tropical maritime region can return their descendants to their summer range as habitats once again become suitable (see Johnson, 1969, Part 5). Because insects have short life spans, seasonal insect migrations more closely resemble the biogeographical migrations of organisms with longer generation times than they do the seasonal migrations of animals such as birds.

The seasonal migrations of birds have captured the attention and imagination of man since the beginning of recorded history, and virtually everyone is at least partially aware of the seasonality of the phenomenon. Several factors are probably responsible for this: (1) birds and their eggs have been a source of food for many north temperate zone peoples, (2) only a very few birds can hibernate to survive the rigors of overwintering at high latitudes, (3) birds have high energetic demands, and (4) birds are generally conspicuous in their activites. Many books have summarized the studies of bird migration (e.g., Bykhovskii, 1974; Dorst, 1962; Griffin, 1974; McClure, 1974; Moreau, 1972; Schüz, 1971), and general books on animal migration have devoted considerable space to birds (e.g., Baker, 1978; Orr, 1970; Schmidt-Koenig, 1975). Not only does the global annual climatic cycle strongly influence the timing and directions of the migrations, but much is known about the influence of meterological factors on the movements once they are under way. In addition to seasonal latitudinal shifts in ranges many species show seasonal vertical migrations in virtually all major mountain ranges in temperate zones of the world. Because of the quantity of available information, only certain aspects of the influence of seasonal climatic factors on the migrations of birds will be emphasized.

Four timing mechanisms that operate on the central control system and are responsible for the timing of events in an individual bird or bird species can be recognized (Farner, 1967): (1) primary timing mechanisms that are related to the annual climatic cycle such as photoperiod or monsoon rainfall, (2) essential supplemental mechanisms (e.g., presence of mate) that fine tune temporal adjustments set in motion by the primary timing mechanism, (3) modifying mechanisms that accelerate or slow down the primary timing mechanism such as weather aspects, and (4) terminal timing mechanisms that terminate or reset the operation of the central control system (e.g., destruction of nest by storm or predator). In the remainder of this section on the seasonal migrations of birds, I will first treat the primary timing mechanisms and then concentrate on the modifying mechanisms.

Because of the relative constancy of the annual climatic cycle, it is not surprising that the seasonal migrations of a wide variety of bird species have been shown to be proximately regulated by endogenous annual or circannual rhythms (Gwinner, 1977). Circannual rhythms guarantee proper adjustment of internal biological activities to external seasonal changes, and in free living birds these rhythms are entrained by the annual climatic cycle (photoperiod). Gwinner (1977) has emphasized that the significance of circannual rhythms is "in the initiation of migration at the appropriate times of the year for long-distance migrants that winter in unpredictable tropical environments," and that such rhythms "can also determine the temporal course and possibly the distance of migration, at least in some long-distance migrants on their first migration."

Circannual rhythms appear to be more important in long-distance migrants than in short-distance migrants (Gwinner, 1977, p. 401). Circannual rhythms in long-distance migrants serve as a proximate mechanism to initiate migration in advance of the seasonal occurrence of ultimate factors. In short-distance migrants migration is more closely tied to weather conditions that may be proximate and ultimate at the same time (see Berthold, 1975). Evidence from field studies of the seasonal timing of migration in a number of species supports the laboratory findings that short- and medium-distance migrants show greater interindividual variablity in the development, duration, total amount, and maximal values of migratory restlessness than do long-distance migrants. A number of studies have shown that arrival dates of migrants early in the spring are more variable than the arrival dates of migrants later in the season (e.g., Slagsvold, 1976; Weydemeyer, 1973; Pinkowski and Bajorek, 1976) and that migrants that arrive early are more influenced by changes in weather conditions

(Slagsvold, 1976). In a 48-year study of spring arrivals of migrants in Montana, Weydemeyer (1973) found that ranges in dates of arrival were greatest during late March and April and least in late May and June. Slagsvold (1976) working in Norway found that for the country as a whole, bird arrival was delayed six days for each 10-day delay in vegetation development. The arrival of migrants at higher latitudes and elevations was faster than the development of vegetation. Slagsvold also found that earlier arriving species varied considerably in arrival date at a particular locality from year to year, but late arriving species had much less variation in arrival time. Pinkowski and Bajorek (1976) examined the spring arrival dates of 29 common or conspicuous migrants and summer residents in southern Michigan over a 7-year period. They concluded that granivorous, omnivorous, and aquatic species tend to arrive earlier than strictly insectivorous species, and that earlier arriving species have a greater variance in date of arrival than later arriving species.

Before leaving the topic of primary timing mechanisms it should be emphasized that photoperiod is but one source of such information; there are several seasonal climatic events that are highly correlated with the various phases of the annual climatic cycle. The movement of mean value isotherms northward in spring and southward in fall may serve as proximate and ultimate factors that regulate the seasonal timing of bird migration in many species. Likewise the rather orderly movement of monsoons and the atmospheric currents that regulate these conditions may be used as proximate or ultimate timers for the phenology of bird migration in other species (Gauthreaux, 1980). Clearly we have just begun to understand the nature of primary timing mechanisms and much of this information has come from birds in the temperate zone where photoperiod shows great seasonal variation. The migrations of birds in the tropics are only poorly known (Moreau, 1972; Nix, 1976; Keast and Morton, 1980), and perhaps in this region primary timing mechanisms are quite different. The importance of fruiting seasons in the tropics to intertropical migrations deserves more attention (see Snow, 1965).

Considerable research has been devoted to modifying mechanisms that accelerate or inhibit the progression of seasonal migrations in birds (Weise, 1974), and most of these can be attributed to stochastic weather conditions previously discussed. An excellent review of the timing and the amount of bird migration in relation to weather has recently appeared (Richardson, 1978). Richardson concludes on the basis of his exhaustive review that peak numbers of birds migrate with synoptic weather conditions such that the pressure gradient falls from the right to the left side of their primary direction and under such conditions the migrants have following winds.

Although many birds migrate in the fair and relatively calm centers of high pressure systems, most fly with tail winds so that birds appear to select wind directions in different portions of the synoptic pattern that are blowing toward their migratory goals. In general, spring migration in north temperate areas takes place with falling pressure, high or rising temperature, and low humidity. In the fall rising pressure, low or falling temperature, and low humidity are most conducive to large movements of birds. Immediately behind a cold front in fall migrants tend to move SE–SSE, but farther behind the front movements to SSW–WSW are more common. Unsettled weather with clouds and precipitation tend to suppress migration, but not eliminate it. There is little evidence that fog reduces the number of migrants aloft. Occasionally migrating birds encounter strong contraseasonal weather conditions in spring and fall, and when these circumstances arise, some migrants perform reverse migrations. The utility of this behavior in spring is obvious, but its appearance in fall is the subject of debate. Although this section has emphasized the deterministic and stochastic influences of climate on the seasonal migrations of birds, many other terrestrial organisms show pronounced seasonal migrations in response to the same climatic cycles and factors, and Baker (1978) and Schmidt-Koenig (1975) have discussed many examples of these movements.

3. Irruptive Migrations

The winter irruptive movements of several boreal species of seed-eating birds have been shown to be generally synchronous and related to variations in the annual climatic cycle. Bagg (1969) showed that the irruptive movements of black-capped chickadees (*Parus atricapillus*) in Ontario occurred approximately every other year from 1951 to 1968 in response to decreases in conifer cone crops. Red-breasted nuthatches (*Sitta canadensis*) and pine siskins (*Spinus pinus*) are also known to invade the south–central United States in alternate years (James, 1967), and Bock and Lepthien (1972) found that the continent-wide pattern of winter irruptions in the red-breasted nuthatch was perfectly synchronized with those of other species such as black-capped chickadees, siskins, and nuthatches, and occur in alternate years. In a later analysis, Bock and Lepthien (1976) found a generally synchronous pattern of winter irruptive movements among eight species of boreal seed-eating birds in response to circumboreally synchronized pattern of seed crop fluctuations in certain high-latitude tree species. Although there was some variability, the movements tended to occur during the fall and winter of odd numbered years. Bock and Lepthien concluded that "the only likely way in which tree species might achieve such a synchrony would be to evolve a common sensitivity to one or a group of regularly fluctuating climatic variables." They pointed out also that Bryson and Dutton (1961) report an approximately alternate year pattern to many tree-ring thickness series. There is of course a quasi-biennial cycle of climatic variability (Mitchell, 1976, and Figs. 1 and 2) that is harmonically related to the annual change in climate. This cycle results from geometric asymmetries in the annual radiation changes as well as a tendency for the climatic system to respond asymmetrically to the changes.

Although the annual climatic cycle is not considered to be very important in the regulation of the demographic cycles of small mammals (Lidicker, 1975), some workers (e.g., Fuller, 1967, 1969) consider weather factors to be important in the timing of population explosions and crashes in microtene rodents. There is one report of a quasi two-year cycle of *Microtus californicus* on an island (Lidicker, 1973), but climatic factors have not been suggested as the causative agent of this cycle. The fact that climatic periodicities of 2.2 and 5.5 years have been indentified commonly (Lamb, 1977) suggest that more work is needed on the influence of climate on demographic cycles in small mammals, before climatic effects can be ruled out entirely. A general treatment of the irruptive migrations of animals can be found in Orr (1970, pp. 4–13).

V. Diel and Semidiel Periodicities and
   Migrational Responses

A. Diel and Semidiel Changes and Biotic Responses

For many organisms the change in the environment during the 24-hour cycle of the earth's rotation on its axis is as important in terms of survival and reproduction as the seasonal climatic cycle is to others. For phytoplanktonic and zooplanktonic organisms the change from light to dark during a 24-hour period represents a climatic cycle equivalent to the seasonal cycle of photoperiod in many tundra plants and animals inhabiting the arctic and antarctic. The relationship between the earth and the moon produces earth tides, and as the earth rotates eastward upon its axis, two tidal centers move westward with respect to the surface of the earth. As each center passes a given point on the surface of the globe near the equator, ocean levels rise to their maximum and halfway between the points the waters fall to a minimum. Thus the typical semidiel tidal rhythm represents an additional external forcing mechanism on the climatic regime of many organisms. Because of the high predictability of these environmental events, a number of organisms have been shown to have endogenous circadian and tidal rhythms that anticipate important daily and semidiel climatic changes in their environments. Aschoff (1965), Bunning (1973), and Neumann (1975) have discussed the nature of circadian, lunar, and tidal clocks at length. The range of biological responses to the climatic changes induced by these periodicities is great, but only migratory responses to these changes will be discussed.

B. DIEL AND SEMIDIEL CLIMATIC CHANGES

The change from light to dark in the daily climatic cycle is a deterministic one in that it is controlled by the speed of rotation of the earth on its axis and by the inclination of the earth's axis in relation to the plane of the ecliptic. The actual climate generated by this cycle is also strongly influenced by internal stochastic processes that have been discussed previously. In general, all weather elements show a daily periodicity when large scale (synoptic) influences are held constant. Synoptic influences nonetheless exist and exert a strong influence on the day-to-day values of weather elements during particular phases of daily cycle. In lakes and oceans environmental variables change far more rapidly in the vertical than in the horizontal plane so that migratory adjustments to daily climatic cycles are usually vertical, but other constraining factors may alter this pattern (Longhurst, 1976).

C. DIEL AND SEMIDIEL MIGRATIONS

1. Aquatic and Marine Migrations

Throughout the oceans of the world and also in fresh water, planktonic organisms show a distinct vertical migration in relation to the phase of the diel light and dark cycle (Cushing, 1951; Longhurst, 1976). At dusk very great numbers of planktonic animals migrate from dimly illuminated depths to near surface waters where they spend the night. At dawn these organisms begin a downward movement so that they spend the daylight hours at dimly illuminated depths. There can be little doubt that light serves as a proximate factor in the timing of the planktonic movements. Sournia (1974) has discussed the circadian periodicities in natural populations of marine phytoplankton, and Enright and Hamner (1967) obtained mixed results in their attempts to discover circadian rhythms in several species of diel migrants. The journeys may cover over 500 m, and movements of 250 m in depth are average. These diel planktonic migrations show seasonal differences at high latitudes. In polar seas during periods of continuous daylight in summer, the vertical migrations are largely suppressed. The adaptive significance of oceanic vertical migrations is the source of considerable speculation (see Enright, 1977b, 1979; Pearre,

1979; Koslow, 1979; McLaren, 1963; Rudjakov, 1970; Enright and Honegger, 1977), and the two most widely held hypotheses relate to predation
avoidance and metabolic or bioenergetic optimization.

The predator avoidance hypothesis implies that the zooplanktonic
organisms escape predation by migrating downward during the day. The
bioenergetic hypothesis stated by McLaren (1963) suggests that vertically
migrating zooplankton realize an energetic advantage by migrating to the
warm surface at night to feed and by remaining at colder depths during the
day to grow. Moreover the grazing of phytoplankton by zooplankton during the night may actually stimulate greater productivity during the day in
the former (McAllister, 1969). Additional discussions of the bioenergetics
of vertical migrations can be found in Kerfoot (1970), Longhurst (1976),
Swift (1976), and Tseitlin (1977), and several studies have concentrated on
the feeding and dietary aspects of vertical migrants (e.g., Hu, 1978; Pearre,
1973). The effects of temperature gradients in the water column have been
considered by Derenbach et al. (1979) and Kamykowski and Zentara (1977)
for phytoplankton and by McLaren (1963) for zooplankton. At high
latitudes thermal barriers may suppress vertical migrations or establish
boundaries to limit the vertical movements as in the case of a sharp
thermocline. Copepods are strongly represented in the diel vertical migrations of zooplankton (Enright, 1977a; McLaren, 1974; Moraitou-
Apostolopoulou, 1971; White et al., 1979), but other groups are also
involved [e.g., hydrozoans (Schmidt, 1973); pelagic shrimps (Barsukov and
Ivanov, 1979; Omori and Gluck, 1979); squid (Belman, 1978); tunicates
(Harbison and Campenot, 1979)]. Vertical migrations of the meiofauna
also occur in estuarine and marine sediments (Dye, 1978a) and these diel
vertical migrations are also strongly influenced by seasonal factors as well
(Dye, 1978b). Interactions between the diel vertical migrations of
organisms and tidal cycles have been recorded frequently (e.g., Al-Adhub
and Naylor, 1975; Bosch and Taylor, 1973; Cronin and Forward, 1979.)

In the study by Bosch and Taylor (1973) estuarine cladocerans, Podon
polyphemoides remain in the Chesapeake Bay despite a net-tidal seaward
current. This is achieved by the pattern of vertical migration shown by the
podonids. During the day the bulk of the population is situated in the upper part of the two-layered estuarine circulation, but at night they migrate
to deeper water where landward advection of the deeper currents helps
counterbalance seaward population loss during the day. Thus the diel
migrations permit this species to remain within the estuary despite the
dynamic estuarine environment.

A diversity of timing mechanisms (e.g., endogenous lunar rhythms, circadian rhythms, and hourglass timers) has been demonstrated in the intertidal midge, Clunio marinus (Neumann, 1975). The length of a generation

in *Clunio* is temperature dependent and ranges in Helgoland from 3–4 months in summer to 7–10 months in winter. In late spring with rising sea temperatures emergence occurs every 15 days, at the times of the full and new moon, resulting in a semilunar periodicity of reproduction. The periodicity of pupation and emergence is semilunar and also circadian, and emergence of adults must be synchronized with the low water at a specific time of day (the imago has a life span of only about 2 hours). The combination of semilunar and diel timing mechanisms thus allows for the temporal coordination of reproduction with the optimum phases of the intertidal environmental cycle. Although *Clunio* does not show migratory behavior, its timing mechanisms have been thoroughly studied and serve as a model for other organisms that have migratory behavior closely coupled with tidal cycles.

VI. Overview

The environment of an organism is constantly changing in time because of periodic climatic changes that affect it directly, or indirectly through other elements of the biota. If the organism cannot tolerate the change, it must change in space, or perish. Certain spatial boundaries are imposed on the movement pattern of the organism because of the velocity of its locomotion in relation to the periodicity of the environmental change. In general, the shorter the periodicity, the more limited the space traversed. For short periodicities (e.g., semidiel and diel), the distances of movement are usually considerably less than 1 km (vertical migration of oceanic plankton) and rarely more than 100 km (cyclic roosting flights of birds). For annual periodicities, the movements may cover distances of 10,000 km (the seasonal migrations of birds). For periodicities greater than annual ones, there is no increase in the spatial boundaries for organisms as mobile as birds, but for more sedentary organisms, the spatial boundaries would increase at climatic cycles of wavelengths greater than a year. This is in keeping with the fact that most climatic variation on time scales less than one year are regional, rather than global in extent (Mitchell, 1976).

Because of the hierarchical organization of climatic variation, the movement patterns of organisms in response to various climatic changes can also be thought of in terms of a hierarchical organization. The total range of a species may change slowly in response to long-term climatic changes (biogeographical migration) while the individuals that constitute the species may change location between breeding and wintering ranges in response to the annual climatic cycle (seasonal migration). On the wintering grounds individuals may change locations between sleeping areas and feeding areas in response to the diel light and dark cycle (roosting movements). Even movements to feeding areas during the day may be dependent on semidiel fluctuations in the availability of food. Consequently the temporal patterns of environmental change that influence the timing and duration of optimal conditions for survival and reproduction (Levins, 1968) have been of the utmost importance in shaping the timing and spatial aspects of the patterns of animal movement.

Certain climatic cycles are controlled externally by deterministic forcing mechanisms that are independent of the state of the climatic system. These climatic cycles influence the *general* patterns of animal movement. Just as climatic variability is dependent substantially on stochastic mechanisms that are internal to the climatic system (Mitchell, 1976), the variability in the general patterns of animal movement is dependent on stochastic mechanisms that are internal to the biotic system (e.g., interactions between the different parts of the system). Additional variability in the general patterns of movement comes directly from climatic variability. Thus stochastic processes internal to the climatic and biotic systems have shaped the *specific* patterns of animal movements within the boundaries imposed by climatic cycles generated by deterministic forcing mechanisms external to the climatic and biotic systems. This perspective of the patterns of animal "migrations" has permitted me to approach the study of climatic influences on the movements more meaningfully, and hopefully, it will have utility for investigations of other aspects of the life history of organisms in their environment.

Acknowledgments

During the preparation of the manuscript I benefited greatly from discussions with my departmental colleagues Jim Schindler and Jack Waide, and their graduate students Paul Giammatteo, John Hains, and Chris Waldron. F. R. Harden Jones and G. P. Harris kindly provided me with copies of manuscripts in press. Donna Dixon and Anne Snider were indispensable in helping me to track down reference material, and Sue McConnell was unbelieveably patient in typing the final draft.

References

Aaby, B. (1976). *Nature (London)* **263**, 281–284.
Ab'Saber, Z. N. (1977). *Paleoclimas* **3**, 1.
Agadjanian, A. K., and Agadzhanyan, A. K. (1977). *Quartär* **27-28**, 111–145.
Al-Adhub, A. H. Y., and Naylor, E. (1975). *J. Mar. Biol. Assoc. U. K.* **55**, 801–810.
Allee, W. C., Emerson, A. E., Park, O., Park, T., and Schmidt, K. P. (1949). "Principles of Animal Ecology." Saunders, Philadelphia, Pennsylvania.

Andersen, S. T. (1973). In "Quaternary Plant Ecology" (H. J. B. Birks and R. G. West, eds.), pp. 108–114. Blackwell, Oxford.

Anderson, E. (1968). Univ. Colo. Stud., Ser. Earth Sci. 6, 1–59.

Anderson, E. (1974). Geol. Surv. Wyo., Rep. Invest. 10, 78–87.

Andrewartha, H. G., and Birch, L. C. (1954). "The Distribution and Abundance of Animals." Univ. of Chicago Press, Chicago, Illinois.

Arnold, G. P. (1974). Biol. Rev. Cambridge Philos. Soc. 49, 515–576.

Arntzen, J. W. (1978). J. Biogeogr. 5, 339–345.

Aschoff, J. (1965). "Circadian Clocks." North-Holland Publ., Amsterdam.

Ashworth, A. C. (1972). Entomol. Scand. 3, 211–224.

Ashworth, A. C. (1980). Quat. Res. 13, 200–212.

Auffenberg, W., and Milstead, W. W. (1965). In "The Quaternary of the United States" (H. E. Wright, Jr. and D. G. Frey, eds.), pp. 557–568. Princeton Univ. Press, Princeton, New Jersey.

Austin, H. M., and Custer, O. (1977). N. Y. Fish Game J. 24, 53–68.

Bagg, A. M. (1969). Audubon Field Notes 23, 4–12.

Baker, R. R. (1978). "The Evolutionary Ecology of Animal Migration." Holmes & Meier, New York.

Banfield, A. W. F. (1962). Natl. Mus. Can. Bull. 177, 1–137.

Bannister, P. (1977). "Physiological Plant Ecology." Wiley, New York.

Barr, T. C. (1969). In "The Distributional History of the Biota of the Southern Appalachians. Part I: Invertebrates" (P. C. Holt, ed.), pp. 67–92. Virginia Polytechnic Institute, Blacksburg.

Barry, R. G., and Chorley, R. J. (1978). "Atmosphere, Weather, and Climate." Halsted Press, Somerset.

Barry, R. G., and Perry, A. H. (1973). "Synoptic Climatology. Methods and Applications." Methuen, London.

Barsukov, V. N., and Ivanov, B. G. (1979). Biol. Morya (Vladivostok) 3, 18–23.

Basedow, T. (1977) Z. Angew. Entomol. 83, 173–183.

Beck, S. D. (1968). "Insect Photoperiodism." Academic Press, New York.

Belman, B. W. (1978). Limnol. Oceanogr. 23, 735–739.

Berenbeim, D. Ya., Dubrovin, I. Ya., and Studenikina, E. M. (1973). J. Ichthyol. (Engl. Transl.) 13, 313–317.

Bernabo, J. C., and Webb, T. (1977). Quat. Res. 8, 64–96.

Berthold, P. (1975). In "Avian Biology" (D. S. Farner, J. R. King, and K. C. Parkes, eds.), Vol. 5, pp. 77–128. Academic Press, New York.

Berti, A. A. (1975). Quat. Res. 5, 591–620.

Bishop, W. W., and Coope, G. R. (1977) In "Studies in the Scottish Late Glacial Environment" (J. M. Gray and J. J. Lowe, eds.), pp. 61–88. Pergamon, Oxford.

Blair, W. F. (1958). In "Zoogeography" (C. L. Hubbs, ed.), Publ. 51, pp. 433–468. Am. Assoc. Adv. Sci., Washington, D. C.

Blair, W. F. (1965). In "The Quaternary of the United States" (H. E. Wright, Jr. and D. G. Frey, eds.), pp. 543–556. Princeton Univ. Press, Princeton, New Jersey.

Bleakney, J. S. (1958). Natl. Mus. Can., Bull. 155, 1–119.

Bligh, J., Cloudsley-Thompson, J. L., and Macdonald, A. G. (1977). "Environmental Physiology of Animals." Halsted Press, Somerset.

Bochenski, Z. (1974). Przegl. Zool. 18, 51–54.

Bock, C. E., and Lepthien, L. W. (1972). Am. Birds 26, 558–561.

Bock, C. E., and Lepthien, L. W. (1976). Am. Nat. 110, 559–571.

Bogorov, B. G. (1960). In "Perspectives in Marine Biology" (A. A. Buzzati-Traverso, ed.), pp. 145–158. Univ. of California Press, Berkeley.

Bonny, A. P. (1972). New Phytol. 71, 393–405.

Bosch, H. F., and Taylor, W. R. (1973). Mar. Biol. 19, 172–181.

Boucher, K. (1976). "Global Climate." Halsted, New York.

Bousfield, E. L. (1958). Can. Field Nat. 72, 55–113.

Bowden, J. (1973). Meded. Fac. Landbouwwet., Rijksuniv. Gent 38, 785–796.

Bowden, J. (1976). J. Entomol. Soc. South Afr. 39, 207–245.

Bowden, J., and Gibbs, D. G. (1973). Bull. Entomol. Res. 62, 571–596.

Bowen, D. Q. (1978), "Quaternary Geology: A Stratigraphic Framework for Multidisciplinary Work." Pergamon, Oxford.

Bowler, J. M., Hope, G. S., Jennings, J. N., Singh, G., and Walker, D. (1976). Quat. Res. 6, 359–394.

Boyd, E. M., and Nunneley, S. A. (1964). Bird-Banding 35, 1–8.

Brattstrom, B. H. (1967). Copeia pp. 188–202.

Briggs, J. C. (1974). "Marine Zoogeography." McGraw-Hill, New York.

Brinkmann, W. A. R. (1976). Quat. Res. 6, 355–358.

Broecker, W. S., and van Donk, J. (1960). Rev. Geophys. Space Phys. 8, 169–198.

Brunnschweiler, D. (1964). Z. Geomorphol. [N. S.] 8, 223–231.

Bryson, R. A. (1966). Geogr. Bull. 8, 228–269.

Bryson, R. A. (1974). Science 184, 753–760.

Bryson, R. A., and Dutton, J. A. (1961). Ann. N. Y. Acad. Sci. 95, 580–604.

Bryson, R. A., and Hare, R. K. (1974). In "Climates of North America" (R. A. Bryson and F. K. Hare, eds.), pp. 1–47. Elsevier, Amsterdam.

Bryson, R. A., and Wendland, W. M. (1967). In "Life, Land and Water" (W. J. Mayer-Oakes, ed.), pp. 271–298. Univ. of Manitoba Press, Winnipeg.

Bryson, R. A., Baerreis, D. A., and Wendland, W. M. (1970). In "Pleistocene and Recent Environments of the Central Great Plains" (W. Dort, Jr. and J. K. Jones, Jr., eds), pp. 53–74. Univ. of Kansas Press, Lawrence.

Budyko, M. I. (1974). "Climate and Life." Academic Press, New York.

Bunning, E. (1973). "The Physiological Clock: Circadian Rhythms and Biological Chronometry," Springer-Verlag, Berlin and New York.

Butzer, K. W. (1976). Geosci. & Man 13, 27–44.

Byers, G. W. (1969). In "The Distributional History of the Biota of the Southern Appalachians. Part I: Invertebrates" (P. C. Holt, ed.), pp. 265–276. Virginia Polytechnic Institute, Blacksburg.

Bykhovskii, B. E. (1974). "Bird Migration: Ecological and Physiological Factors." Wiley, New York.

Campion, D. G., Bettany, B. W., McGinnigle, J. B., and Taylor, L. R. (1977). Bull. Entomol. Res. 67, 501–522.

Chavin, W. (1973). "Responses of Fish to Environmental Changes." Thomas, Springfield, Illinois.

Clark, N. E., Blasing, T. J., and Fritts, H. C. (1975). Nature (London) 256, 302–305.

Coe, M. J. (1967). "The Ecology of the Alpine Zone of Mount Kenya." Junk, The Hague.

Coetzee, J. A. (1967). In "Palaeoecology of Africa III" (E. M. van Zinderen Bakker, ed.), pp. 1–146. Balkema, Cape Town.

Colbert, E. H. (1953). In "Climatic Change. Evidence, Causes, and Effects" (H. Shapley, ed.), pp. 249–271. Cambridge Univ. Press, London and New York.

Cooch, F. G. (1963) Proc. In. Ornithol. Congr. 13th, 1962 pp. 1182–1194.

Cooke, H. B. S. (1972). *Q. Rev. Biol.* **43**, 234–264.

Coope, G. R. (1968). *Philos. Trans. R. Soc. London, Ser. B* **254**, 425–456.

Coope, G. R. (1975). *In* "Ice Ages Ancient and Modern" (A. E. Wright and F. Moseley, eds.), Geol. J. Spec. Issue No. 6, pp. 153–168. Gallery Press, Liverpool, England.

Coope, G. R. (1977a). *In* "British Quaternary Studies" (F. W. Shotton, ed.), pp. 55–68. Oxford Univ. Press (Clarendon), London and New York.

Coope, G. R. (1977b). *Philos. Trans. R. Soc. London, Ser. B* **280**, 313–340.

Coope, G. R., and Angus, R. B. (1975). *J. Anim. Ecol.* **44**, 365–391.

Cooper, R. A., and Uzmann, J. R. (1971). *Science* **171**, 288–290.

Corliss, B. H., (1975). *Palaeogeogr. Palaeoclimatol., Palaeoecol.* **18**, 45–61.

Cremer, K. W. (1977). *Aust. For.* **7**, 225–228.

Cresswell, G. R. (1971). *CSIRO Aust. Div. Fish. Oceanogr. Rep. No. 50.*

Crisp, D. J. (1959). *Geogr. J.* **125**, 1–19.

Crisp, D. J. (1965). *In* "The Biological Significance of Climatic Changes in Britain" (C. G. Johnson and L. P. Smith, eds.), Inst. Biol. Symp. No. 14, pp. 63–77. Academic Press, New York.

Cronin, T. W., and Forward, R. B., Jr. (1979). *Science* **205**, 1020–1022.

Cross, F. B. (1970). *In* "Pleistocene and Recent Environments of the Central Great Plains" (W. Dort, Jr. and J. K. Jones, Jr., eds.), pp. 241–257. Univ. of Kansas Press, Lawrence.

Curray, J. R. (1960). *In* "Recent Sediments, Northeast Gulf of Mexico" (F. P. Shepard, ed.), pp. 221–226. Am. Assoc. Pet. Geol., Tulsa, Oklahoma.

Cushing, D. H. (1951). *Biol. Rev. Cambridge Philos. Soc.* **26**, 158–192.

Cushing, D. H. (1969). *FAO Fish. Tech. Pap.* **84**, 1–40.

Cushing, D. H. (1971). *J. Cons., Cons. Int. Explor. Mer.* **33**, 340–362.

Cushing, D. H. (1975). "Marine Ecology and Fisheries." Cambridge Univ. Press, London and New York.

Cushing, D. H. (1976a). *Geogr. J.* **142**, 216–227.

Cushing, D. H. (1976b). *In* "The Ecology of the Seas" (D. H. Cushing and J. J. Walsh, eds.), pp. 317–340. Saunders, Philadelphia, Pennsylvania.

Cushing, D. H. (1978). *Rapp. P.-V. Reun., Cons. Int. Explor. Mer* **173**, 107–116.

Cushing, D. H., and Dickson, R. R. (1976). *Adv. Mar. Biol.* **14**, 1–122.

Dabrowski, M. J. (1971). *Bull. Acad. Pol. Sci., Ser. Sci. Terre* **19**, 29–36.

Dadswell, M. J. (1974). *Natl. Mus. Can., Publ. Zool.* **11.**

Dalquest, W. W. (1962). *J. Paleontol.* **36**, 568–582.

Dalquest, W. W. (1964). *Trans. Kans. Acad. Sci.* **67**, 499–505.

Dalquest, W. W. (1965). *J. Paleontol.* **39**, 63–79.

Danilevskii, A. S. (1965). "Photoperiodism and Seasonal Development of Insects." Oliver & Boyd, Edinburgh.

Danthanarayana, W. (1976). *Oecologia* **23**, 271–282.

Daubenmire, R. (1978). "Plant Geography with Special Reference to North America." Academic Press, New York.

Davis, M. B. (1969). *Am. Sci.* **57**, 317–332.

Davis, M. B. (1976). *Geosci. & Man* **13**, 13–26.

Deevey, E. S., Jr. (1949). *Geol. Soc. Am. Bull.* **60**, 1315–1416.

Degerbøl, M. (1964). *J. Anim. Ecol.* **33**, Suppl., 71–85.

Demarcq, G., and Mourier-Charvire, C. (1976). *Geobios.* **9**, 125–141.

Denton, G. H., and Karlen, W. (1973). *Quat. Res.* **3**, 155–205.

Derenbach, J. B., Astheimer, H., Hansen, H., Leach, H. P., and Leach, H. (1979). *Mar. Ecol.* **1**, 187–194.

de Veen, J. F. (1978). *Neth. J. Sea Res.* **12**, 115–147.

DeWorms, G. G. M. (1958). *Proc. Int. Congr. Entomol., 10th, 1956* Vol. 1, pp. 737–740.

Dillon, L. S. (1956). *Science* **123**, 167–176.

Dingle, H. (1972). *Science* **175**, 1327–1335.

Dingle, H. (1978). "Evolution of Insect Migration and Diapause." Springer-Verlag, Berlin and New York.

Dingle, H. (1979). *In* "Movement of Highly Mobile Insects: Concepts and Methodology in Research" (R. L. Rabb and G. G. Kennedy, eds.), pp. 64–87. North Carolina State University Graphics, Raleigh.

Donner, J. J. (1964). *Am. J. Sci.* **262**, 355–376.

Dorst, J. (1962). "The Migrations of Birds." Houghton, Boston, Massachusetts.

Dort, W., Jr., and Jones, J. K. Jr., eds. (1970). "Pleistocene and Recent Environments of the Central Great Plains." Univ. of Kansas Press, Lawrence.

Dowdy, W. W. (1944). *Ecology* **25**, 449–460.

Dreimanis, A., and Raukas, A. (1975). *R. Soc. N. Z., Bull.* **13**, 109–120.

Duviard, D. (1977). *Bull. Entomol. Res.* **67**, 185–204.

Dye, A. H. (1978a). *Zool. Afr.* **13**, 201–205.

Dye, A. H. (1978b). *Zool. Afr.* **13**, 207–212.

Dyer, T. G. J., and Curtis, B. (1978). *Afr. J. Sci.* **74**, 176–178.

Ekman, S. (1953). "Zoogeography of the Sea." Sidgwick & Jackson, London.

Elliott, J. A. (1976). *Fed. Proc., Fed. Am. Soc. Exp. Biol.* **35**, 2339–2346.

Emiliani, C., Gartner, S., Lidz, B., Eldridge, D., Elvey, D. K., Huang, T. C., Stipp, J. J., and Swanson, M. F. (1975). *Science* **189**, 1083–1088.

Enright, J. T. (1976). *Oecologia* **24**, 295–310.

Enright, J. T. (1977a). *Limnol. Oceanogr.* **22**, 118–125.

Enright, J. T. (1977b). *Limnol. Oceanogr.* **22**, 856–872.

Enright, J. T. (1979). *Limnol. Oceanogr.* **24**, 788–791.

Enright, J. T., and Hamner, W. M. (1967). *Science* **157**, 937–941.

Enright, J. T., and Honegger, H. W. (1977). *Limnol. Oceanogr.* **22**, 873– 886.

Ericson, D. B., and Wollin, G. (1970). *Science* **167**, 1483–1485.

Ericson, D. B., Ewing, M., Wollin, G., and Heezen, B. C. (1961). *Geol. Soc. Am. Bull.* **72**, 193–286.

Ericson, D. B., Ewing, M., and Wollin, G. (1964). *Science* **146**, 723–732.

Etheridge, R. (1958). *Copeia* pp. 94–101.

Etheridge, R. (1960). *Pap. Mich. Acad. Sci., Arts Lett.* **45**, 113–117.

Etherington, J. R. (1975). "Environment and Plant Ecology." Wiley, New York.

Faegri, K. (1963). *In* "North Atlantic Biota and Their History" (A. Love and D. Love, eds.), pp. 221–232. Macmillan, New York.

Farner, D. S. (1967). *Proc. Int. Ornithol. Congr., 14th, 1966* pp. 107–133.

Farner, D. S., and Follett, B. K. (1979). *In* "Hormones and Evolution" (E. J. W. Barrington, ed.), Vol. 2, pp. 829–872. Academic Press, New York.

Farner, D. S., and Lewis, R. A. (1971). *Photophysiology* **6**, 325–370.

Farner, D. S., Lewis, R. A., and Darden, T. R. (1973). *In* "Biology Data Book" (P. L. Altman and D. S. Dittmer, eds.), Vol. II, pp. 1047–1052. Fed. Am. Soc. Exp. Biol., Bethesda, Maryland.

Farrand, W. R. (1971). *In* "The Late Cenozoic Glacial Ages" (K. K. Turekian, ed.), pp. 493–529. Yale Univ. Press, New Haven, Connecticut.

Findley, J. S. (1969). *In* "Contributions in Mammalogy" (J. K. Jones, Jr., ed.), Publ. Mus. Nat. Hist. No. 51. pp. 113–128. University of Kansas, Lawrence.

Firbas, F. (1949). "Spat-und Nacheiszeitliche Waldgeschichte Mitteleuropas Nordlich der Alpen." Fischer, Jena.

Flint, R. F., and Deevey, E. S., Jr. (1951). *Am. J. Sci.* **249**, 257–300.

Flint, R. W. (1977). *Am. Midl. Nat.* **97**, 280–292.

Frakes, L. A. (1979). "Climates Throughout Geologic Time." Am. Elsevier, New York.

Frenzel, B. (1967). "Die Klimasch Wankungen des Eiszeitalters." Vieweg, Braunschweig.

Frenzel, B. (1968). *Science* **161**, 637–648.

Frenzel, B. (1973). "Climatic Fluctuations of the Ige Age." Case Western Reserve Univ. Press, Cleveland, Ohio.

Frey, D. G. (1958). *Arch. Hydrobiol.* **54**, 209–275.

Frey, D. G. (1961). *Int. Assoc. Limnol. Proc.* **14**, 271–278.

Frey, D. G. (1964). *Arch. Hydrobiol., Suppl. Ergeb. Limnol.* **2**, 1–114.

Frey, D. G. (1965). *In* "The Quaternary of the United States" (H. E. Wright, Jr. and D. G. Frey, eds.), pp. 613–631. Princeton Univ. Press, Princeton, New Jersey.

Fritts, H. C. (1976). "Tree Rings and Climate." Academic Press, New York.

Fritts, H. C. (1978). *Naturwissenschaften* **65**, 48–56.

Fuji, N. (1973). *Proc. Jpn. Acad.* **49**, 737–748.

Fujii, T. (1975). *Mem. Fac. Fish., Hokkaido Univ.* **22**, 99–192.

Fuller, W. A. (1967). *Terre Vie* **114**, 97–115.

Fuller, W. A. (1969). *Ann. Zool. Fenn.* **6**, 113–144.

Garrod, D. J., and Colebrook, J. M. (1978). *Rapp. P.-V. Reun., Cons. Int. Explor. Mer* **173**, 128–144.

Gates, D. M. (1979). "Biophysical Ecology." Springer-Verlag, Berlin and New York.

Gates, D. M., and Schmerl, R. (1975). "Perspectives of Biophysical Ecology." Springer-Verlag, Berlin and New York.

Gauthreaux, S. A., Jr. (1980). *Proc. Int. Ornithol. Congr., 17th 1978* (in press).

Gauthreaux, S. A., Jr., and LeGrand, H. E. (1974). *Am. Birds* **28**, 771–778.

Glemarec, M. (1979). *Oceanol. Acta* **2**, 365–372.

Godwin, H. (1956). "The History of the British Flora." Cambridge Univ. Press, London and New York.

Godwin, H. (1975). "The History of the British Flora," 2nd ed. Cambridge Univ. Press, London and New York.

Golding, F. D. (1928). *Emp. Cotton Grow. Rev.* **5**, 128–133.

Good, R. (1953). "The Geography of the Flowering Plants." Longmans, Green, New York.

Grace, J. (1978). "Plant Response to Wind." Academic Press, New York.

Graham, R. W. (1976). *Paleobiology* **2**, 343–350.

Grant, V. (1977). "Organismic Evolution." Freeman, San Francisco, California.

Gribbin, J. (1978). "Climatic Change." Cambridge Univ. Press, London and New York.

Griffin, D. R. (1974). "Bird Migration." Dover, New York.

Griggs, G. B., Kulm, L. D., Duncan, J. R., and Fowler, G. A. (1970). *Palaeogeogr., Palaeoclimatol., Palaeoecol.* **7**, 5–12.

Gromov, V. I., Nikiforova, K. V., and Kahlke, H. D. (1978). *Schriftenr Geol. Wiss.* **9**, 145–160.

Gruger, E. (1972a). *Quat. Res.* **2**, 217–231.

Gruger, E. (1972b). *Geol. Soc. Am. Bull.* **83**, 2715–2734.

Gruger, J. (1973). *Geol. Soc. Am. Bull.* **84**, 239–250.

Guelpen, L. V., and Davis, C. C. (1979). *Trans. Am. Fish. Soc.* **108**, 26–37.

Guilday, J. E. (1962). *Ann. Carnegie Mus.* **36**, 87–122.

Guilday, J. E. (1967). *Am. Antiq.* **32**, 231–232.

Guilday, J. E. (1971). *In* "The Distributional History of the Biota of the Southern Appalachians. Part III: Vertebrates" (P. C. Holt, ed.), pp. 232–262. Virginia Polytechnic Institute, Blacksburg.

Guilday, J. E., and Adam, E. K. (1967). *Tebiwa* **10**, 26–36.

Guilday, J. E., and Parmalee, P. W. (1972). *Quat. Res.* **2**, 170–175.

Guilday, J. E., Hamilton, H. W., and Adam, E. K. (1967). *Contrib. Geol.* **6**, 97–99.

Gwinner, E. (1977). *Annu. Rev. Ecol. Syst.* **8**, 381–405.

Haffer, J. (1974). *Bonn. Zool. Beitr.* **25**, 87–117.

Haggis, M. J. (1971). *East Afr. Agric. For. J.* **37**, 100–108.

Hall, E. R. (1958). *In* "Zoogeography" (C. L. Hubbs, ed.), Publ. 51, pp. 371–373. Am. Assoc. Adv. Sci., Washington, D. C.

Hamilton, P. V. (1978). *Mar. Biol.* **46**, 49–58.

Hamm, F. (1951). *Beitr. Naturkd. Neidersachs.* **4**, 1–14.

Handley, C. O., Jr. (1971). *In* "The Distributional History of the Biota of the Southern Appalachians. Part III: Vertebrates" (P. C. Holt, ed.), pp. 263–303. Virginia Polytechnic Institute, Blacksburg.

Harbison, G. R., and Campenot, R. B. (1979). *Limnol. Oceanogr.* **24**, 1081–1091.

Harden Jones, F. R. (1968). "Fish Migration." Arnold, London.

Harden Jones, F. R. (1977). *In* "Fisheries Mathematics" (J. H. Steele, ed.), pp. 145–170. Academic Press, New York.

Harden Jones, F. R. (1980) *In* "Fish Behavior and its Use in the Capture and Culture of Fishes" (J. E. Bardoch, J. J. Magnuson, R. C. May, and J. M. Reinhart, eds.), International Center for Living Aquatic Resources Management, Manila (in press).

Harden Jones, F. R., Walker, M. G., and Arnold, G. P. (1978). *In* "Advances in Oceanography" (H. Charnock and G. Deacon, eds.), pp. 185–207. Plenum, New York.

Harden Jones, F. R., Arnold, G. P., Walker M. G., and Scholes, P. (1980). *J. Cons., Cons. Int. Explor. Mer,* (in press).

Harmsworth, R. V. (1968). *Ecol. Monogr.* **38**, 223–241.

Harris, A. (1964). *Weather* **19**, 70–79.

Harris, G. P. (1980). *Can J. Fish. Aquat. Sci.* **37**, 877–900.

Harris, G. P., and Smith, R. E. H. (1977). *Limnol. Oceanogr.* **22**, 887–899.

Haskell, E. (1972). "Full Circle: The Moral Force of Unified Science." Gordon & Breach, New York.

Haury, L. R., McGowan, J. A., and Wiebe, P. H. (1978). *In* "Spatial Pattern in Plankton Communities" (J. H. Steele, ed.), pp. 277–328. Plenum, New York.

Hay, O. P. (1923). *Carnegie Inst. Washington, Publ.* **322**, 1–499.

Hay, O. P. (1924). *Carnegie Inst. Washington, Publ.* **322A**, 1–385.

Hay, O. P. (1927). *Carnegie Inst. Washington, Publ.* **322B**, 1–346.

Haynes, J. M., Gray, R. H., and Montgomery, J. C. (1978). *Trans. Am. Fish. Soc.* **107**, 275–280.

Hays, J. D. (1967). *Prog. Oceanogr.* **4**, 117–131.

Henriksen, K. (1933). *Vidensk. Medd. Dan. Naturhist. Foren.* **96**, 77–355.

Herman, J. R., and Goldberg, R. A. (1978) *NASA Spec. Publ.* **426.**

Hibbard, C. W. (1949). *Geol. Soc. Am. Bull.* **60**, 1417–1428.

Hibbard, C. W. (1955). *Contrib. Mus. Paleontol. Univ. Mich.* **12**, 47–96.

Hibbard, C. W. (1970). *In* "Pleistocene and Recent Environments of the Central Great Plains" (W. Dort, Jr. and J. K. Jones, Jr., eds.), pp. 395–433. Univ. of Kansas Press, Lawrence.

Hibbard, C. W., and Taylor, D. W. (1960). *Contrib. Mus. Paleontol., Univ. Mich.* **16**, 1–223.

Hibbard, C. W., Ray, D. E., Savage, D. E., Taylor, D. W., and Guilday, J. E. (1965). *In* "The Quaternary of the United States" (H. E. Wright, Jr. and D. G. Freys, eds.), pp. 509–525. Princeton Univ. Press, Princeton, New Jersey.

Hill, R. W. (1976). "Comparative Physiology of Animals." Harper, New York.

Hobbs, H. H., Jr. (1969). *In* "The Distributional History of the Biota of the Southern Appalachians. Part I: Invertebrates;" (P. C. Holt, ed.), pp. 93–178. Virginia Polytechnic Institute, Blacksburg.

Hoffmann, R. S., and Jones, J. K., Jr. (1970). *In* "Pleistocene and Recent Environments of the Central Great Plains" (W. Dort, Jr. and J. K. Jones, Jr., eds.), pp. 355–394. Univ. of Kansas Press, Lawrence.

Holman, J. A. (1963). *J. Grad. Res. Cent.* **31**, 152–167.

Holman, J. A. (1964). *Herpetologica* **20**, 73–83.

Holman, J. A. (1976). *Herpetologica* **32**, 290–295.

Hope, G. S., Peterson, J. A., Radok, U., and Allison, I. (1976). "The Equitorial Glaciers of New Guinea." Balkema, Rotterdam.

Hu, V. J. H. (1978). *Limnol. Oceanogr.* **23**, 296–306.

Hubbard, J. P. (1971). *In* "The Distributional History of the Biota of the Southern Appalachians. Part III: Vertebrates" (P. C. Holt, ed.), pp. 197–232. Virginia Polytechnic Institute, Blacksburg.

Hubbs, C. L. (1948). *J. Mar. Res.* **7**, 459–482.

Hubbs, C. L. (1960). *Syst. Zool.* **9**, 134–147.

Hughes, M. K., Gray, B., Pilcher, J., Baillie, M., and Leggett, P. (1978). *Nature (London)* **272**, 605–606.

Humphrey, S. R. (1974). *BioScience* **24**, 457–462.

Hustich, I. (1952). *Fennia* **75**, 97–105.

Imbrie, J., and Imbrie, J. Z. (1980). *Science* **207**, 943–953.

Imbrie, J. C., and Kipp, N. G. (1971). *In* "The Late Cenozoic Glacial Ages" (K. K. Turekian, ed.), pp. 71–182. Yale Univ. Press, New Haven, Connecticut.

Ingram, M. J., Underhill, D. J., and Wigley, T. M. L. (1978). *Nature (London)* **276**, 329–333.

Irving, L. (1972). "Arctic Life of Birds and Mammals, Including Man." Springer-Verlag, Berlin and New York.

Itoh, S., Ogata, K., and Yoshimura, H. (1972). "Advances in Climatic Physiology." Springer-Verlag, Berlin and New York.

Iversen, J. (1973). *Dan. Geol. Unders. [Afh.], Raekke 5* **7C**, 1–126.

James, F. C. (1967). *Audubon Field Notes* **21**, 426–430.

Jenkins, R. E., Lochner, E. A., and Schwartz, F. J. (1971). *In* "The Distributional History of the Biota of the Southern Appalachians. Part III: Vertebrates" (P. C. Holt, ed.), pp. 43–117. Virginia Polytechnic Institute, Blacksburg.

Johnson, C. G. (1969). "Migration and Dispersal of Insects by Flight." Methuen, London.

Johnson, C. G., and Smith, L. P., eds. (1965). "The Biological Significance of Climatic Changes in Britain," Inst. Biol. Symp. No. 14. Academic Press, New York.

Johnson, D. A., and Knoll, A. H. (1974). *Quat. Res.* **4**, 206–216.

Johnson, D. A., and Knoll, A. H. (1975). *Quat. Res.* **5**, 99–110.

Johnson, S. E. (1976). *Int. J. Biometeorol.* **20**, 207–217.

Johnson, T., and Muller, K. (1978). *Aquilo, Ser. Zool.* **18**, 57–61.

Jones, R. (1973). *Rapp. P.-V. Ren., Cons. Int Explor. Mer* **164**, 156–173.

Jopson, H. G. M. (1971). *In* "The Distributional History of the Biota of the Southern Appalachians. Part III: Vertebrates" (P. C. Holt, ed.), pp. 189–196. Virginia Polytechnic Institute, Blacksburg.

Joyce, R. J. V. (1976). *Symp. R. Entomol. Soc. London* **7**, 135–155.

Kaesler, R. L., and Mulvany, P. S. (1977). *Proc. Int. Symp. Ostracods, 6th, 1976* pp. 33–43.

Kalela, O. (1949). *Bird-Banding* **201**, 77–103.

Kamykowski, D., and Zentara, S.-J. (1977). *Limnol. Oceanogr.* **22**, 148–151.

Kanciruk, P., and Herrnkind, W. (1978). *Bull. Mar. Sci.* **28**, 601–623.

Kapp, R. O. (1965). *Contrib. Mus. Paleontol., Univ. Mich.* **19**, 167–255.

Kapp, R. O., and Gooding, A. M. (1964). *J. Geol.* **72**, 307–326.

Karlstrom, E. L. (1962). *Univ. Calif., Berkeley, Publ. Zool.* **62**, 1–104.

Keany, J., and Kennett, J. P. (1972). *Deep-Sea Res.* **19**, 529–548.

Keast, A., and Morton, E., eds. (1980). "Migrant Birds in the Neotropics." Smithsonian Institution Press, Washington, D. C.

Kendall, R. L. (1969). *Ecol. Monogr.* **39**, 121–176.

Kennedy, J. S. (1975). *In* "Insects, Science, and Society" (D. Pimentel, ed.), pp. 103–119. Academic Press, New York.

Kerfoot, W. B. (1970). *Am. Nat.* **104**, 529–546.

Kerney, M. P. (1968). *Symp. Zool. Soc. London* **22**, 273–291.

Kerney, M. P. (1976a). *J. Conchol.* **29**, 26–28.

Kerney, M. P. (1976b). *J. Conchol.* **29**, 47–50.

Kerney, M. P. (1977). *In* "British Quaternary Studies" (F. W. Shotton, ed.), pp. 31–42. Oxford Univ. Press (Clarendon), London and New York.

Keve, A. (1963). *Proc. Int. Ornithol. Congr., 13th, 1962* pp. 1124–1127.

Kisimoto, R. (1976). *Ecol. Entomol.* **1**, 95–109.

Knox, A. S. (1962). *Pollen Spores* **4**, 357–358.

Koslow, J. A. (1979). *Limnol. Oceanogr.* **24**, 783–784.

Kowalski, K. (1967). *In* "Pleistocene Extinctions: The Search for a Cause" (P. S. Martin and H. E. Wright, eds.), pp. 349–364. Yale Univ. Press, New Haven, Connecticut.

Krebs, C. J. (1978). "Ecology: The Experimental Analysis of Distribution and Abundance." Harper, New York.

Kurten, B. (1968). "Pleistocene Mammals of Europe." Weidenfeld & Nicolson, London.

Kurten, B., and Anderson, E. (1980). "Pleistocene Mammals of North America." Columbia Univ. Press, New York.

Kutzbach, J. E. (1976). *Quat. Res.* **6**, 471–480.

Kutzbach, J. E. (1978). *IEEE Trans. Geosci. Electron.* **ge-16**.

LaBar, G. W., McCleave, J. D., and Fried, S. M. (1978). *J. Cons., Cons. Int. Explor. Mer* **38**, 257–269.

LaMarche, V. C., Jr. (1978). *Nature (London)* **276**, 334–338.

Lamb, H. H. (1972). "Climate," Vol. 1. Methuen, London.

Lamb, H. H. (1977). "Climate: Present, Past and Future," Vol. 2. Methuen, London.

Larcher, W. (1980). "Physiological Plant Ecology." Springer-Verlag, Berlin and New York.

Laurs, R. M., and Lynn, R. J. (1977) *Fish. Bull.* **75**, 795–822.

Lecoq, M. (1978). *C. R. Hebd. Seances Acad. Sci., Ser. D* **286**, 419–422.

Leggett, W. C. (1977a). *J. Fish. Res. Board Can.* **34**, 1422–1426.

Leggett, W. C. (1977b). *Annu. Rev. Ecol. Syst.* **8**, 285–308.

Leonard, A. B. (1953). *Am. J. Sci.* **251**, 369–376.

Levins, R. (1968). "Evolution in Changing Environments." Princeton Univ. Press, Princeton, New Jersey.

Levitt, J. (1972). "Responses of Plants to Environmental Stresses." Academic Press, New York.

Libby, L. M., and Pandolfi, L. J. (1977). *Nature (London)* **266**, 415–417.

Lidicker, W. Z., Jr. (1973). *Ecol. Monogr.* **43**, 271–302.

Lidicker, W. Z., Jr. (1975). *In* "Small Mammals: Their Productivity and Population Dynamics" (F. B. Golley, K. Petrusexicz, and L. Ryszkowski, eds.), I.B.P. 5, pp. 103–128. Cambridge Univ. Press, London and New York.

Lieth, H. (1974). "Phenology and Seasonality Modeling." Springer-Verlag, Berlin and New York.

Livingstone, D. A. (1975). *Annu. Rev. Ecol. Syst.* **6**, 332–341.

Longhurst, A. R. (1976). *In* "The Ecology of the Seas" (D. H. Cushing and J. J. Walsh, eds.), pp. 116–137. Saunders, Philadelphia, Pennsylvania.

Löppenthin, B. (1963). *Proc. Int. Ornithol. Congr., 13th, 1962* pp. 1128–1133.

Lowry, W. P. (1969). "Weather and Life: An Introduction to Biometeorology." Academic Press, New York.

Lozek, V. (1972). *Quat. Res.* **2,** 327–334.

Lundelius, E. L., Jr. (1967). *In* "Pleistocene Extinctions: The Search for a Cause" (P. S. Martin and H. E. Wright, Jr., eds.), pp. 287–319. Yale Univ. Press, New Haven, Connecticut.

Lundelius, E. L., Jr. (1976). *Geosci. & Man* **13,** 45–59.

McAllister, C. D. (1969). *J. Fish Res. Board Can.* **26,** 199–220.

McClure, H. E. (1974). "Migration and Survival of the Birds of Asia." U.S. Army Medical Component, SEATO Medical Project, Bangkok, Thailand.

McIntyre, A. (1967). *Science* **158,** 1314.

McIntyre, A., Ruddiman, W. F., and Jantzen, R. (1972). *Deep-Sea Res.* **19,** 61–77.

Mackintosh, N. A. (1937). *'Discovery' Rep.* **16,** 365–412.

McLaren, I. A. (1963). *J. Fish Res. Board Can.* **20,** 685–727.

McLaren, I. A. (1974). *Am. Nat.* **108,** 91–102.

McManus, M. L. (1979). *In* "Radar, Insect Population Ecology, and Pest Management" (C. R. Vaughn, W. Wolf, and W. Klassen, eds.), NASA Conf. Publ. 2070, pp. 3–16. NASA Wallops Island, Virginia.

Maloiy, G. M. O., ed. (1972). "Comparative Physiology of Desert Animals," Symp. Zool. Soc. London, No. 31. Academic Press, New York.

Marshall, S. M., and Orr, A. P. (1960). *J. Mar. Biol. Assoc. U. K.* **39,** 135–147.

Martin, P. S. (1958). *In* "Zoogeography" (C. L. Hubbs, ed.), Publ. 51, pp. 375–420. Am. Assoc. Adv. Sci., Washington, D.C.

Mason, B. J. (1976). *Endeavour* **35,** 51–57.

Maxwell, J. A., and Davis, M. B. (1972). *Quat. Res.* **2,** 506–530.

May, R. M. (1979). *Nature (London)* **281,** 177–178.

Mayhew, D. F. (1975). The quaternary history of some British rodents and lagomorphs. Ph.D. Thesis, University of Cambridge.

Mecham, J. S. (1959). *Southwest. Nat.* **3,** 17–27.

Mere, R. M. (1961). *Proc. R. London Entomol. Nat. Hist. Soc., 1961* pp. 63–73.

Meyer, H. J., and Appleby, J. E. (1973). *Ann. Entomol. Soc. Am.* **66,** 505–508.

Michel, R., and Albrecht, F. O. (1978). *Behaviour* **67,** 208–216.

Miller, A. H. (1937). *Condor* **39,** 248–252.

Miller, A. H. (1940). *Proc. Pac. Sci. Congr., 6th, 1939* pp. 807–810.

Miller, R. R. (1958). *In* "Zoogeography" (C. L. Hubbs, ed.), Publ. 51, pp. 187–222. Am. Assoc. Adv. Sci., Washington, D.C.

Miller, R. R. (1965). *In* "The Quaternary of the United States" (H. E. Wright, Jr. and D. G. Frey, eds.), pp. 569–581. Princeton Univ. Press, Princeton, New Jersey.

Milstead, W. W. (1967). *Copeia* pp. 168–179.

Mitchell, J. M., Jr. (1963). *Arid Zone Res.* **20,** 161–181.

Mitchell, J. M., Jr. (1971). *In* "Man's Impact on the Climate" (W. H. Matthews, W. N. Kellogg, and G. D. Robinson, eds.), pp. 133–140. MIT Press, Cambridge, Massachusetts.

Mitchell, J., Jr. (1976). *Quat. Res.* **6,** 481–493.

Momot, W. T., and Gowing, H. (1972). *Ecology* **53,** 479–483.

Monod, T. (1964). *In* "African Ecology and Human Evolution" (F. C. Howell, and F. Bourliere, eds.), pp. 117–229. Methuen, London.

Monteith, J. L., ed. (1975). "Vegetation and the Atmosphere," 2 Vols. Academic Press, London.

Mooney, H. A. (1975). *In* "Physiological Adaptation to the Environment" (F. J. Vernberg, ed.), pp. 19–36. Intext Educational Publishers, New York.

Moore, P. D. (1978). *Nature (London)* **272**, 578–579.

Moore, T. C., Jr. (1973). *Quat. Res.* **3**, 99–109.

Moraitou-Apostolopoulou, M. (1971), *Mar. Biol.* **9**, 92–98.

Moreau, R. E. (1955). *Proc. Zool. Soc. London* **125**, 253–295.

Moreau, R. E. (1963). *Proc. Zool Soc. London* **141**, 395–421.

Moreau, R. E. (1969). *J. Zool.* **158**, 39–61.

Moreau, R. E. (1972). "The Palaearctic-African Bird Migration System" (J. F. Monk. ed.). Academic Press, New York.

Morel, G. J., and Morel, M.-Y. (1978). *Cah. ORSTOM, Ser. Biol* **13**, 347–358.

Morgan, A. (1973). *Boreas* **2**, 173–212.

Mourer-Chauvire, C. (1974). *Anthropologie* **78**, 37–48.

Mourer-Chauvire, C. (1975). *Geobios* **8**, 333–352.

Mourer-Chauvire, C. (1976). *Arch. Mus. Hist. Nat. Lyon* **14**, 41–42.

Munn, R. E. (1970). "Biometeorological Methods." Academic Press, New York.

Munro, I. S. R. (1975). *In* "National Prawn Seminar" (P. C. Young, ed.), pp. 60–78. Australian Government Publishing Service, Canberra.

Neumann, D. (1975). *In* "Physiological Adaptation to the Environment" (F. J. Vernberg, ed.), pp. 451–463. Intext Educational Publishers, New York.

Neves, R. J., and Depres, L. (1979). *Fish. Bull.* **77**, 199–212.

Newell, R. C. (1976). "Adaptation to Environment: Essays on the Physiology of Marine Animals." Butterworth, London.

Newell, R. E. (1979). *Am. Sci.* **67**, 405–416.

Nichols, H. (1974). *In* "Arctic and Alpine Environments" (J. D. Ives and R. G. Barry, eds.), pp. 637–667. Methuen, London.

Nigrini, C. A. (1971). *In* "The Micropaleontology of Oceans" (B. M. Funnel and W. R. Riedel, eds.), pp. 443–461. Cambridge Univ. Press, London and New York.

Nix, H. A. (1976). *Proc. Int. Ornithol. Congr., 16th, 1974* pp. 272–305.

Norton, P. E. P. (1977). *In* "British Quaternary Studies" (F. W. Shotton, ed.), pp. 43–53. Oxford Univ. Press (Clarendon), London and New York.

Omori, M., and Gluck, D. (1979). *Fish. Bull.* **77**, 183–198.

Orr, R. T. (1970). "Animals in Migration." Macmillan, New York.

Ortmann, A. E. (1913). *Proc. Am. Philos. Soc.* **52**, 287–395.

Osborne, P. J. (1974). *Quat. Res.* **4**, 471–486.

Parker, R. H. (1960). *In* "Recent Sediments, Northwest Gulf of Mexico" (F. P. Shepard, ed.), pp. 302–337. Am. Assoc. Pet. Geol., Tulsa, Oklahoma.

Parker, R. H. (1964a). *Vidensk. Medd. Dansk. Naturh. Foren.* **126**, 1–178.

Parker, R. H. (1964b). *In* "Marine Geology of the Gulf of California" (T. H. van Andel and G. C. Shor, Jr., eds.), Mem. 3, pp. 331–376. Am. Assoc. Pet. Geol., Tulsa, Oklahoma.

Parmalee, P. W. (1968). *Univ. Ill. Coll. Agric., Agric. Exp. Stn., Spec. Publ.* **14**, 104–113.

Pavshtiks, E. A. (1959). *U. S., Fish Wildl. Serv., Spec. Sci. Rep.—Fish.* **327**, 104–139.

Peabody, F. E., and Savage, J. M. (1958). *In* "Zoogeography" (C. L. Hubbs, ed.), Publ. 51, pp. 159–186. Am. Assoc. Adv. Sci., Washington, D.C.

Pearre, S., Jr. (1973). *Ecology* **54**, 300–314.

Pearre, S., Jr. (1979). *Limnol. Oceanogr.* **24**, 781–782.

Pearson, R. (1978). "Climate and Evolution." Academic Press, New York.

Pengelley, E. T., ed. (1974). "Circannual Clocks, Annual Biological Rhythms," Academic Press, New York.

Pigott, C. D. (1975). *Weather* **30**, 82–90.

Pinkowski, B. C., and Bajorek, R. A. (1976). *Jack Pine Warbler* **54**, 62–68.

Pittock, A. B., Frakes, L. A., Jenssen, D., Peterson, J. A., and Zillman, J. W. (1978). "Climatic Change and Variability: A Southern Perspective." Cambridge Univ. Press, London and New York.

Platt, T. (1978). *In* "Spatial Pattern in Plankton Communities" (J. H. Steele, ed.), pp. 73–84. Plenum, New York.

Pocklington, R. (1972). *J. Geophys. Res.* **77**, 6604–6607.

Pocklington, R. (1978). *Nature (London)* **273**, 407.

Porter, D. M. (1976). *Nature (London)* **264**, 745–746.

Porter, K. R. (1964). *Herpetologica* **19**, 229–247.

Porter, K. R. (1972). "Herpetology." Saunders, Philadelphia, Pennsylvania.

Rabb, R. L., and Kennedy, G. G., eds. (1979). "Movement of Highly Mobile Insects: Concepts and Methodology in Research." North Carolina State University Graphics, Raleigh.

Rabb, R. L., and Stinner, R. E. (1979). *In* "Radar, Insect Population Ecology, and Pest Management" (C. R. Vaughn, W. Wolf, and W. Klassen, eds.), NASA Conf. Publ. 2070, pp. 3–16. NASA Wallops Island, Virginia.

Rainey, R. C. (1951). *Nature (London)* **168**, 1057–1060.

Rainey, R. C. (1973). *Weather* **28**, 224–239.

Rainey, R. C. (1974). *Annu. Rev. Entomol.* **19**, 407–439.

Rainey, R. C. (1976). *Symp. R. Entomol. Soc. London* **7**, 75–112 and 272–273.

Rainey, R. C. (1978). *In* "Evolution of Insect Migration and Diapause" (H. Dingle, ed.), pp. 33–48. Springer-Verlag, Berlin and New York.

Rainey, R. C. (1979). *In* "Movements of Highly Mobile Insects: Concepts and Methodology in Research" (R. L. Rabb and G. G. Kennedy, eds.), pp. 109–119. North Carolina State University Graphics, Raleigh.

Ramsay, A. T. S., ed. (1977). "Oceanic Micropaleontology," Vol. 2. Academic Press, New York.

Reichle, D. E., O'Neil, R. V., and Harris, W. F. (1975). *In* "Unifying Concepts in Ecology" (W. H. van Dobben and R. H. Lowe-McConnell, eds.), pp. 27–43. Junk, The Hague.

Reid, J. L., Brinton, E., Fleminger, A., Venrick, E. L., and McGowan, J. A. (1978). *In* "Advances in Oceanography" (H. Charnock and G. Deacon, eds.), pp. 65–130. Plenum, New York.

Richardson, W. J. (1978). *Oikos* **30**, 224–272.

Riley, D., and Spolton, L. (1974). "World Weather and Climate." Cambridge Univ. Press, London and New York.

Riley, J. R., and Reynolds, D. R. (1979). *Proc. R. Soc. London, Ser. B* **204**, 67–82.

Riordan, A. J. (1979). *In* "Movement of Highly Mobile Insects: Concepts and Methodology in Research" (R. L. Rabb and G. G. Kennedy, eds.), pp. 120–132. North Carolina State University Graphics, Raleigh.

Robertshaw, D. (1974). "Environmental Physiology." University Park Press, Baltimore, Maryland.

Robertshaw, D. (1977). "Environmental Physiology. II," Int. Rev. Physiol., Vol. 15. University Park Press, Baltimore, Maryland.

Robinson, A. R. (1976). *Oceanus* **19**, 2–17.

Rose, D. J. W., and Law, A. B. (1976). *J. Entomol. Soc. South Afr.* **39**, 125–130.

Rosenberg, N. J. (1974). "Microclimate: The Biological Environment." Wiley, New York.

Ross, H. H. (1965). *In* "The Quaternary of the United States" (H. E. Wright, Jr. and D. G. Frey, eds.), pp. 583–596. Princeton Univ. Press, Princeton, New Jersey.

Ross, H. H. (1970). *In* "Pleistocene and Recent Environments of the Central Great Plains" (W. Dort, Jr. and J. K. Jones, Jr., eds.), Spec. Publ. 3, pp. 225–240. Univ. of Kansas Press, Lawrence.

Ruddiman, W. F. and McIntyre, A. (1976). *Mem., Geol. Soc. Am.* **145**, 111–146.

Ruddiman, W. F., Sancetta, C. D., and McIntyre, A. (1977). *Philos. Trans. R. Soc. London, Ser. B* **280**, 119–142.

Rudjakov, J. A. (1970). *Mar. Biol.* **6**, 98–105.

Sachs, H. M. (1973a). *Quat. Res.* **3**, 73–88.

Sachs, H. M. (1973b). *Quat. Res.* **3**, 89–98.

Salisbury, E. (1976). *Proc. R. Soc. London, Ser. B.* **192**, 323–329.

Salomonsen, F. (1948). *Dan. Ornithol. Foren. Tidsskr.* **42**, 85–99.

Saunders, D. S. (1976). "Insect Clocks." Pergamon, Oxford.

Savage, J. M. (1960). *Syst. Zool.* **9**, 184–212.

Schaefer, G. W. (1976). *Symp. R. Entomol. Soc. London* **7**, 157–197.

Schaefer, G. W. (1979). *Philos. Trans. R. Soc. London, Ser. B.* **287**, 459–465.

Schmidt, H.-E. (1973). *Mar. Biol.* **18**, 61–68.

Schmidt, K. P. (1938). *Ecology,* **19**, 396–407.

Schmidt-Koenig, K. (1975). "Migration and Homing in Animals." Springer-Verlag, Berlin and New York.

Schneider, F. (1962). *Annu. Rev. Entomol.* **7**, 223–242.

Schopf, T. J. M. (1980). "Paleoceanography." Harvard Univ. Press, Cambridge, Massachusetts.

Schultz, C. B. (1972). *Quat. Res.* **2**, 337–340.

Schultz, C. B., and Hillerud, J. M. (1977). *In* "Quaternary Glaciation of the Northern Hemisphere," Rep. 4, pp. 218–233. Int. Union Quat. Res., Stuttgart.

Schultz, C. B., and Martin, L. D. (1970). *In* "Pleistocene and Recent Environments of the Central Great Plains" (W. Dort, Jr. and J. K. Jones, Jr., eds.), pp. 341–353. Univ. of Kansas Press, Lawrence.

Schultz, C. B., Lueninghoerner, G. C., and Frankforter, W. D. (1951). *Bull. Univ. Nebr. State Mus.* **3**, 1–41.

Schultz, G. E. (1967). *In* "Pleistocene Extinctions: The Search for a Cause" (P. S. Martin and H. E. Wright, Jr. eds.), pp. 321–336. Yale Univ. Press, New Haven, Connecticut.

Schultz, G. E. (1969). *Geol. Soc. Am., Spec. Pap.* **105**, 1–85.

Schüz, E. (1971). "Grundriss der Vogelzugskunde." Parey, Berlin.

Seddon, B. (1971). "Introduction to Biogeography." Harper, New York.

Selander, R. K. (1965). *In* "The Quaternary of the United States" (H. E. Wright, Jr. and D. G. Frey, eds.), pp. 527–542. Princeton Univ. Press, Princeton, New Jersey.

Serventy, D. L. (1958). *Proc. Pac. Sci. Congr., 8th, 1953* Vol. 3, pp. 461–487.

Serventy, D. L. (1960). *In* "Biology and Comparative Physiology of Birds" (A. J. Marshall, ed.), Vol. 1, pp. 95–126. Academic Press, New York.

Serventy, D. L. (1977). *Emu* **77**, 162–166.

Sharp, G. D., and Dizon, A. E., eds. (1978). "The Physiological Ecology of Tunas." Academic Press, New York.

Shotton, F. W. (1977). "British Quaternary Studies: Recent Advances." Oxford Univ. Press (Clarendon), London and New York.

Shutts, G. J., and Green, J. S. A. (1978). *Nature (London)* **276**, 339–341.

Simon, H. A. (1973). *In* "Hierarchy Theory" (H. H. Patte, ed.), pp. 1–28. Braziller, New York.

Simpson, B. B. (1974). *Science* **185**, 698–700.

Simpson, G. G. (1953). "The Major Features of Evolution." Columbia Univ. Press, New York.

Sissons, J. B. (1979). *Nature (London)* **280**, 199–203.

Slagsvold, T. (1976). *Norw. J. Zool.* **24**, 161–173.

Slaughter, B. H., and Hoover, B. R. (1963). *J. Grad. Res. Cent.* **31**, 132–148.

Slaughter, B. H., and Ritchie, R. (1963). *J. Grad. Res. Cent.* **31**, 117–131.

Smith, G. R., and Fisher, D. R. (1970). *In* "Pleistocene and Recent Environments of the Cen-

tral Great Plains" (W. Dort, Jr. and J. K. Jones, Jr., eds.), pp. 259–277. Univ. of Kansas
  Press, Lawrence.
Smith, K. (1975). "Principles of Applied Climatology." Halsted Press, New York.
Smith, P. E. (1978). Rapp. P.-V. Reun., Cons. Int. Explor. Mer 173, 117–127.
Smith, P. W. (1957). Ecology 38, 205–218.
Smith, R. L. (1976). In "The Ecology of the Seas" (D. H. Cushing and J. J. Walsh, eds.), pp.
  23–58. Saunders, Philadelphia, Pennsylvania.
Snow, D. W. (1965). Oikos 15, 274–281.
Snow, J. W., and Copeland, W. W. (1969). U.S., Dep. Agric., Prod. Res. Rep. 110, 1–9.
Solbreck, C. (1975). Oikos 27, 134–143.
Sømme, J. D. (1933). Am. Nat. 67, 33–34 and 42–43.
Sorensen, A. E. (1978). Nature (London) 276, 174–176.
Sournia, A. (1974). Adv. Mar. Biol. 12, 325–389.
Southwood, T. R. E. (1962). Biol. Rev. Cambridge Philos. Soc. 37, 171–214.
Southwood, T. R. E. (1975). In "Insects, Science, and Society" (D. Pimentel, ed.), pp.
  151–191. Academic Press, New York.
Southwood, T. R. E. (1977). J. Anim. Ecol. 46, 337–365.
Sparks, B. W. (1961). Proc. Linn. Soc. London 172, 71–80.
Sparks, B. W., and West, R. G. (1972). "The Ice Age in Britain." Methuen, London.
Staples, D. J. (1979). Aust. J. Mar. Freshwater Res. 30, 143–157.
Staplin, F. L. (1963). J. Paleontol. 37, 758–797 and 1164–1203.
Starkel, L. (1977). Philos. Trans. R. Soc. London, Ser. B 280, 351–372.
Steele, J. H., and Frost, B. W. (1977). Philos. Trans. R. Soc. London, Ser. B. 280, 285–534.
Steuerwald, B. A., and Clark, D. L. (1972). J. Palaeontol. 46, 573–580.
Strickland, A. H. (1947). J. Anim. Ecol. 16, 1–10.
Stuart, A. J. (1974). Biol. Rev. Cambridge Philos. Soc. 49, 225–266.
Stuart, A. J. (1976). Philos. Trans. R. Soc. London, Ser. B 276, 221–250.
Stuart, A. J. (1977a). In "British Quaternary Studies" (F. W. Shotton, ed.), pp. 69–81. Oxford
  Univ. Press (Clarendon), London and New York.
Stuart, A. J. (1977b). Philos. Trans. R. Soc. London, Ser. B. 280, 295–312.
Suggate, R. P. (1965). N. Z., Geol. Surv., Rep. 77.
Sutcliffe, A. J., and Kowalski, K. (1976). Bull. Br. Mus. (Nat. Hist.), Geol. 27, 33–147.
Swift, M. C. (1976). Ecology 57, 900–914.
Talbot, G. B., and Sykes, J. E. (1958). Fish. Bull. 58, 473–490.
Tauber, M. J., and Tauber, C. A. (1976). Int. J. Biometeorol. 20, 218–222.
Taylor, D. W. (1965). In "The Quaternary of the United States" (H. E. Wright, Jr. and D. G.
  Frey, eds.), pp. 597–611. Princeton Univ. Press, Princeton, New Jersey.
Taylor, L. R. (1974). J. Anim. Ecol. 43, 225–238.
Taylor, L. R., and Taylor, R. A. J. (1977). Nature (London) 265, 415–421.
Terasmaë, J. (1960). Geol. Surv. Can., Mem. 56, 1–41.
Theobald, N. (1952). Geol. Rundsch. 40, 89–92.
Thunnell, R. C, and Lohmann, G. P. (1979). Nature (London) 281, 211–213.
Tihen, J. A. (1962). Am. Midl. Nat. 68, 1–50.
Tseitlin, V. B. (1977). Okeanologiya 17, 532–539.
Tsukada, M. (1966). Anthropology 55, 543–549.
Tsukada, M. (1967). Am. J. Bot. 54, 821–831.
Ullrich, W. G., and Coope, G. R. (1974). J. Entomol., Ser. B 42, 207–212.
U.S. Committee for GARP (1975). "Understanding Climatic Change." United States Commit-
  tee for the Global Atmospheric Research Program, National Research Council, National
  Academy of Sciences, Washington, D.C.

Uspenskii, S. M. (1969). *Bird-Banding* **41**, 161–163.

Valentin, P. C. (1976). *U.S., Geol. Surv., Prof. Pap.* **916**, 1–47.

Van der Hammen, T. (1974). *J. Biogeogr.* **1**, 3–26.

Van der Hammen, T., Wijmstra, T. A., and Zagwijn, W. H., (1971). *In* "The Late Cenozoic Glacial Ages" (K. K. Turekian, ed.), pp. 391–424. Yale Univ. Press, New Haven, Connecticut.

Van der Pijl, L. (1972). "Principles of Dispersal in Higher Plants." Springer-Verlag, Berlin and New York.

Van Devender, T. R., and Spaulding, W. G. (1979). *Science* **204**, 701–710.

Vaughn, T. A. (1978). "Mammalogy." Saunders, Philadelphia, Pennsylvania.

Vermeij, G. J. (1978). "Biogeography and Adaptation: Patterns of Marine Life." Harvard Univ. Press, Cambridge, Massachusetts.

Vernberg, F. J., ed. (1975a). "Physiological Adaptation to the Environment." Intext Educational Publishers, New York.

Vernberg, F. J., ed. (1975b). "Physiological Ecology of Estuarine Organisms." Univ. of South Carolina Press, Columbia.

Vernberg, W., and Vernberg, F. J. (1972). "Environmental Physiology of Marine Animals." Springer-Verlag, Berlin and New York.

Veronis, G. (1978). *J. Mar. Res.* **36**, 1–44.

von Uexküll, J. (1909). "Umwelt and Innenwelt der Tiere." Springer-Verlag, Berlin and New York.

Vuilleumier, B. S. (1971). *Science* **173**, 711–780.

Wagner, F. J. E. (1977). *Can. J. Earth Sci.* **14**, 1305–1323.

Walker, D. (1978). *In* "Climatic Change and Variability: A Southern Perspective" (A. B. Pittock, L. A. Frakes, D. Jenssen, J. A. Peterson, and J. W. Zilman, eds.), pp. 82–97. Cambridge Univ. Press, London and New York.

Walker, E. M. (1957). *In* "Changes in the Fauna of Ontario" (F. A. Urquhart, ed.), pp. 4–12. R. Ont. Mus. Zool. Paleontol. Contrib., Toronto.

Walker, M. G., Harden Jones, F. R., and Arnold, G. P. (1978). *J. Cons., Cons. Int. Explor. Mer* **38**, 58–86.

Wall, R. C. (1977). *Nautilus* **91**, 47–51.

Walter, H. (1979). "Vegetation of the Earth: In Relation to Climate and the Eco-Physiological Conditions," 2nd ed. Springer-Verlag, Berlin and New York.

Watts, W. A. (1973). *Quat. Res.* **3**, 257–268.

Webb, S. D. (1974). *In* "Pleistocene Mammals of Florida" (S. D. Webb, ed.), pp. 5–31. Univ. of Florida Press, Gainesville.

Weise, C. M. (1974). *In* "Phenology and Seasonality Modeling" (H. Lieth, ed.), pp. 139–147. Springer-Verlag, Berlin and New York.

Wellington, W. G. (1979). *In* "Movements of Highly Mobile Insects: Concepts and Methodology in Research" (R. L. Rabb and G. G. Kennedy, eds.), pp. 104–108. North Carolina State University Graphics, Raleigh.

West, R. G. (1977). "Pleistocene Geology and Biology," 2nd ed. Longmans, Green, New York.

Wetmore, A. (1959). *Wilson Bull.* **71**, 178–182.

Wetmore, A. (1962). *Smithson. Misc. Collect.* **145**, 1–17.

Wetmore, A. (1967). *Bull. Ga. Acad. Sci.* **25**, 151–153.

Weydemeyer, W. (1973). *Condor* **75**, 400–413.

Weyl, P. K. (1978). *Science* **202**, 475–481.

White, H. H., Heaton, J. S., and Schmitz, K. B. (1979). *Estuaries* **2**, 61–63.

Whitehead, D. R. (1965). *In* "The Quaternary of the United States;" (H. E. Wright, Jr. and D. G. Frey, eds.), pp. 417–432. Princeton Univ. Press, Princeton, New Jersey.

Whittaker, R. H. (1975a). *In* "Unifying Concepts in Ecology" (W. H. van Dobben and R. H. Lowe-McConnell, eds.), pp. 169–181. Junk, The Hague.

Whittaker, R. H. (1975b). "Communities and Ecosystems." Macmillan, New York.

Wijmstra, T. A. (1969). *Acta Bot. Neerl.* **18**, 511.

Wijmstra, T. A. (1978). *In* "Climatic Change" (J. R. Gribbin, ed.), pp. 25–45. Cambridge Univ. Press, London and New York.

Williams, C. B. (1958). "Insect Migration." Collins, London.

Williams, D. F., and Johnson, W. (1975). *Quat. Res.* **5**, 237–250.

Williamson, K. (1975). *Bird Study* **22**, 143–164.

Williamson, K. (1976). *Weather* **31**, 362–384.

Wilson, A. T., Hendy, C. H., and Reynolds, C. P. (1979). *Nature (London)* **279**, 315–317.

Woillard, G. (1974). Exposé de recherches palynologiques sur le Pleistocène dans les Vosges Lorraines. Thèse doctoral, Université Catholique deLouvain, Louvain.

Wright, H. E., Jr. (1971). *In* "The Late Cenozoic Glacial Ages" (K. Turekian, ed.), pp. 425–464. Yale Univ. Press, New Haven, Connecticut.

Wright, H. E., Jr. (1976). *Quat. Res.* **6**, 581–596.

Wright, H. E., Jr., and Frey, D. G., eds. (1965). "The Quaternary of the United States." Princeton Univ. Press, Princeton, New Jersey.

Wright, H. P. (1932). *Ecol. Monogr.* **2**, 233–259.

Wulff, E. V. (1943). "An Introduction to Historical Plant Geography." Chronica Botanica, Waltham, Massachusetts.

Wurtman, R. J. (1975). *Annu. Rev. Physiol.* **37**, 467–483.

Yousef, M. K., Horvath, S. M., and Bullard, R. W., eds. (1972). "Physiological Adaptations: Desert and Mountain." Academic Press, New York.

Zagwijn, W. H. (1974). *Boreas* **3**, 75–97.

Zagwijn, W. H. (1975). *In* "Ice Ages: Ancient and Modern" (A. E. Wright and F. Moseley, eds.), pp. 137–152. Seel House, Liverpool.

Zaslavsky, V. A. (1976). "Photoperiodism in Animals and Plants." USSR Academy of Sciences, Zoological Institute, Leningrad.

Zeuner, F. E. (1958). "Dating the Past." Methuen, London.

Zeuner, F. E. (1959). "The Pleistocene Period: Its Climate, Chronology, and Faunal Successions," 2nd ed. Hutchinson, London.

Zinsmeister, W. J., and Emerson, W. K. (1979). *Veliger* **22**, 32–40.

3 The Energetics of Migration

Charles R. Blem

I. Introduction

Natural selection favors any organismal movement in which the benefits of changing location exceed the risks of travel. In temperate zone species, such movements are often of a seasonal nature and alternate between reproductive sites and wintering grounds or feeding–maturation areas. In highly mobile organisms capable of storing relatively great amounts of energy or able to forage over a wide range of habitat, these movements may become extensive, culminating in the annual 40,000 km cycle of travel by the Arctic tern, *Sterna paradisaea* (Berger and Hart, 1974).

Energetic strategies (*strategy* here being used to describe a possible or realized selective pattern in an organism's life history) are closely related to

Animal Migration, Orientation, and Navigation

temporal patterns of migration, distances traveled, and barriers to be traversed. In general, short-distance migrants or species that feed as they migrate, traveling short distances each day over a long period of time, do not often demonstrate extensive metabolic or energetic specializations. Therefore, the bulk of this chapter deals with long-range migrants that generally deposit and expend a relatively large amount of energy in a short period of time. This account is largely based upon the research literature dealing with insects and birds. This bias is due to the extensive work done with migratory species within these taxa (*insects*: see Williams, 1958; Johnson, 1965, 1969; Dingle, 1972; Crabtree and Newsholme, 1975; *birds*: see Odum, 1960a; King and Farner, 1965; Orr, 1970; Berthold, 1975; Blem, 1976), the extensive movements made by some species and the degree of adaptation and importance of migration in the life histories of many birds and insects. A fragmented literature exists for bats (see Griffin, 1970; also Allen, 1939; Davis and Hitchcock, 1966; Cockrum, 1969), sea turtles (Carr, 1964; Carr and Goodman, 1970), and fishes (Harden-Jones, 1968; Brett, 1972); therefore only occasional mention will be made of these forms. Relatively little relevant work appears to have been done with migratory cetaceans (but see Brodie, 1975) or long-distance terrestrial migrants (see Orr, 1970).

Analyses of the energetics of migrating organisms must include the costs of locomotion and the patterns of storage and utilization of energy reserves. The former involve the energetics and biophysics of locomotion, circulatory and respiratory adaptations, and the size and life form of the organism. The latter include strategies of energy accumulation, patterns of fuel storage in relation to distances traveled and barriers to be crossed, and kinds of fuels and biochemical pathways of synthesis and catabolism. Additional adaptations or limitations such as loss of water during long-range flight are related to maintenance of appropriate body temperatures and biochemistry of fuel utilization and must also be considered.

Energetic strategies of migration are similar to life history patterns of storage and utilization of energy reserves for any stressful period. Proximate cues stimulate mechanisms that determine the timing of collection of energy reserves and possibly the amounts to be accumulated. Ultimate factors involving energy demands select those individuals having appropriate timing and the ability to accumulate sufficient reserves to support energy expenditures necessary for survival of extreme conditions encountered during migration (see Berthold, 1975). A mathematical model incorporating some of the benefits and disadvantages of migration has been developed by Cohen (1967).

This chapter is intended to be a review of the physiological and morphological adaptations that promote sufficient energy supplies for migration as well as those modifications that encourage efficient use of limited energy reserves. In many instances, great gaps are apparently present in the research literature. In some instances, these blank areas are apparently due to lack of interest in the phenomena, but in other cases there has been little or no progress apparently because of the difficulties in finding or obtaining particular organisms before, during, and after migration. Where it is relevent, such gaps will be indicated.

II. Energy Reserves

A. METABOLIC SUBSTRATE

Although little or no research has been done of many migratory forms, the available evidence from studies of muscle physiology and from more detailed examinations of insects, birds, and bats indicates the lipid provides the major (but not exclusive) source of energy during migration. The words *lipid* and *fat* are used interchangeably in the following account (and in many other publications) to describe those materials, primarily triglycerides and fatty acids, extractable in nonpolar, organic solvents, particularly petroleum ether and/or chloroform (see Blem, 1976). Lipids provide a relatively light-weight, energy-rich source of fuel and are therefore advantageous for flying migrants. Lipids have a higher energy yield (9.0–9.5 kcal/gm) than carbohydrates (4.2–4.5 kcal/gm) and may often be stored without the addition of water. For these reasons, Johnson (1969, after Weis-Fogh, 1952a) suggests that in some flying insects lipid is about eight times more productive of energy than the same weight of glycogen. The available evidence indicates that lipids provide similar advantages for birds and bats. The concept that fat may be a major fuel for vigorous muscular activity is relatively new. As recently as 1960, Drummond and Black, in a review of the fuels of muscle metabolism, stated that fat is generally considered an inferior source of energy because it is less readily mobilized and could only be broken down aerobically.

Among insects, several fuels may be used to support flight, including carbohydrates, lipids, and certain amino acids (Sacktor, 1965, 1970, 1974; Beenakkers, 1969; Bailey, 1975). Furthermore, some species may use different fuels as flight becomes prolonged. Glycogen and trehalose (a nonreducing disaccharide which hydrolyzes to glucose) provide sources for glucose used directly in metabolic pathways of some long-range migrants including some diptera and lepidoptera (Bücher and Klingenberg, 1958; Clegg and Evans, 1961; Gilby *et al.*, 1967; Wyatt, 1967; Stevenson, 1968; Johnson, 1969; Van Handel and Nayar, 1972; Crabtree and Newsholme,

1975). In some insects whose flight muscles depend upon carbohydrates, there is little indication that lipids can be metabolized (Sacktor, 1965). Other insects use relatively small amounts of glycogen during the early part of migratory flights, but change over to fats as local glycogen supplies are depleted (Krogh and Weis-Fogh, 1951; Weis-Fogh, 1952b; Cockbain, 1961a). During the first half-hour of continuous flight in laboratory tests of the desert locust, *Schistocerca gregaria*, the respiratory quotient (carbon dioxide produced/oxygen consumed) was approximately 1.0, indicating that carbohydrate is probably the principal fuel source (Krogh and Weis-Fogh, 1951). After this time, the respiratory quotient decreased to near 0.7, suggesting a shift to the use of lipids. Lipid reserves declined during prolonged flight in this species and provided about 85–90% of the energy expended (Weis-Fogh, 1952b). Similar findings have been reported for the migratory locust, *Locusta migratoria* (Jutsum and Goldsworthy, 1976). Zebe (1954) demonstrated that the respiratory quotient in lepidoptera also is approximately 0.7. Lipid reserves also decline with prolonged flight in the beet leafhopper, *Euttetix fenellus*, (Fulton and Romney, 1940), monarch butterfly, *Danaus plexippus* (Beall, 1948), and aphid *Aphis fabae* (Cockbain, 1961a), douglas fir beetle, *Dendroctonus pseudotsugae* (Atkins, 1966, 1969), a dragonfly, *Pantala flavescens* (Kallapur and George, 1973), and tettigoniid cricket, *Homorocorhyphus nitidulus* (Karuhize, 1972). Enzyme activity also indicates that lipids are oxidized by flight muscles of some insects, although research of lipid metabolism in insects is not extensive compared to studies of carbohydrates and nitrogenous compounds (see George *et al.*, 1958; Chefurka, 1965ab; Sacktor, 1974, 1975). Crabtree and Newsholme (1972), in a survey of lipase activity in a variety of insects, found greater activities in those species using fat as fuel during sustained flights (also see Tietz, 1965).

Suggestions that some lepidoptera convert the charbohydrates of nectar, on which they feed as adults, to lipid energy stores (Zebe, 1953, 1954) have been disputed because of the presence of high activities of enzymes that break down carbohydrates in some lepidopteran flight muscles (Crabtree and Newsholme, 1975) and because of the specific activities of carbon dioxide and glycerides of moths fed radioactive glucose (Van Handel and Nayar, 1972). However, Brown and Chippendale (1974) suggest that female monarch butterflies can rapidly incorporate glucose into abdominal glycerides.

In birds, glucose and glycogen are used as subtrate for muscular activity (George and Berger, 1966), but the accumulation and utilization of lipids necessary for long-range migration has been widely demonstrated (it is known from at least forty families of birds) and reviewed (Farner, 1960;

Odum, 1960a, 1965; King and Farner, 1965; King, 1972; Berthold, 1975; Blem, 1976; and many others).

Nonesterified fatty acids provide fuel for avian muscular contraction (George and Vallyathan, 1964; George and Berger, 1966), particularly when exercise becomes prolonged. In the few experimental tests in which carbon dioxide production and oxygen consumption have been simultaneously measured in birds (Tucker, 1968b, 1972a; Berger and Hart, 1972), respiratory quotients ranged from 0.66 to 0.85, suggesting that lipids are being used. In some experiments, however, respiratory quotients rose with prolonged flight; either early oxygen debts or carbon dioxide retention were implicated. LeFebvre's (1964) simultaneous measurements of carbon dioxide production and lipid utilization during flight demonstrate that combustion of lipid is the sole source of energy.

Migration of bats is poorly known in comparison with knowledge of the phenomenon in birds and insects. Indeed, it is not clear whether some species migrate. It appears that extensive movements of marked individuals are relatively uncommon (see Davis, 1966, for a summary of records). The substrate that supports metabolism of flight is not well known because of a paucity of studies of the respiration of flying bats, particularly migratory forms. Bats are known to utilize lipids in hibernation (see Dodgen and Blood, 1956) and circumstantial evidence is strong that lipids provide the major source of energy for migration. Although the importance of the reserves in migration is not known, late summer lipid deposition suitable to support long-range flight has been reported for the evening bat, *Nycticieus humeralis* (Baker et al., 1968), Mexican free-tailed bat, *Tadarida brasiliensis* (Herreid, 1963a; Pagels, 1975), big brown bat, *Eptesicus fuscus* (Weber and Findley, 1970), fringed myotis, *Myotis thysanodes*, and Yuma myotis, *M. yumanensis* (Ewing et al., 1970), gray bat, *Myotis grisescens* (Krulin and Sealander, 1972), and several species of bats in Poland (Krzanowski, 1961). However, experimental evidence from tests of flying bats indicates that carbohydrates may be an important source of energy, particularly in the initial minutes of flight. In studies of the nonmigratory, tropical bats *Phyllostomus hastatus and Pteropus gouldii* (Thomas and Suthers, 1972; Thomas, 1975), mean respiratory quotients after about 1 min of flight ranged from 1.04–1.12; values from 15-min flights were less (0.88–0.93) and declined in subsequent tests performed on the same day. This may indicate a shift from carbohydrate catabolism to lipid utilization with prolonged exercise as has been demonstrated in other mammals (e.g., Issekutz et al., 1967).

The metabolic pathway by which lipids appear to be used as fuel in insects, fish, birds, and mammals is most likely a direct oxidation of fatty acids (β oxidation). In this process, fatty acids are broken down into two-carbon units, converted to acetyl-CoA which then enters the tricarboxylic acid cycle (see Farner et al., 1961; Sacktor, 1965; George and Berger, 1966; Bilinski, 1969; Hochachka, 1969; Crabtree and Newsholme, 1975; Hazelwood, 1976). Lipid is usually stored in the form of triglyceride (three fatty acid molecules attached to a glycerine molecule), but may be found in monoglyceride or diglyceride form during lipogenesis or lipid catabolism. Diglyceride has been identified as the form in which lipid is transported from fat bodies to wing muscles in some insects (Gilbert, 1967; Tietz, 1967; Mayer and Candy, 1969). Lipid apparently cannot be broken down anaerobically, therefore systems utilizing this fuel must have efficient oxygen delivery systems (see below).

The energetic cost of lipid deposition is not well known. Although the energy equivalent of extractable lipid in birds and bats ranges from 8.2 to 9.5 kcal/gm (Rogers and Odum, 1964; Odum et al., 1965; Johnston, 1970; Ewing et al., 1970), the energy involved in premigratory fattening or in lipid utilization during migration is difficult to calculate from live weight changes. Changes in body weight do not reflect addition or removal of lipid alone, even though water content and lean dry weights may remain relatively constant in birds and some bats (Odum et al., 1964; Rogers and Odum, 1964; Hicks, 1967; Ewing et al., 1970). Caloric equivalents of body weight changes in birds vary from 1.8 to 6.0 kcal/gm (Helms, 1963, 1968; Dolnik and Blyumental, 1964; Dolnik, 1968; Brisbin, 1969; Kendeigh et al., 1969; Dolnik and Gavrilov, 1973), and change with season, physiological condition, activity, time of day, and possibly may vary between species.

B. Composition of Lipid Reserves

Although the energy content of fatty acids increases slightly with chain length and degree of saturation, premigratory fat deposits generally are composed of triglycerides containing a mixture of fatty acids of various chain lengths and degrees of saturation (Hilditch and Williams, 1964; Ewing et al., 1970; Johnston, 1973; Bailey, 1975; see Table I). An appreciable amount of data on lipid composition is available for insects (see Fast, 1964; Gilbert, 1967), fish (Hilditch and Williams, 1964; Sinnhuber, 1969), and bats (Ewing et al., 1970), but by far the great majority of studies of premigratory lipid composition have been done with birds (see Blem, 1976). Lipids in all migratory birds tested so far were comprised mainly of 16- or 18-carbon fatty acids, in some cases more than 90% of the total, and the majority were unsaturated (Nakamura, 1963, 1964; Walker, 1964; West and Meng, 1968; Caldwell, 1973; Johnston, 1973; Blem, 1976). The major fatty acid components of insects, and the fatty acids that appear to be used

TABLE I

Fatty Acid Composition of Lipid Reserves in Some Selected Migrants [a]

Species	Carbon atoms:double bonds							Reference
	14:0	16:0	16:1	18:0	18:1	18:2	18:3	
Douglas fir beetle	0.8	13.2	23.7	1.8	59.8	1.1		Thompson and Bennett, 1971
Migratory locust	1.0	24.5	2.1	7.3	12.4	35.1	17.3	Hilditch and Williams, 1964
Migratory locust	11.0	42.0		18.0	14.0	5.0	10.0	Beenakkers, 1965
Coho salmon[b]	8.4	27.9	17.6	11.2	36.7	22.1	1.8	Krueger et al., 1968
Ruby-throated hummingbird[c]	1.0	17.6	5.2	20.9	33.4	1.8	3.7	Caldwell, 1973
Starling	1.3	19.3	6.3	5.8	45.3	14.9	2.2	Caldwell, 1973
Philadelphia vireo	1.0	21.7	5.7	2.6	38.5	13.9	16.6	Caldwell, 1973
Blackpoll warbler	3.6	25.8	6.8	6.4	39.1	8.7	8.8	Caldwell, 1973
Nashville warbler[d]	19.9	18.1	7.8	2.6	31.0	12.0	3.3	Caldwell, 1973
Tennessee warbler	5.8	22.8	8.3	3.6	37.4	7.8	12.4	Caldwell, 1973
Blackburnian warbler	2.6	20.3	6.8	2.6	40.9	17.3	9.3	Caldwell, 1973
Gray-cheeked thrush		30.2	8.4	3.6	53.2	3.2		Caldwell, 1973
Bobolink		13.0	8.0	26.0	19.0	6.0		Walker, 1964
Northern oriole[e]	7.5	15.1	5.0	2.6	36.0	27.3	1.1	Caldwell, 1973
Rose-breasted grosbeak[f]	3.9	21.7	2.3	9.7	34.3	15.8	3.8	Caldwell, 1973
Indigo bunting		23.2	4.5	6.7	41.0	23.5		Caldwell, 1973
White-crowned sparrow[g]	21.6	19.9	3.4	9.3	23.9	15.4	1.9	Morton and Liebmann, 1974
Little brown bat	2.3	12.3	16.9	3.5	31.3	18.6	5.3	Ewing et al., 1970
Yuma myotis	1.4	18.3	14.5	2.5	45.0	9.8	5.2	Ewing et al., 1970
Fringed myotis	0.7	14.9	9.1	1.5	49.1	17.0	6.2	Ewing et al., 1970

[a] Fatty acids comprising less than 1% of total lipids generally have been omitted.
[b] Not including 14:1 (2.0), 16:2 (2.6), 18:4 (7.9), and eight fatty acids of C_{20} or greater.
[c] Not including 20:4 (10.6).
[d] Not including 12:0 (2.2).
[e] Not including 10:0 (1.4), 12:0 (2.8).
[f] Not including 10:0 (5.9), 12:0 (2.6).
[g] Not including 12:0 (2.1), 14:1 (1.7).

most often, are the C_{16} and C_{18} saturated and unsaturated acids (Thompson and Bennett, 1971; Bailey, 1975). A similar relationship holds for the few bats tested (Ewing *et al.*, 1970). Highly unsaturated fat stores may have increased mobility and therefore may be more available during the metabolic demands of migration (Johnston, 1973). Nakamura (1963) found a decrease in melting point of lipids obtained from the Eastern great reed warbler, *Acrocephalus arundinaceus*, in the premigratory period. However, Hicks (1967) and McGreal and Farner (1956) cite exceptions and I have been unable to detect trends in saturation of total lipids in a survey of the studies of composition of migratory and nonmigratory birds (Blem, 1976).

There is no apparent consistent difference in composition of lipids in migrating birds when compared to nonmigrants, or migrants in the non-migratory period, except that oleic/linoleic ratios (18:1/18:2, carbon atoms:double bonds) tend to be greater than one in migratory species and less than one in nonmigrants. The functional significance of this difference, if any, remains to be determined. Variations in the composition of depot lipid between species appear to be due either to differences in diet (but see Hazelwood, 1972) or to species differences (see Blem, 1976, for a review), but not to stage of migratory preparation (Berthold, 1975). However, some migratory species show significant seasonal variations in fatty acid composition (Bower and Helms, 1968; West and Meng, 1968; Morton and Liebmann, 1974).

C. LIPID STORAGE SITES

Fat-body tissues in the thorax and abdomen are the major sites of lipid storage in insects. These organs may also function as the primary site of lipogenesis (Zebe and McShan, 1959; Tietz, 1961; Bade, 1964; Sun and Brookes, 1968; Thomas, 1974). Although a variety of lipids may be present there, the major component in the fat-body tissues of all species examined is triglyceride (Fast, 1964; Bailey, 1975; Downer and Matthews, 1976).

Although some lipids are stored within avian locomotor muscles (George and Jyoti, 1957; Hartman and Brownell, 1961; George and Berger, 1966), in birds the majority of lipid deposition occurs at specific anatomical sites in a fairly precise sequence. For example, the white-crowned sparrow, *Zonotrichia leucophrys*, has fat bodies in 15 different regions (McGreal and Farner, 1956). Subcutaneous layers of lipid associated with the feather tracts appear in the first stages of fat deposition. Subsequent fattening is evidenced by greater amounts of subcutaneous fat, particularly in the furcular region (the claviculocoracoid fat organ). Further increases in lipid deposition results in intraabdominal deposits. In the fattest birds, subcutaneous deposits are composed of extensive masses of fat; the interfurcular region and the abdominal cavity are bulging with fat deposits. During premigratory fattening, most organs and areas of the body demonstrate increases in lipid content except the heart (Odum and Perkinson, 1951; Vallythan and George, 1964).

Depot fat in birds contains little lipid other than triglyceride (Johnston, 1973). There seems to be no information on differences in the fatty acid composition of the triglycerides deposited at each site, even though one might expect variations in chain length and saturation, and therefore ease in mobilization, between warmer abdominal sites and cooler subcutaneous deposits.

Avian fat bodies generally increase in lipid content without corresponding changes in lean dry weight or relative water content (Connell et al., 1960; Turĉek, 1960; Johnston, 1964, 1973; Odum et al., 1964; Rogers and Odum, 1964; Odum, 1965; King and Farner, 1965; Helms et al., 1967; Hicks, 1967; Child and Marshall, 1970; Page and Middleton, 1972; but see King et al., 1965; Zimmerman, 1965b; Child, 1969). However, decreases in lean body mass in several species may represent adaptive reduction of wing loading that may reduce energetic costs of flight (Dolnik, 1963; Dolnik and Blyumental, 1964, 1967; King et al., 1965; but see aerodynamic discussion below). Lipid seems to be added to vacuoles of adipocytes, specialized cells for lipid storage, without an increase in number of cells (but see Dolnik, 1972). Histological studies verify this observation (Hicks, 1967; Johnston, 1973). The adipocytes apparently do not synthesize significant amounts of fatty acids but obtain them from plasma triglycerides (Goodridge and Ball, 1967a; Leveille et al., 1968). The liver is the major site of fatty acid synthesis and regulation of lipogenesis in birds (Fisher and Bartlett, 1957; George and Naik, 1962; Goodridge, 1964; Goodridge and Ball, 1967a,b,c; Farner et al., 1969; Hazelwood, 1972).

Storage of lipids in migratory bats appears to occur in much the same fashion as in birds, even though it is known that some mammals deposit lipids not only within exisiting adipocytes, but also by the production of more fat cells (Behnke, 1962; Dole, 1962). Although considerable subcutaneous lipid deposition has been observed, there is relatively little variation in the lean dry weight of the evening bat (Baker, et al., 1968), Mexican free-tailed bat (Pagels, 1975), and fringed myotis (Ewing et al., 1970). However, the weight of nonfat, dry components in the Yuma myotis increases with fat deposition (Ewing et al., 1970).

It is unfortunate that most students of the composition of bats have not critically examined the relationship between fat deposition and lean dry weight to determine if new cells are being created rather than the simple filling of preexisting cells. Expected increases in lean dry material during adipocyte construction are small and probably amount to 7–8% or less

(Helms *et al.*, 1967). Slight increases in lean dry weight might be concealed by the variation within samples, but should be detectable in regression analyses of lean dry weight versus fat content. These analyses are apparently not available. Furthermore, histological examinations of depot fats of chiropterans also appear to be lacking.

Lipid storage is not particularly well known in other forms. Although it is virtually certain that lipids, particularly non-esterified fatty acids (Bilinski, 1963) provide the major metabolic substrate in fishes, there are few studies of lipid storage in migratory fishes. Major lipid storage sites are in connective tissue or in mesentaries of intestines (Lagler *et al.*, 1977), skeletal muscle and liver (Vague and Fenasse, 1965), or in head, skin, and tail (Idler and Bitners, 1959).

D. Sources of Energy for Premigratory Reserves

The energy necessary to deposit lipid reserves for long-range migration must be collected at the same time the energy demands of normal daily activities are being satisfied. Several mechanisms have been postulated that may provide energy reserves in the premigratory period. These include (1) temporary decreases in basal energy requirements, (2) increasing ambient temperature in the vernal premigratory period, therefore reducing the costs of thermoregulation and providing an increased amount of surplus or productive energy, (3) increased assimilation efficiency, (4) changes in locomotor activity, (5) shifts to high-energy food, and/or (6) hyperphagia. As described above, some insects demonstrate biochemical modifications at the time of lipid deposition which enhance the synthesis of lipid from carbohydrate. However, I have found only weak indications that the major source of energy for premigratory fuel storage in any organism is anything other than hyperphagia (for reviews, see King and Farner, 1956, 1965; Johnson, 1969; Farner, 1970).

Premigratory hyperphagia has been most thoroughly studied in birds (King and Farner, 1956, 1959, 1963, 1965; Kendeigh *et al.*, 1960; Odum, 1960b; Dolnik, 1970; Whittow, 1976b), and relatively little evidence of the phenomenon exists for insects, bats, or other forms. During vernal and autumnal premigratory periods, hyperphagia results in hyperlipogenesis and may produce extreme levels of lipid deposition in a few days (King and Farner, 1956, 1965; Farner, 1960; Odum, 1960b). However, Merkel (1958), in studies of the European robin, *Erithacus rubecula*, and whitethroat, *Sylvia communis*, was not convinced that hyperphagia accounted for all of the autumnal fat deposition.

During hyperphagia, metabolized energy levels increase by 20% in white-crowned sparrows (King, 1961) and 25–30% in bobolinks, *Dol-*

ichonyx oryzivorous (Gifford and Odum, 1965). Conversely, there is little evidence for decreased rates of standard metabolism in several species tested (Wallgren, 1954; Merkel, 1958; King, 1961). However, Pohl (1971) found a lower rate of oxygen consumption in bramblings, *Fringilla montifringilla*, caged outdoors during the migratory period. Likewise, there is no apparent increase in efficiency of assimilation (King, 1961; Gifford and Odum, 1965; but see Zimmerman, 1965a). There appears to be no decrease in premigratory activity that would provide an energy savings (Farner *et al.*, 1954; Weise, 1956; King and Farner, 1963). In fact, *Zugunruhe* possibly provides the opposite effect. Nocturnal activity is more intense in captive long-range migrants than those that normally make shorter migratory flights (Engels, 1962; Weise, 1963; Gifford and Odum, 1965; Evans, 1968; Berthold, 1973). Kendeigh *et al.* (1960) postulated that energy demands of *Zugunruhe* in the tree sparrow, *Spizella arborea*, not only were replaced by food intake during the day, but that the birds overcompensated in their feeding, resulting in a weight (lipid?) increase. Kontogiannis (1968) was able to duplicate this effect in caged individuals of the nonmigratory house sparrow, *Passer domesticus*, by subjecting them to forced nighttime exercise. However, he was not able to produce greater weights in the migratory white-throated sparrow, *Zonotrichia albicollis*.

   Zugunruhe may occur even if fattening is prevented (King and Farner, 1963; Lofts *et al.*, 1963); however, Dolnik and Blyumental (1967) and Evans (1968) found the reverse to be true. During the premigratory period, a starved bird quickly regains fat reserves when refed. Under natural conditions, Morton (1967) found that white-crowned sparrows fed more intensively, but without increased feeding time, in the middle of the day a few days before migration and probably accumulated surplus energy for lipid reserves. There is evidence of shifts in food selection during the premigratory period in a few species (Dolnik and Blyumental, 1964, 1967; Bower and Helms, 1968; Hintz and Dyer, 1970). However, Odum and Major (1956) found that lipid levels of the diet had little effect on the rate or final level of fat deposition in caged white-throated sparrows maintained under stimulatory photoperiods. However, insectivorous migrants may shift to foods that are high in carbohydrates or lipids before migration (Dolnik and Blyumental, 1964; Fry, 1967; Hintz and Dyer, 1970), thus promoting the production of premigratory fat. Deposition of sufficient quantities of lipid may be an important stimulus for the onset of migration (Blyumental, 1962, 1963; Dolnik, 1963; Dolnik *et al.*, 1963a).

   The rate at which exhausted fat reserves may be replenished following a long migratory flight may determine whether migration will continue (see Mueller and Berger, 1966; Gauthreaux, 1971), but migrants are sometimes grounded by unfavorable weather conditions even though they have sufficient energy stores (e.g., Pearson and Backhurst, 1976). Reaccumulation rates are not well known, but it appears that some migrants can deposit approximately 0.5–1.0 gm lipid/day under favorable weather conditions (Dolnik and Blyumental, 1964, 1967; Helms, 1968).

Bats have a possible alternative strategy for accumulating premigratory reserves; hypothermia or lowered body temperature during part of the day decreases energy requirements by more than 40% (Ewing *et al.*, 1970; Pagels and Blem, 1973). Krzanowski (1961), Ewing *et al.*, (1970), and O'Farrell and Studier (1970) have suggested that at least part of the energy necessary for autumnal fat deposition is provided by these savings.

E. QUANTITY AND PATTERN OF LIPID STORAGE

1. Insects

Little attention has been given to lipid levels among most migratory insects (but see Fast, 1964; Gilbert, 1967), and the available data are given in absolute terms or as percentage of dry or fresh weights (Table II) and are difficult to compare to fat indices of birds and bats (see below). The only recognizable classifications of insect migrants with respect to amounts of lipid deposited and distances traveled are (1) short-distance migrants that travel overland and feed as they go, and (2) long-distance migrants that generally deposit large amounts of lipid and may make long, nonstop flights over barriers such as oceans (some moths more that 3200 km; see Johnson, 1969).

TABLE II

Lipid Reserves in Migratory Insects [a]

Species	Distance of migration (km)	Lipid (%)	Reference
Migratory locust	Occasionally over 2500	35–69	Osman and Schmidt, 1961
Desert locust[b]	Occasionally over 1000	8–15	Weis-Fogh, 1952a,b
Monarch butterfly[c]	Up to 3200	125	Beall, 1948
Monarch butterfly		42.5	Brown and Chippendale, 1974
Noctuid moth[d]	Up to 200	5–15	Koerwitz and Pruess, 1964
Aphid[e]	15 or less	31.4	Cockbain, 1961a
Beet leafhopper	640–720	32	Johnson, 1969

[a] Amounts of lipid are given as precentage of total dry weight of the insect.
[b] Lipid given as percentage of fresh weight of insect.
[c] Lipid given as percentage of lean weight of thorax and abdomen.
[d] *Chorizagrotis auxiliaris*; lipid is percentage of fresh weight of insect.
[e] *Aphis fabae*.

The monarch butterfly in North America is probably the best known example of a long-range insect migrant. In Ontario, it deposits up to 125% of its lean dry weight in lipid (an increase of 95% over emergence levels) before flying across the Great Lakes (Beall, 1948). Butterflies that drowned during storms *en route* over the lakes had depleted lipid reserves to about 30%, while those captured in New Orleans after a total migration of more than 1600 km had only 2% of this lipid remaining. Monarch butterflies found more than 1100 km out over the ocean (Walker, 1886; in Johnson, 1969) provide evidence that much of the journey could be completed without stopping. Longer migration routes leading into southwestern United States and Mexico may be followed. The trip may be relatively rapid; one marked individual covered nearly 1700 km in 18 days (Urquhart, 1960). Lipid reserves possibly adequate for the northward spring migration (Cenedella, 1971) are accumulated overwinter and Urquhart (1960) suggests that little or no nectar is ingested in the northward trip.

In the douglas fir beetle, the aphid, *Aphis fabae*, and a tettigoniid cricket, *Homorocorhyphus nitidulus*, there is a positive correlation between lipid reserves and the tendency for flight (Atkins, 1966; Cockbain, 1961a; Karuhize, 1972). The desert locust begins to rapidly produce lipid from dietary carbohydrate 6–8 days after adult emergence; triglycerides are stored in the fat body (Hill *et al.*, 1968; Walker and Bailey, 1970; Walker *et al.*, 1970). This lipogenic activity occurs at the time when adult flight muscles become completely developed (Brosemer *et al.*, 1963), presumedly in preparation for migratory flights. Migration is relatively rapid and takes place over several days with occasional stops for feeding (Johnson, 1969). The ability to produce lipid is retained for up to 20 days so as to produce flight fuels for individuals in the migratory swarm (Downer and Matthews, 1976).

2. Birds

Temperate zone migrants typically demonstrate annual lipid cycles in which peaks of lipid deposition occur in spring and autumn (Fig. 1). While the timing may be different, the fattening of some tropical migrants is similar to that of the temperate zone species (Ward, 1963, 1964a; McNeil and de Itriago, 1968).

A minimum of four patterns of premigratory fat deposition have been recognized: (1) short-range intracontinental migrants that begin migration before peak lipid deposition, (2) short-range intracontinental migrants that begin migration after peak lipid deposition, (3) medium- or long-range migrants that begin migration before maximum lipid deposition, and (4) long-range migrants that accumulate an extremely great lipid reserve before

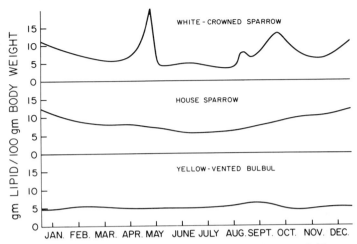

Fig. 1. Seasonal lipid cycles in three species of birds (from Blem, 1976). The curve for the white-crowned sparrow, *Zonotrichia leucophrys*, is drawn from data in King and Farner (1959, 1965) and King *et al.* (1963, 1965). Data for the house sparrow, *Passer domesticus*, are taken from Barnett (1970) and Blem (1973). The yellow-vented bulbul, *Pycnonotus goiavier*, curve is from data in Ward (1969). The white-crowned sparrow is a temperate zone migrant, the house sparrow, a temperate zone nonmigrant, and the yellow-vented bulbul, a tropical nonmigrant.

a flight over a barrier such as a desert or ocean (Odum, 1960a; Odum *et al.*, 1961; Johnston, 1964). Helms and Smythe (1969) recognize only (1) intracontinental migrants that migrate at speeds and over distances directly correlated with fat reserves, and (2) intercontinental migrants that are indistinguishable from the former over land, but build up reserves as they approach barriers (also see Caldwell *et al.*, 1963).

Birds that feed *en route* to migratory destinations may migrate relatively slowly. They do not need to accumulate large lipid reserves and so avoid severe wing loading. They may also stop or turn back along migration routes under conditions of severe weather (Duvall, 1966). Very obese birds tend to migrate more rapidly, partly because of reduced foraging time, but also because of aerodynamic considerations (Pennycuick, 1969). Pennycuick's analyses, based upon general mechanical principles, demonstrate that the addition of weight, as in premigratory fat deposition, increases the power needed to fly, and also results in an increased range of flight at a higher speed. For example, white-crowned sparrows have heavier lipid reserves and a more rapid migration in spring than in autumn (King, 1963; King *et al.*, 1963; also see Blyumental, 1968; King, 1972). Selection undoubtedly favors the faster vernal rate because early arrivals on the breeding grounds may be more successful in establishing territories and ac-quiring mates. The larger vernal reserves may also be beneficial during harsh weather conditions and may supplement energy demands early in the breeding season when foraging conditions are poor (King and Wales, 1965; King et al., 1965; Berthold, 1975).

Well-known extremes of nonstop, long-range migration include the Pacific golden plover, *Pluvialis dominica fulva*, which migrates 3900 km or more over ocean from the Aleutian Islands to Wake Island (Johnston and McFarlane, 1967), the blackpoll warbler, *Dendroica striata*, which flies nonstop from New England to Bermuda (1300 km) and probably continues to Central America without a stop for a total flight of approximately 2600 km (Nisbet et al., 1963; but see Murray, 1965), and the ruby-throated hummingbird, *Archilochus colubris*, which crosses the Gulf of Mexico, a distance of 800 km. These species demonstrate the interaction between body weight and fuel reserves in flying migrants. Larger birds generally deposit smaller proportions of lipid reserve than small birds (e.g., Johnston, 1966; Johnston and McFarlane, 1967; Pearson et al., 1970; Thomas and Dartnall, 1970). In comparison of the three species mentioned above, the 130-gm plover may deposit 20–25% of its weight in lipid, the warbler (weighing 10–20 gm, depending upon the amount of lipid) and hummingbird (3–5 gm) up to 50%. In general, with increased body weight, flight speed increases (Pennycuick, 1969, 1975; Berger and Hart, 1974; Tucker, 1974), and the power input/weight ratio decreases (Schaefer, 1968; Schmidt-Nielsen, 1972a; Tucker, 1975a). The result is that flight duration and range of flight increase with body weight and percentage lipid composition (see Berger and Hart, 1974; and discussion of flight ranges below), but larger birds need to increase lipid reserves less than small birds in order to attain similar velocities, durations, and ranges.

Long-distance migrants deposit considerable amounts of lipid (Table III), particularly those about to cross wide barriers. Lipid indices (gm lipid/gm lean dry weight) of nonmigrants range from 0.05 to 0.43, short- and medium-range migrants (in migration) from 0.17 to 0.77, and long-range migrants, 0.45 to 4.36. Some species deposit lipid depots equivalent to 50% of their total weight. The highest levels are known from species which fly across the Gulf of Mexico (e.g., ruby-throated hummingbird, North American wood warblers, vireos, scarlet tanager, *Piranga olivacea* and others; for a comparative series of values see Odum and Connell, 1956, Odum, 1960a; Caldwell et al., 1963; Johnston, 1964; Odum et al., 1965; Rogers, 1965; Zimmerman, 1965b; Helms et al., 1967; Brisbin, 1968; Newton, 1969; Helms and Symthe, 1969; Evans, 1969; Barnett, 1970; Yarbrough, 1970; McNeil, 1971; Blem, 1973, 1976) or the Sahara Desert (see Ward, 1963, 1964b; Ludlow, 1966; Smith, 1966; Fry et al., 1970; Pearson, 1971; Fogden, 1972).

TABLE III

Lipid Indices (gm lipid/gm lean dry weight) in Selected Species of Vertebrate Migrants

Species	Approximate weight (gm)	Lipid indices			Reference
		Nonmigratory	Spring migration	Autumn migration	
Short-range migrants					
Yuma myotis (*Myotis yumanensis*)	6.8	0.30($\female$)	—	0.87($\female$)	Ewing *et al.*, 1970
Little brown bat (*Myotis lucifugus*)	8.9	0.18 ($\male$) 0.58 ($\female$)	—	0.38 ($\male$) 0.69 ($\female$)	Ewing *et al.*, 1970
Fringed myotis (*Myotis thysanodes*)	10.0	0.16 ($\female$)	—	0.68–0.73 ($\female$)	Ewing *et al.*, 1970
Yellow-rumped warbler (*Dendroica coronata*)	12.5	0.18–0.68	0.55–0.84	—	Yarbrough and Johnston, 1965
Tree sparrow (*Spizella arborea*)	18.0	0.17–0.43	0.34	0.20	Helms and Smythe, 1969
Savannah sparrow (*Passerculus sandwichensis*)	21.0	0.33	—	0.77	Odum, 1960b
Dark-eyed junco (*Junco hyemalis*)	21.0	0.18–0.61	0.41–0.44	—	Helms *et al.*, 1967; Holmes, 1976

White-throated sparrow (*Zonotrichia albicollis*)	28.0	0.17–0.85	—	0.20–0.68	Caldwell *et al.*, 1963; Odum, 1960a,b
Long-range migrants					
Ruby-throated hummingbird (*Archilochus colubris*)	3.0	—	—	3.13–3.50	Odum, 1960b, 1965
American redstart (*Setophaga ruticilla*)	10.0	0.18–0.26	0.34–0.43	0.73–1.29	Caldwell *et al.*, 1963; Holmes, 1976
Northern waterthrush (*Seiurus noveboracensis*)	18.0	0.13	0.45	0.86–1.51	Caldwell *et al.*, 1963; Holmes, 1976; Odum and Connell, 1956
Blackpoll warbler (*Dendroica striata*)	24.0	—	—	3.42	Odum, 1965
Red-eyed vireo (*Vireo olivaceus*)	25.0	0.15–0.28	0.54–0.67	0.74–4.36	Caldwell *et al.*, 1963; Holmes, 1976; Odum, 1960a, 1965; Odum and Connell, 1956
Dickcissel (*Spiza americana*)	28.0	0.12–0.66	—	0.37–0.53	Zimmerman, 1965b
Scarlet tanager (*Piranga olivacea*)	28.0	0.08–0.22	—	1.17–2.16	Caldwell *et al.*, 1963; Holmes, 1976; Odum, 1960b; Odum and Connell, 1956; Odum *et al.*, 1965
Bobolink (*Dolichonyx oryzivorus*)	45.0	—	0.87–0.97	2.73	Caldwell *et al.*, 1963

3. Bats

In bats known or suspected to be migratory, lipid deposition generally begins in late summer or early autumn. Lipid indices may increase from lows of 0.08–0.40 to highs of 0.70–1.70 (Baker *et al.*, 1968; Ewing *et al.*, 1970; Krulin and Sealander, 1972; Pagels, 1975). There is no apparent information on speed of migration (but see Villa and Cockrum, 1962) or distances traveled in single flights. It is known that some species (e.g., evening bat, Humphrey and Cope, 1968; little brown bat, Davis and Hitchcock, 1966) make relatively short (100–300 km) flights between breeding sites and hibernaculae, while others make relatively long trips between northern breeding areas and wintering grounds in the south (e.g., hoary bat, *Lasiurus cinereus*, Findley and Jones, 1964; Mexican free-tailed bat, Twente, 1956; Villa and Cockrum, 1962).

4. Other Taxa

Some individuals of the green turtle, *Chelonia mydas*, make annual migrations of more than 4800 km from the Brazilian coast to breeding grounds on Ascension Island in the south Atlantic Ocean. During this time they are generally traveling over abyssal waters where food is presumedly not available and additionally individual turtles may continue to fast for 6–8 weeks after reaching their destination (Carr and Goodman, 1970). Part of the population does not travel as extensively from Brazil, and are individually smaller than the Ascension Island migrants. Comparative studies of lipid deposition of migratory individuals with the more sedentary forms of this species have not been made, but it appears that migrating turtles are considerably fatter than nonmigrants and that females taken late in the nesting season are lean (Carr, 1964; Carr and Goodman, 1970). It has been suggested that adult body size of green turtles is influenced more by size at maturation and by factors involved with migratory movement than by growth after maturation. The large size of the Ascension Island turtles provides for storage of larger amounts of fat and greater locomotor efficiency to assist in their long trip (Carr and Goodman, 1970).

Estimates of the fatness of green turtles range from less than 1 to over 4%, excluding muscle which contains 7.6% lipid (Rao and Dutt, 1965; Prange, 1976). Estimates of energy utilization during migration indicate that these estimates are probably low and do not represent animals about to migrate (Prange, 1976).

Although many anadromous or catadromous fishes make annual migrations, little attention has been given to patterns of premigratory energy storage. It is known that lipid reserves in the coho salmon, *Oncorhynchus kisutch*, decrease with forced exercise (Krueger *et al.*, 1968), and that herring deposit large amounts of lipid which are used to move from feeding grounds to breeding sites (Orr, 1970). However, few complete analyses of storage and utilization of lipids in fishes are available. The most comprehensive model may be found in studies of the sockeye salmon, *Oncorhynchus nerka*, by Idler and his associates (Idler and Tsuyuki, 1958; Idler and Bitners, 1958, 1959; Idler and Clemens, 1959). Migrating adult sockeye salmon accumulate considerable fat reserves which are largely used on migratory trips to upstream breeding sites, a distance often over 1200 km (Idler and Bitners, 1960). During the migration, the lipid content of sockeye salmon decreases more than 70% and protein content of the flesh decreases approximately 40% (Idler and Bitners, 1958, 1959). Sockeye salmon in other studies consume 91–96% of their fat reserves and 53% of their protein during the course of migration (Idler and Clemens, 1959). The major site of lipid storage is not in the flesh, but in head, tail, and skin (Idler and Bitners, 1959).

III. Energetic Costs of Migration

Energetic characteristics of long-range migrants include relatively high rates of metabolism, great endurance, and a large metabolic scope (ratio of peak metabolic rate to standard or resting metabolic rate). Other factors affecting the ability to travel long distances include the size of the organism and/or its speed of locomotion.

A. THE ENERGETICS OF LOCOMOTION

The number of publications dealing with energetic costs of locomotion and physical theory of propulsion of all sorts of organsims has literally exploded in the past ten years. The following highly simplified account is an attempt to synthesize the available literature as it relates to migration.

The strategy of long-distances migration is relatively restricted to flying and swimming forms and to a few terrestrial organisms generally above 1 kg in weight (Tucker, 1975a). Per unit weight, swimming requires less energy over a given distance than flight, and both are energetically cheaper than all natural forms of terrestrial locomotion (Table IV; Schmidt-Neilsen, 1972a; Gold, 1973; Tucker, 1975a). In addtion, the cost of locomotion is an inverse function of body weight (Tucker, 1970); migration is therefore relatively expensive in small organisms. Long-distance travel in small terrestrial or aquatic organisms (unaided by currents) is a rarity; energy expenditures are prohibitively large in relation to potential energy stores.

Most estimates of rates of energy expenditure during locomotion have been based upon indirect calorimetry, changes in body weight and/or com-

TABLE IV

Costs of Transport of Some Migrants

Species	Weight (gm)	Energy expended (kcal/kg-km)	Reference
Mosquito	3.2×10^{-3}	18.9	Hocking, 1953
Mosquito	8.2×10^{-3}	35.0	Hocking, 1953
Desert locust	2.0	5.0	Weis-Fogh, 1952a
Migratory locust	2.0	4.9	Jutsum and Goldsworthy, 1976
Hummingbird	3.0	4.0	Lasiewski, 1963
Eurasian siskin	12.5	3.8	Dolnik and Blyumental, 1967
Chaffinch	20–22	3.5–4.2	Dolnik and Blyumental, 1967; Dolnik and Gavrilov, 1973; Dolnik, 1974
Brambling	23.0	3.4	Dolnik and Gavrilov, 1973
Common bullfinch	29.0	3.3	Dolnik and Gavrilov, 1973
Purple martin	50.5	2.5	Utter and LeFebvre, 1970
Laughing Gull	310.0	1.5	Tucker, 1969
Lemming	61.0	40.9	Hart and Heroux, 1955
Sockeye salmon	3–1500	0.5–1.1	Brett, 1964; Schmidt-Nielsen, 1972a

position (material balance studies), correlations between rate of energy expenditure and other physiological variables, and aerodynamic considerations (insects: see Krogh and Weis-Fogh, 1951; Weis-Fogh, 1952b, 1956, 1961, 1964b, 1967a,b, 1972, 1976; Pringle, 1957; Keister and Buck, 1964; Johnson, 1969; Casey, 1976a; fish: Brett, 1964, 1965a,b, 1971, 1972; Fry and Hochachka, 1970; Fry, 1971; Webb, 1971a,b, 1975; Brett and Glass, 1973; sea turtles: Prange, 1976; birds: Brown, 1961; Dolnik and Blyumental, 1964; Tucker, 1966, 1968b, 1969, 1971, 1972a, 1973a,b, 1974, 1975a,b; Dolnik, 1969a, 1971c; Pennycuick, 1969, 1972, 1975; Farner, 1970; Dolnik and Gavrilov, 1971; Berger and Hart, 1974; Berthold, 1975; bats: Thomas and Suthers, 1970, 1972; Suthers et al., 1972; Thomas, 1975). The best estimates of the costs of locomotion in migratory forms, or in forms that might provide relevant information, are from indirect calorimetry and material balance studies and are discussed below. Aerodynamic factors are considered below in relation to range of migration.

1. Indirect Calorimetry

Measurements of energy utilization by means of indirect calorimetry involve quantifying the rate of oxygen uptake and/or carbon dioxide production and converting these measurements to energy terms on the basis of known or estimated respiratory quotients.

a. Insects. Indirect calorimetry is occasionally used to measure energy utilization in flying insects. Because of the small absolute amounts of gaseous exchange, small closed systems are generally used. Respiration rates of flying insects are quite large in relation to resting rates, and rates of oxidation can go from resting levels to 50 to 100 times resting rates in a short period of time (Sacktor, 1965, 1975). On a weight-specific basis, the metabolic rate of many insects approximates or exceeds the rates of hovering hummingbirds, flying birds or any organism, including mammals, that utilize any of the forms of terrestrial locomotion (see Tucker, 1970; Schmidt-Nielsen, 1972a). In fact, oxygen consumption in many flying insects exceeds all other forms of continuous energy expenditure (Weis-Fogh, 1961; Heinrich, 1974).

Weight-specific costs of flight are quite high in insects, but are in line with quantitative relationships established by Tucker (1970) and Schmidt-Nielsen (1972a) for birds and insects in which the cost of flight is an inverse exponential function of body weight. In comparison with birds and bats, insects have relatively low wing loads (Weis-Fogh, 1976). Theoretical analyses of fast forward flight in the Desert Locust indicate that Pennycuick's (1969) model of power input and flight speed of birds may be applicable to insects (Weis-Fogh, 1976). The power required to fly at intermediate speeds is less than that needed at lower and higher velocities (see discussion of Pennycuick's model below). A study of flight costs of the desert locust support this analysis (Jensen, 1956). No other complete experimental study appears to be available for insect migrants (see Weis-Fogh, 1976). Calculated minimum cost velocities for the desert locust approximate those found in prolonged flights in the laboratory. The speed at which minimum cost of transport occurs is considerably higher. This is because of the interaction between flight costs and velocity at flight speeds above minimum cost velocities (i.e., costs do not increase as rapidly as flight velocity during part of the curve). Therefore, while minimum costs of flight of the locust were found at speeds of 3.5. m/sec, minimum costs of transport were found at 4.5–5.5 m/sec. Measurements of flight speeds of naturally migrating desert locusts have ranged mostly between 4–6 m/sec (Weis-Fogh, 1976).

b. Fishes. Respiration rates of active fishes, particulary salmon, have generally been measured in closed systems containing flowing water (e.g., see Brett, 1964). The energetic cost of swimming increases with water velocity or swimming speed in an exponential fashion up to the maximum rate of oxygen consumption which represents the top sustained speed of the fish (Brett, 1965a,b, 1971, 1972; Brett and Sutherland, 1965; see Phillips, 1969; Fry and Hochachka, 1970; Fry, 1971). Maximum sustained metabolic rates in migratory forms such as salmon, when simultaneously affected by both high ambient temperatures and activity, may be more than 20 times standard metabolic rates measured at low temperature (Brett, 1964). Active metabolic rates at 15°C may be 10–12 times standard metabolic rates at the same temperature. The effects of ambient temperature upon costs of locomotion are difficult to assess, as are those of insects, because standard rates of metabolism increase with temperature. Rao (1968) found higher metabolic costs with increased temperature in trout. Brett (1967, 1971, 1972) demonstrated that absolute levels of oxygen uptake, rates of increase of oxygen demand with speed, and maximum oxygen uptake rates vary with temperature in sockeye salmon. For example, above 24%C the ratio between active and standard metabolic rates (metabolic scope) decreases greatly, with the maximum occurring at 15%C.

Fish may swim in bursts and incur oxygen debts that may be replaced during periods of decreased activity. This partially offsets the handicap of respiration in water. High demands for oxygen during vigorous activity in aquatic habitat requires more respiratory effort to obtain sufficient oxygen and is difficult to satisfy because of low diffusion rates of oxygen in water (see Irving, 1964; Brett, 1972). Maximum sustained cruising speeds may be related to maximum sustainable rates of oxygen transfer from water to muscle (Kutty and Saunders, 1973; Wardle, 1977).

In general, migration appears to be limited to larger fishes (see Bainbridge, 1960). Fish in the size range of most small migratory birds (3–30 gm) can expend energy on a sustained basis at a rate approximately 1% of that of the birds (Brett, 1972). Additionally, maximum and cruising speeds of swimming are related to length of the fish (Wardle, 1977). It is possible that small fishes cannot swim rapidly enough and store or expend sufficient energy to travel long distances. Small fishes such as young eels may, however, make part or all of a long migration by drifting with prevailing currents (Lagler et al., 1977).

c. Birds. Measurements of the energetics of avian flight by indirect calorimetry include those by Pearson (1950, 1964), Lasiewski (1963), LeFebvre (1964), Tucker (1966, 1968b, 1972a), Hainsworth and Wolf (1969, 1972), Teal (1969), Berger et al. (1970b), Utter and LeFebvre (1970), and Berger and Hart (1972). Techniques used include measurements of carbon dioxide production and/or oxygen utilization of birds held in closed chambers, or by using birds wearing masks attached to collecting or sampling devices while flying in wind tunnels (see Tucker, 1971), or flying free (see Hart and Roy, 1966). A second, more subtle, technique involves the use of doubly labeled water (D_2O^{18}), which makes it possible to quantify the carbon dioxide produced in unrestrained birds (see Lifson et al., 1955; LeFebvre, 1964; Utter and LeFebvre, 1970; Mullen, 1973). Because oxygen in body water is in equilibrium with oxygen in respiratory carbon dioxide, loss of labeled oxygen (through evaporation and carbon dioxide production) exceeds hydrogen loss through evaporation only. The difference in rates of loss of hydrogen and oxygen can be used to estimate carbon dioxide production. These estimates can then be converted to energy terms in the same fashion as respirometry measurements.

Both aerodynamic theory and empirical data indicate that power input varies with flight speed in birds (Pennycuick, 1969; Tucker, 1968b, 1971, 1973a, 1974, 1975b,c). The energy required to fly at low and high speeds is relatively great (see Fig. 2), as the power needed to support the bird's weight in the air (induced power) declines with velocity and the power required to overcome friction and drag (parasite power) increases with velocity. Power requirements for overcoming the drag of flapping wings (profile power) remains proportional to the sums of induced and parasite power. At intermediate speeds the power required is at a minimum since the sums of the power components are minimized (see discussion of insect flight costs above.).

d. Bats. Bats trained to fly while wearing a collecting mask attached to an oxygen analyzer provide the few measurements of the energetics of bat flight that have been done (Thomas and Suthers, 1970, 1972; Thomas, 1975). Metabolic rates of flying Phyllostomus hastatus and Pteropus gouldii were more than 20 to 30 times basal rates and approximated values for flying birds of similar size (Thomas and Suthers, 1972; Thomas, 1975). These levels are significantly greater than values reported for exercised terrestrial mammals. The effects of flight velocity upon energetics are similar to those of birds; minimum metabolic rates (and costs of transport) occur at intermediate velocities which approximate best estimates of flight speed in those species (Thomas, 1975).

e. Other Taxa. Little information is available on energy utilization in other migratory forms. However, Prange (1976) provides an interesting analysis of the costs of swimming in the green turtle which indicates that oxygen consumption during maximum swimming speeds may be 3–4 times resting rates. The absolute costs for long-distance travel in this species is quite low, possibly in the range of the costs of locomotion of fishes (Prange, 1976).

2. Material Balance Studies

Material balance measurements of the costs of locomotor activity generally involve measurement of lipid levels (or sometimes other reserves) of fasting individuals before and after stress. In the course of nonstop, continuous exercise such as the migratory flights of insects, birds, and probably bats, the composition and weight of the migrant changes considerably (see Ward, 1963; Ash, 1969; Salomonsen, 1969). Quantification of the energy used may involve comparison of control animals with experimentals, or individuals collected at the beginning and end of a migratory flight. Often utilization of reserves may be estimated from weight loss during a flight of known length.

a. *Insects.* Material balance studies are frequently used to measure energy utilization of insects tethered in flight mills (e.g., Weis-Fogh, 1952b). In such tests the insect flies for a known period of time and weight changes can be precisely measured and related to compositional changes. Depletion of reserves during such tests frequently give estimates of the cost of flight that are comparable with respirometry estimates.

b. *Fish.* Material balance techniques occasionally have been applied to fish. Analyses of the bodies of two species of migratory salmon, forced to swim nearly 900 km in laboratory apparatus, provided measurements of utilized energy that compare favorably with rates of energy use under similar conditions in a respirometer (Brett, 1965b; see Idler and Clemens, 1959; Brett, 1972).

c. *Birds.* Avian energy reserves may be greatly depleted in at least some individuals (Kuroda, 1964; Johnston, 1968; Diamond et al., 1977) and energy demands may be so extensive that nonfat tissues (and possibly muscle tissue) are partially used as fuel (Rogers and Odum, 1964; Evans, 1969). Lipid and glycogen levels are depleted and weight decreases are correlated with the amount of flight work performed (Berger and Hart, 1974). Weight losses resulting from respiratory evaporation may also be significant (see below, and Berger and Hart, 1974).

Estimates of the energy expended in avian migration frequently have been obtained from measurements of weight loss during migration. The loss is either assumed to represent lipid and is therefore convertable to energy terms by multiplying the loss by the caloric equivalent of fat (see Johnston, 1970), or assumed to represent the dissipation of a combination of materials and therefore is converted to calories by the use of the caloric equivalent of weight loss as determined empirically (Dolnik and Blyumental, 1967; Keskpaik, 1968; Dolnik and Gavrilov, 1973). The nonfat components of a migrating bird may remain relatively unchanged so long as lipid deposits are not exhausted (Odum et al., 1964, 1965), but severely stressed birds may show decreases in lean dry weight that represent losses in gut contents, glycogen stores, and perhaps the use of the muscle mass as fuel (Rogers and Odum, 1966; Newton, 1968; Evans, 1969). Pennycuick

(1975) discusses the strategy of the use of flight muscles as fuel reserves. He points out that if body proportions are appropriate at the beginning of a migratory flight (i.e., flight muscles should be about equal to total body mass to the 1.5 power), 5–10% of the energy normally derived from fat might be obtained from muscle without loss of flight efficiency. Goldspink (1965), working with fasted mice, has shown that utilization of muscle as an energy source is possible.

Although none of the material balance studies produces extremely precise results, in several studies (but not all) there is a remarkable agreement between such estimates and those obtained from indirect calorimetry. For example, LeFebvre (1964) measured both carbon dioxide production and lipid depletion in domestic pigeons and obtained similar estimates of flight energy. Material balance studies have been applied to avian migrants more often than any other group. This is because other techniques were more appropriate for nonavian migrants, but also because capture of sufficient individuals at migratory intervals is often difficult or the energy demands of migration are supplemented by daily feeding and body composition and weight are not representative of distances traveled.

B. Adaptations Related to the Energetics of Migration

The vigorous muscular activity attendant with locomotion over wide distances in migration obviously is favored by adaptations that decrease unnecessary energy expenditures, permit sufficient energy stores to be accumulated and carried, provide oxygen to muscles, and supply or conserve water. The ability to maintain a relatively high rate of energy expenditure is necessary for long-distance locomotion. Since this process is almost always fueled by lipid and therefore is an aerobic process, long-range migrants generally have the ability to obtain and utilize oxygen readily.

1. Muscles and Respiration

Insect flight muscles demonstrate the highest rate of oxygen consumption of any known tissue (Sacktor, 1975), and may account for about 99% of total metabolism during flight (Weis-Fogh, 1964a). Adaptive to this very high rate of oxidation is the high degree of development of the tracheal system, the mechanism of respiratory gas exchange of insects. Terminal tracheoles penetrate individual flight muscle fibers in all but a few insects (Elder, 1975), and may be in close juxtaposition with individual mitochondria (Weis-Fogh, 1964a). Flight muscles in migratory insects are mostly of the nonfibrillar, synchronous type (see Elder, 1975), but the type of muscle is more of a phylogenetic function than one specifically adaptive to long-range flight. Mitochondria in these muscles are large and numerous and are probably associated with high levels of aerobic metabolism (Sacktor, 1965).

Flying birds also have relatively high metabolic rates and metabolic scope (up to 14.5 times standard metabolic rates, see Farner, 1970; Calder and King, 1974). The major flight muscles (pectoralis) of birds are variously composed of white and/or red skeletal muscle fibers. White muscle fibers contain high glycogen levels and low amounts of lipid, have few mitochondria, and little myoglobin (George and Talesara, 1961; George and Berger, 1966). Red fibers are narrower, have less glycogen and more mitochondria, and myoglobin and lipid levels are higher in red than in white fibers. Red fibers appear to be highly adapted for sustained aerobic respiration. White fibers are not capable of sustaining vigorous activity, but can undergo an oxygen debt. It has been suggested that avian pectoral muscles undergo hypertrophy in the premigratory period (Evans, 1969; Fry et al., 1970), but the evidence for this phenomenon is equivocal (see Baggott, 1975; Berthold, 1975). Avian pectoral muscles are either composed entirely of red fibers (hummingbird, sparrow), or of a combination in which either red ("red-mixed": domestic pigeon) or white ("white-mixed": domestic fowl) fibers predominate. No examples of "white-unmixed" pectoral muscles are known (George and Berger, 1966).

Fishes also possess two types of muscle fibers. The red fibers contain more lipid, respire aerobically, give a slow, long-lasting contraction, and may be more active in prolonged swimming (Bilinski, 1969). White fibers are not well adapted for lipid metabolism, may respire anaerobically, and are used for quick movements. An oxygen debt may be temporarily incurred during intensive exercise and satisfied later during periods of reduced activity. The similarity between avian and piscine muscle fibers may be only superficial. Fish muscle fibers may possess a variety of further specializations (see Bilinski, 1969). For example, glycogen storage may be mostly a function of red muscle. Red muscle provides the lipid storage site in many fish, rather than fat bodies (see Tashima and Cahill, 1965). Also, thermogenesis is a primary function of red muscle in some warm-bodied fishes (Carey and Teal, 1966; Gordon, 1968).

During avian flight, wing beats and respiration rates are often coordinated (Tomlinson and McKinnon, 1957; Lord et al., 1962; Hart and Roy, 1966; Berger et al., 1970a; Lasiewski, 1972; Berger and Hart, 1974), the assumption being that mechanical efficiency would be increased if the two actions are synchronized. Wing-beat frequency in stable flight is generally equal to ventilation frequency in the domestic pigeon, Columba livia, common crow, Corvus brachyrhynchos, and the spear-nosed bat,

Phyllostomus hastatus (Hart and Roy, 1966; Berger *et al.*, 1970a; Suthers *et al.*, 1972). Several other species exhibit coordination in which wing rates exceed respiration rates (Berger *et al.*, 1970a). Respiration may not be coordinated with wing beat in the budgerigar, *Melopsittacus undulatus*, depending on flight speed (Tucker, 1968b) or ambient temperature (Aulie, 1975), or in hummingbirds (Berger and Hart, 1974). There appear to be no obligatory interactions between respiratory movements and wing beat, although it appears that maximal contraction of the pectoral muscles as in takeoff or wing downstroke may interfere with inspiration (see Berger and Hart, 1974).

Heart weights of birds may be 1.4–2.0 times greater in birds than in mammals of similar weight (Berger and Hart, 1974) and may be related to the ability to maintain high rates of gaseous exchange. Heart rates of small birds in flight may reach 7/sec to 10/sec or greater, and although there seems to be no obligatory link between heartbeat and activity of pectoral muscles, in some species the heart may not contract at the same time as the pectoral muscles (Aulie, 1971a, 1972). There does not appear to be any great difference between heart rates of flying birds and flying bats (Studier and Howell, 1969; Thomas and Suthers, 1970; Berger and Hart, 1974) or nonflying mammals undergoing vigorous exercise.

Flying vertebrates may also have blood that is highly adapted for oxygen transport. Hematocrit values in birds and bats are higher than those of most other terrestrial vertebrates, except diving organisms and those living at high altitudes. Oxygen capacity of the blood is related to hematocrit level and is also correspondingly high (Thomas and Suthers, 1972).

2. Body Temperature during Flight

Many insects demonstrate elevations in temperature during flight, and it has been found that a relatively high thoracic temperature (up to 40°–45°C is some species) is a prerequisite for flight in larger insects (Heinrich, 1974). Heat production is brought about by activation of the power muscles for flight without appreciable wing movement and results in apparent shivering. Insects producing heat in this fashion include a wide variety of forms including a number of migratory species within the orders Orthoptera and Lepidoptera (Church, 1960a; Heath and Adams, 1967; Kammer, 1970; Casey, 1976b). Behavioral thermoregulation may assist in this process, as demonstrated by the migratory desert locust (Heinrich, 1974). The body temperature of flying orthopterans, butterflies, moths, locusts, and other insects may be 5°–32°C above ambient temperature (Krogh and Zeuthen, 1941; Sotavalta, 1954; Church, 1960a,b; Adams and Heath, 1964; Bartholomew and Heinrich, 1973; Heinrich, 1974), and some insects may regulate their thoracic temperature within a few degrees variation during flight. The energetic cost of heat production is apparently not great, and may be only a fraction of the cost of flight (Cloudsley-Thompson, 1970).

Body temperatures of birds may become elevated during flight, even in the absence of high ambient temperature (Dawson and Hudson, 1970; Berger and Hart, 1974; Whittow, 1976a). For example, body temperatures of flying domestic pigeons stabilize at about 44.5°C, or 2°C above resting levels, after about 2 min of flight (Hart and Roy, 1967). If the flight is continued, at 10 min body temperatures reach 43.6°–44.7°C, the pigeon refuses to fly further, and appears to be overheated (Aulie, 1971b). The budgerigar becomes hyperthermic while flying less than 20 min at ambient temperatures of 36°–37°C and will not continue to fly (Tucker, 1968b). At lower ambient temperatures (25°–29°C), budgerigars stabilize at body temperatures of 42.1°C and continue to fly (Aulie, 1971b). Torre-Bueno (1976) found that body temperatures of flying starlings, *Sturnus vulgaris*, rose to 42.7°–44.0°C. The upper temperatures are only a few degrees below lethality. In starlings, the increase is independent of ambient temperature and therefore does not represent heat storage or adjustment to heavy exercise. It appears that the flight muscles may operate more efficiently at higher temperatures, therefore providing increased power output and allowing birds to fly with smaller, lighter muscle mass (Torre-Bueno, 1976). This finding, in the light of the contention of several authors that overheating may occur in flight at elevated temperatures, demonstrates the need for further analysis of the effects and/or benefits of elevated temperatures in flying birds.

Some birds such as the poorwill, *Phalaeoptilus nuttallii*, may fly with fairly low body temperatures (e.g., 34.0°–37.5°C, see Berger and Hart, 1974), and bats apparently do not demonstrate extreme body temperatures during flight. O'Farrell and Bradley (1977), in a survey of eleven species of bats, found the average body temperature at which nine species could initiate and maintain flight ranged from 19.6° to 26.3°C. The remaining species flew at temperatures up to 31.6°C. Studier and O'Farrell (1972) present a list of minimum body temperatures required for initiation of flight in 19 species of bats. Values range from 21.3° to 33.0°C (also see Reeder and Cowles, 1951; Herreid, 1963b; Bradley and O'Farrell, 1969; Kulzer *et al.*, 1970). Conversely, Thomas and Suthers (1972) found that deep rectal temperatures of the spear-nosed bat, a large (70–110 gm) neotropical species, ranged from 41.2° to 42.1°C after flights of about 3 min or less. These temperatures are very near upper lethal temperatures (42°–43°C). Berger and Hart (1974) speculate that relatively low body temperatures are possible in some bats and birds having low wing loads and low rates of wingbeat, while forms having high wing loads and rapid wingbeats such as the hummingbird (and possibly the spear-nosed bat) require higher body temperatures for flight. This idea achieves even greater significance when one considers that a positive relationship between wing loading and flight temperatures has been demonstrated within several families of moths (Dorsett, 1962; Bartholomew and Heinrich, 1973).

Perhaps it is also significant that some of the long-range migrants among fishes, particularly the bluefin tuna (Mather, 1962), demonstrate a remarkable ability to maintain relatively high body temperatures above the ambient temperature of the surrounding water (Barrett and Hester, 1964; Carey and Teal, 1966, 1969a,b; Carey et al., 1971). Undoubtedly, high body temperatures in such fishes (occasionally over 30°C) not only permit greater activity in cold waters, but also probably have a similar function in permitting more efficient muscular activity as in birds and perhaps insects (see Neville and Weis-Fogh, 1963).

3. Water Loss

It appears that low rates of water loss protect migratory insects from desiccation (see Weis-Fogh, 1967a; Heath et al., 1971), rather than any mechanisms for storing water or producing great quantities of metabolic water. The tracheal system appears to lose less water during respiration than the lungs of terrestrial vertebrates. In insects, metabolic rates may increase 50 times or more during flight (Sacktor, 1965, 1975), but water loss increases only by perhaps 6 times (Bursell, 1964). Because oxidation of one gram of lipid of average saturation theoretically yields about 1.1 gm of water, or about twice as much as may be derived from carbohydrate, water balance in insect migrants should not generally be a severe problem. Heat loss through evaporation may not be necessary in some insects such as the desert locust (e.g., see Church, 1960a). In *Aphis* (Cockbain, 1961b) and in the desert locust (Weis-Fogh, 1956), water losses are balanced by metabolic production of water even at low humidities.

During premigratory fattening, the water content of birds probably remains relatively constant (McGreal and Farner, 1956; Odum et al., 1964; Helms and Smythe, 1969), although a few authors have found that it varies considerably (e.g., Moreau and Dolp, 1970), or may decrease (McNeil, 1970; Dolnik, 1973). The contradictory findings may be a result of variation in the materials from which the lipids must be synthesized. If lipid reserves are synthesized from carbohydrate or protein, the materials must undergo hydrogenation during transformation to fat. Water may be used in the process, and the composition of the body may be modified, depending upon the composition of the diet. Rarely does the water content of the body increase with fattening (Fry et al., 1970).

The available evidence indicates that water requirements rather than energy reserves in some instances may limit flight range (see Bartholomew, 1972; Hart and Berger, 1972; Berthold, 1975; but see Odum *et al.*, 1964). It has long been assumed that long-range migrants unable to obtain fresh water may have a severe tendency to dehydrate (Yapp, 1956, 1962; Fogden, 1972; but see Schmidt-Nielsen, 1972b). Flight requires extremely high rates of energy expenditure and probably results in production of heat that is several times greater than that of resting birds (Aulie, 1971b). Excess heat is partly dissipated by evaporative water loss from the respiratory tract. This loss may account for a large amount of the water loss by flying birds, but since significant amounts (up to nearly 63%) are lost through the skin of resting birds, it is reasonable to assume that this is also a major site of water loss in flying birds (Smith and Suthers, 1969; Bernstein, 1971a,b; Lasiewski *et al.*, 1971). At intermediate temperatures, 13–22% of heat production is dissipated by evaporation in flying domestic pigeons and budgerigars and the remainder is lost through radiation and convection (LeFebvre, 1964; Hart and Roy, 1966; Tucker, 1968b; Lasiewski, 1972). Although most of the experimental work done concerning water loss in birds has not involved species known to be strong migrants, there is no evidence that migratory species would behave differently or have additional mechanisms that promote storage or conservation of water.

Tucker (1968b), working with flying budgerigars, has shown that metabolic production of water partly offsets moisture losses through respiration. However, a net water loss of 1.1% of body weight per hour occurred in his birds. Assuming that a budgerigar survives water losses up to 15% of body weight, it could fly for approximately 14 hours and travel 490 km (Tucker, 1968b). Some birds apparently fly for much longer periods without drinking, perhaps up to 50 or more hrs (Moreau, 1961; Nisbet *et al.*, 1963; Johnston and McFarlane, 1967). Furthermore, in some instances, there is little indication that water content of long-range migrants is particularly low at the end of the flight (Nisbet *et al.*, 1963; Johnston, 1968; Child, 1969; but see Searle, 1956; Moreau and Dolp, 1970; Forgden, 1972). Compounding the enigma is the observation that evaporative water loss exceeds metabolic production in most species tested (LeFebvre, 1964; Hart and Roy, 1967; Dolnik, 1969a; Berger *et al.*, 1971). In resting birds, water loss is reduced at low temperatures (see Shilov, 1973) and water homeostasis (balance between water loss and production of metabolic water) occurs in several species of small birds at ambient temperatures of 14°–18°C. Flying birds show a smaller net loss of water at low ambient temperatures (LeFebvre, 1964; Hart and Roy, 1967; Tucker, 1968b; Berger *et al.*, 1971) and a hummingbird, the glittering-throated emerald, *Amazilia fimbriata*, reaches water homeostasis in flight at 0°C (Berger and Hart, 1974). Additionally, water loss may be reduced by countercurrent heat exchangers within the nasal passages of some birds (Schmidt-Nielsen *et al.*, 1970; also see Berger *et al.*, 1971). Countercurrent or crosscurrent exchangers in the avian lung may increase the extraction ratio of oxygen from inspired air and may result in lower ventilation rates which should also aid in conservation of respiratory water loss during flight (Schmidt-Nielsen *et al.*, 1969; Scheid and Piiper, 1970; Tucker, 1972b).

Although preferred altitudes are not well known, migrating passerines may fly at altitudes of 0.3 to 6.8 km (Meinertzhagen, 1955; Lack, 1960; Phelps, 1961; Nisbet, 1963a; Bellrose, 1971; Blokpoel and Burton, 1975; Richardson, 1976). The best indications are that the majority of small passerines migrate in the lower half of this range. Lower ambient temperatures at higher altitudes may increase cooling rates of birds without as much evaporative water loss (average air temperatures decrease 6.6°C/km; Blokpoel and Burton, 1975). At the upper altitudinal extremes, it is possible that reductions in body temperature (see Tucker, 1968a) might aid in water conservation. Because of lower ambient temperatures and higher relative humidities, nocturnal migration may also be adaptive in the same fashion. Speed of flight may also increase with altitude of migration as the density of air decreases with elevation. Flight times as well as rates of water loss therefore should be reduced at higher altitudes. Further work must be done on migratory birds to obtain realistic, quantitative estimates of water loss under natural, or near natural, conditions of ambient temperature, solar radiation, humidity, and altitude before the question of water balance in migrants will be resolved.

In the few analyses of flying bats, water loss rates were slightly greater than comparable avian rates (Carpenter, 1968, 1969), although most of the heat produced during the vigorous exercise of flight appears to be lost from nonrespiratory sources (Thomas and Suthers, 1972). Low body temperatures during flight may assist in conserving water (see below) and it is likely that migration routes and patterns of travel allow bats to drink *en route*, thus obviating the necessity of selective strategies for water conservation.

4. Behavioral Adjustments

A variety of behavioral adjustments make it possible for insects to reduce the high energetic costs of sustained flight. Aphids and locusts may be able to use updrafts or thermal currents (Rainey, 1958, 1976; Pringle, 1965). The butterfly *Ascia monuste* flies low within the boundary layer of shelter and avoids the effects of wind (Taylor, 1958; Nielsen, 1961). The volume of migration of lepidoptera, dragonflies, and other insects may be greater in periods when wind movements are from appropriate directions (see Johnson, 1969). However, Pringle (1965) contends that long migrations such as those of the monarch butterfly and the butterfly *Pyrameis atalanta* are made by active wing movements sustained by metabolic energy. Monarch butterflies have been observed flying into winds up to 10 miles/hr. Wind direction affects the movements of flying desert locusts, but their orientation to wind is much more complex than simply flying downwind (see Johnson, 1969; Schaefer, 1976).

Birds may save energy by means of formation flight (Lissaman and Shollenberger, 1970; but see Gould and Heppner, 1974), although most small passerines appear to migrate singly (Gauthreaux, 1972; Balcomb, 1977). Some migrants practice thermal soaring (e.g., Alerstam, 1975) which may reduce costs of travel (see Pennycuick, 1975).

Flight speed, duration, and the tendency for migration to occur in insects and birds are known to depend at least partially on wind direction and velocity. The longest nonstop flights may be possible only if birds are aided or at least not greatly hindered by adverse winds, and it appears that a variety of birds sometimes migrate downwind (Gauthreaux and Able, 1970; Able, 1973; Alerstam and Ulfstrand, 1974; Blokpoel and Burton, 1975) or may reorient after flying downwind (Able, 1977). For example, radar studies of nocturnal migration of North American passerines indicate that the largest volume of migrants is usually observed on nights when the wind is blowing from an appropriate direction (Bellrose, 1971; Able, 1973). Other investigators, however, have found that passerine and other small avian migrants rarely fly downwind (Evans, 1970; Nisbet, 1971; Richardson, 1972, 1974, 1976; Steidinger, 1972; Alerstam, 1975; Williams *et al.*, 1977). Tucker and Schmidt-Koenig (1971) observed that some birds appear to fly faster in head winds, but caution that this may be due to imprecise measurements of wind velocities. However, evidence exists that migrants adjust their air speed according to wind velocity (Bellrose, 1967; Bruderer, 1971) and Pennycuick's (1969, 1975) analyses point out the adaptiveness of this phenomenon. It is obvious that the relationships of wind direction and avian migratory flight are variable and complex and remain an area for future development and synthesis of theory.

C. Migration Ranges

In general, investigations of migration ranges have emphasized the migratory flights of birds, and to a lesser extent, insects. Terrestrial migrants tend to migrate relatively short distances, generally do not travel so rapidly, and are capable of obtaining food *en route*. Few species adaptations are necessary for this sort of movement. The 600-km migration by the North American caribou, *Rangifer tarandus*, probably constitutes the longest distance regularly traveled by a land mammal (Irving, 1972).

The theory of the aerodynamics and power of migratory flight have generally focused upon birds (but see Weis-Fogh and Jensen, 1956; Weis-Fogh, 1956, 1961, 1964b, 1967b, 1976, 1977; Kokshaysky, 1977), and for that reason the following discussion largely emphasizes findings derived from avian considerations. A large number of authors have discussed potential flight ranges of birds (e.g., see Odum et al., 1961; Lasiewski, 1962; Yapp, 1962; Dolnik et al., 1963b; Nisbet, 1963b, 1970; Nisbet et al., 1963; Raveling and LeFebvre, 1967; Dolnik, 1969b; Pennycuick, 1969, 1975; Tucker, 1974; Greenewalt, 1975). Migratory range in species that make long, nonstop flights is a function of fuel stores, rate of fuel utilization (metabolic costs of travel), and the speed of flight (Pennycuick, 1969, 1972, 1975). Potential flight ranges have been calculated traditionally on the basis of the best estimates of these variables and the probable energy yield of fuel stores (usually 9.0–9.5 kcal/gm, see Odum et al., 1961, 1965; Farner, 1970; Johnston, 1970). Known fuel reserves are taken to be the amounts of extractable lipid as determined from individuals thought to be ready to begin migration. Glycogen stores probably make only small contributions (see above). However, not all the lipid is available for use as fuel. A small amount of residual lipid is associated with membrane structure or tissue of the nervous system. Sometimes called "reserve fat" or "tissue fat," it has been estimated to amount to 1.0 gm in the indigo bunting, *Passerina cyanea* (Johnston, 1965), and 0.5 gm in other small passerines (Odum, 1960a). Actual measurements of mimimum residual lipid in small passerines which have been severely stressed range from 0.05 to 0.20 gm (Helms et al., 1967; Johnston, 1968; Evans, 1969), or about 0.02 to 0.10 gm lipid/gm lean dry weight. If the initial fuel load, corrected for unusable residual lipid, is known, then flight range may be calculated if the rate of lipid utilization or the metabolic rate of the migrant is known. Early attempts to estimate flight ranges (or conversely to estimate the costs of flight) were based upon measurements of rates of weight loss (see Nisbet, 1963b, 1967; Nisbet et al., 1963; Raveling and LeFebvre, 1967; Hussell, 1969). Later attempts involved knowledge or estimates of the actual energetic costs of flight. In such cases, migration range is generally determined from fuel load multiplied by the caloric content of the fuel and the result divided by the energetic costs of flight. This produces the expected length of flight in units of time. Flight speed must be known in order to predict the flight range.

The magnitude of fuel reserve is limited by wing loading. If the weight of lipid reserves approaches the limits of carrying ability, flying may be so inhabited that costs of flight become prohibitive or the ability to maneuver or evade predation may be seriously affected (see Blem, 1975). Some migratory birds, bats, and insects may differ in physical proportions from related, more sedentary forms, In most instances, strong migrants have longer (sometimes pointed) wings, higher aspect ratios, and possibly faster flight (see Chapman, 1940; Meinertzhagen, 1951; Struhsaker, 1961; Johnson, 1963; Ward, 1964a; Blyumental and Dolnik, 1970; Findley et al., 1972). The adaptive significance of increased wing length may be in decreasing wing loading problems, but the interaction of flight speed, wing beat frequency, and costs of flight are very complex. Although little quantitative work has been done on ths limits of wing loading, Pennycuick's (1969, 1975) analyses are instructive.

Recent developments in the aerodynamic theory of flying organisms (Pennycuick, 1968, 1969, 1972, 1975; Tucker, 1974, 1977; Greenewalt, 1975; Weis-Fogh, 1976; Nachtigall, 1976; Lighthill, 1977; Oehme, 1977) have advanced knowledge of the limitations upon migration and have produced some interesting predictions about the performance of flying organisms. These models incorporate considerations of weight of the migrant and mass of potential fuel stores with aerodynamic theories of the power required for flight in such a way that testable predictions have been produced. Developed largely from analytic techniques derived from aerodynamic engineering, Pennycuick's (1969, 1975) and Tucker's (1974) models generally fit empirically determined power requirements for flight quite well. The basis of these methods for estimating flight performance in birds is the calculation of curves relating mechanical power requirements to forward velocities. Their equations take into account the bird's weight, wingspan (wing disk area), air density (altitude), and functional diameter of the bird's cross section (flat-plate area), and other factors in a manner too complex to describe here in complete detail. The models describe power curves which are U-shaped with respect to forward velocity (Fig. 2.), and resemble actual power requirements as measured in laboratory tests of flight (Tucker, 1968b, 1971, 1972a, 1974; also see Bernstein et al., 1973). These curves predict that some speeds intermediate between slow flight speeds and relatively fast flying rates will require minimum power requirements. Because of the shape of the power curve, minimum power speeds are not those velocities yielding maximum range. At higher velocities (minimum cost of transport or maximum range speeds), greater distances are covered per unit of energy expenditure. Minimum flight and minimum transportation costs are predicted to increase with body mass (Fig. 2) and likewise the associated velocities are greater. [See Pennycuick (1969, 1975) and Tucker (1974) for excellent discussions of the aerodynamic costs of flight.] Larger migrants should therefore fly faster than small ones. The power/weight ratio decreases with size (also see Schaefer, 1968) and larger birds have greater migratory ranges than small birds if the proportion of fuel is similar. Increased weight of premigratory fat deposition will

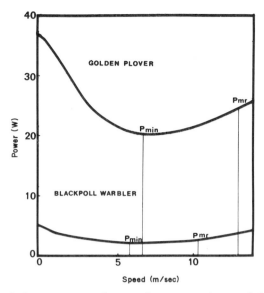

Speed (m/sec)

Fig. 2. Theoretical power curves for two long-range migrants flying at 1000 m, as calculated using Pennycuick's (1969, 1975) methods. Power is in watts, speed in meters per second. Lines for minimum power required for flight (P_{min}) extend to minimum cost flight speeds; lines for power required for maximum ranges (P_{min}) extend to velocities that produce maximum migratory ranges.

result in greater power requirements, and greater maximum range speeds (see Kokshaysky, 1970), but both decline during long-range migratory flights as fuel loads diminish. The analyses of Pennycuick and Tucker also indicate that altitude of flight may be affected early in migration by fuel loads. For example, maximum range speeds and the power required for flight increases with altitude, but as fuel loads are utilized, the velocity and power required for maximum range flight decrease at any given height.

When these aerodynamic models are used to predict potential migratory ranges, results similar to those obtained from material balance studies are generally obtained. Knowledge of the lipid content of the migrant at the beginning of the flight (and sometimes the end of the flight) are needed to calculate potential migration distance. Reasonable estimates of flight range of many migrants may be obtained from both Pennycuick's and Tucker's models. For example, my interpretation of Pennycuick's method for calculating flight range (Fig. 3) produced reasonable estimates for small passerines such as the blackpoll warbler, but underestimates the distances traveled by some larger migrants such as the Pacific golden plover.

Charles R. Blem

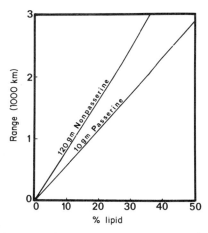

Fig. 3. Nonstop flight ranges for two avian migrants as a function of lipid content, calculated using Pennycuick's (1975) method. Lipid content of the migrants is expressed as percentage of total body weight.

D. Energetic Advantages of Migration

Many migrants obtain an energetic advantage from migration in that periodic, heterogeneous food resources may be exploited. The energetic benefit of long photoperiods and higher ambient temperatures at southern latitudes in winter may be a major selective factor for autumnal migration in a variety of organisms, but is best known in birds (Seibert, 1949). In some species of birds, there is sexual dimorphism in body size and distance of southward migration. For example, Ketterson and Nolan (1976) found that average male dark-eyed juncos, *Junco hyemalis*, are larger, winter farther northward than females, and that sex ratios vary clinally. Similar findings have been reported for more than a dozen other species of birds, most of them strong migrants. The significance of the variation in sex ratios may lay in greater endurance of fasting during inclement winter conditions by larger individuals (often males; Ketterson and Nolan, 1976; Ketterson and King, 1977; also see Blem, 1973).

The costs of migration may be less than or equal to the energy saved in warmer wintering grounds with longer photoperiods (Dolnik, 1971a, 1974) and overall may not be significantly more costly than molt or breeding in caloric expenditure per day (Dolnik, 1971b). Likewise, breeding at higher latitudes having very long summer days may allow birds to accumulate greater amounts of energy for the production of eggs, the feeding of young or both. Bats may obtain a similar energetic advantage by migrating between southern hibernation or wintering sites and northern breeding areas.

In some cases there may not be a direct energetic advantage in migration (e.g., West, 1960), as birds may be migrating to avoid low temperature extremes (Seibert, 1949; Olsen, 1965). Insects and other migrants may move to suitable breeding sites where substrate, climate, or lack of predation may be advantageous, or migration may occur for a variety of other ecological reasons, including avoidance of competition (see Dorst, 1962; Orr, 1970; Dingle, 1972). In some species, migration may replace thermoregulatory acclimatization or be necessitated by the lack of it (Gelineo, 1964). In a variety of species that have traditionally been strong migrants, the appearance of surplus food in the summer range has served to allow at least some individuals to overwinter. The reverse is also true in some sedentary forms.

IV. Summary

This brief review has demonstrated that the storage and utilization of migratory fuel is well known, but only in a relatively small number of species within any of the major taxonomic units discussed above (also see King and Farner, 1965; Johnson, 1969; Berthold, 1975). Further analyses of lipid storage and utilization in a wider diversity of insects, nonpasserine birds (e.g., waterfowl, caprimulgiformes, and swifts), bats, cetaceans, and sea turtles should be fruitful. Specifically, the proximate cues and mechanisms that ensure that appropriate amounts of lipid are deposited and that stimulate continental migrants to deposit greater amounts of fuel as they approach barriers are poorly known. The factors that affect composition of lipid reserves are likewise problematical. New approaches may be necessary to clarify the problem of water balance in long-range avian migrants. It is possible that more complex laboratory tests of the physiology of migration may be necessary. These might incorporate the effects of altitude, humidity, and temperature on migratory birds in the premigratory state. It is unfortunate that, so far, little of the fine quantitative laboratory analysis of the energetics of flight has dealt with birds that are actually long-range migrants. Migrants such as some bats, fishes, sea turtles, and cetaceans may continue to remain poorly known because of difficulties in obtaining data from suitable numbers of migrating individuals. Hopefully, students of these forms will direct their attention, whenever possible, to the problems of energy storage and utilization during the migratory period.

The area of migratory physiology has passed through a period, within the last decade or so, of remarkable development. I am sure that the next decade will provide even more sophisticated analyses and exciting results. Hopefully, when these are integrated with knowledge of life history strategies and the remarkable ability of many migrants to navigate, a comprehensive viewpoint of migration will emerge.

V. Addendum

Several recent publications are relevent to the subject of the energetics of migration. These may be divided into two broad groups: theoretical and empirical analyses of transportation costs and studies of the physiology of locomotion.

Among papers dealing with invertebrates, Savage et al. (1979) present a physical explanation of the role of vortices in the hovering flight of dragonflies and propose a novel alternative to the lift-generating mechanisms previously proposed by Weis-Fogh (1972). May (1979) measured energy metabolism of dragonflies at rest and during preflight warm-up. Cloupeau et al. (1979) directly measured instantaneous lift in desert locusts flying in a wind tunnel and compare their results to Jensen's (1956) measurements on detached wings.

Several important papers have appeared which may be useful in the theoretical analysis of the energetic demands of avian flight and migration. Pennycuick (1978) outlines 15 predictions about avian flight that may be expected from his previously published analyses and discusses how these might be tested. Rayner (1977, 1979) uses a new approach, vortex theory, to analyze the same aspects of flight considered by Pennycuick in earlier papers. Ellington (1978) examines the aerodynamics of hovering. Torre-Bueno and Larochelle (1978) measured metabolic costs of long flights by unrestrained birds in a sealed wind tunnel (described by Linderoth, 1975). The approach is particularly useful since oxygen consumption and carbon dioxide production are measured simultaneously and respiratory quotients are therefore directly obtainable. This increases the precision of the estimates of the cost of flight as well as eliminating the restraint of the equipment used in the series of tests by Tucker (see literature cited this chapter). One important aspect of Torre-Bueno and Larochelle's work that undoubtedly will stimulate further work is their finding that only a slight and nonsignificant increase in metabolic rate was observed with increasing speed of flight. Hails (1979) used the $D_2{}^{18}O$ technique to measure the costs of flight in two species of swallows. His results suggest that low costs of flight in swallows and swifts as compared to other birds are due to behavioral and morphological differences and not to metabolic adaptations. Flynn and Gessaman (1979) provide an important test of telemetric measurement of the metabolism of unrestrained birds as evaluated from heart rates. Oehme *et al.* (1977), in one paper of a series on the biophysics and physiology of avian flight, examine the energetics of flight.

Among the papers on physiology of locomotion, Freadman (1979) provides an analysis of the swimming energetics of two migratory fishes, the striped bass (*Morone saxatilis*) and the bluefish (*Pomatomus saltatrix*). Among his findings is the discovery that the mode of gill ventilation changes with velocity of swimming speed. The energetics of tuna are analyzed by Hochachka *et al.* (1978). George and Stevens (1978) and Hulbert *et al.* (1979) examined fine muscle structure of tuna and found significant differentiation of both red and white fibers. The fibers show differences similar to those of red and white fibers of birds, although white muscle has an ample capillary supply, significant amounts of intracellular lipid, and unusual numbers of mitochondria. Guppy *et al.* (1979) investigated enzyme and metabolite profiles of white and red fibers of tuna in relation to the heat generated during activity and discuss the production of heat leading to the endothermy of some species.

Torre-Bueno (1978) found that evaporative cooling of starlings (*Sturnus vulgaris*) during long flights in a wind tunnel did not result in dehydration at temperatures below 7°C. He suggests that long-range migrants (which apparently do not dehydrate) maintain water balance by flying in colder air at higher elevations in which convection carries away most of the heat produced.

Finally, a voluminous analysis of the evolution and ecology of migration by Baker (1978) includes scattered references to the energetics of locomotion and premigratory fat deposition.

Acknowledgments

This chapter is a tribute to the spiritual support and tolerance of my wife Leann. I thank her for reading earlier versions of the manuscript. Preparation of the figures was supported by Department of Biology, Academic Division, Virginia Commonwealth University.

References

Able, K. P. (1973). *Ecology* 54, 1031–1041.
Able, K. P. (1977). *Auk* 94, 320–330.
Adams, P. A., and Heath, J. R. (1964). *Nature (London* 201, 20–21.
Alerstam, T. (1975). *Ibis* 117, 489–495.
Alerstam, T., and Ulfstrand, S. (1974). *Ibis* 116, 522–542.
Allen, G. M. (1939). "Bats." Harvard Univ. Press, Cambridge, Massachusetts.
Ash, J. S. (1969). *Ibis* 111, 1–10.

Atkins, M. D. (1966). *Can. Entomol.* **98,** 953–991.

Atkins, M. D. (1969). *Can. Entomol.* **101,** 164–165.

Aulie, A. (1971a). *Comp. Biochem. Physiol. A* **38,** 91–97.

Aulie, A. (1971b). *Comp. Biochem. Physiol. A* **39,** 173–176.

Aulie, A. (1972). *Comp. Biochem. Physiol. A* **41,** 43–48.

Aulie, A. (1975). *Comp. Biochem. Physiol. A* **52,** 81–84.

Bade, M. L. (1964). *J. Insect Physiol.* **10,** 333–341.

Baggott, G. K. (1975). *J. Zool.* **175,** 299–314.

Bailey, E. (1975). *In* "Insect Biochemistry and Function" (D. J. Candy and B. A. Kilby, eds.), pp. 89–176. Chapman & Hall, London.

Bainbridge, R. (1960). *J. Exp. Biol.* **37,** 129–153.

Baker, R. R. (1978). "The Evolutionary Ecology of Animal Migration." Holmes & Meier Publ., Inc., New York.

Baker, W. W., Marshall, S. G., and Baker, V. B. (1968). *J. Mammal.* **49,** 314–317.

Balcomb, R. (1977). *Auk* **94,** 479–488.

Barnett, L. B. (1970). *Comp. Biochem. Physiol.* **33,** 559–578.

Barrett, I. and Hester, F. J. (1964). *Nature (London)* **203,** 96–97.

Bartholomew, G. A. (1972). *Proc. Int. Ornithol. Congr., 15th, 1970* pp. 237–254.

Beall, G. (1948). *Ecology* **29,** 80–94.

Beenakkers, A. M. T. (1965). *J. Insect Physiol.* **11,** 879–888.

Beenakkers, A. M. T. (1969). *J. Insect Phiol.* **15,** 353–361.

Behnke, A. R. (1962). *In* "Fat as a Tissue" (K. Rodahl and B. Issekutz, Jr., eds.), pp. 285–313. McGraw-Hill, New York.

Bellrose, F. C. (1967). *Proc Int. Ornithol. Congr., Proc. 14th, 1966* pp. 281–309.

Bellrose, F. C. (1971). *Auk* **88,** 397–424.

Berger, M., and Hart, J. S. (1972). *J. Comp. Physiol.* **81,** 363–380.

Berger, M., and Hart, J. S. (1974). *In* "Avian Biology" (D. S. Farner and J. R. King, eds.), Vol. 4, pp. 416–477. Academic Press, New York.

Berger, M., Roy, O. Z., and Hart, J. S. (1970a). *Z. Vergl. Physiol.* **66,** 190– 200.

Berger, M., Hart, J. S., and Roy, O. Z. (1970b). *Z. Vergl. Physiol.* **66,** 201– 214.

Berger, M., Hart, J. S., and Roy, O. Z. (1971). *Can. J. Zool.* **49,** 767–774.

Bernstein, M. H. (1971a). *Condor* **73,** 468–469.

Bernstein, M. H. (1971b). *Comp. Biochem. Physiol. A* **38,** 611–617.

Bernstein, M. H., Thomas, S. P., and Schmidt-Nielsen, K. (1973). *J. Exp. Biol.* **58,** 401–410.

Berthold, P. (1973). *Ibis* **115,** 594–599.

Berthold, P. (1975). *In* "Avian Biology" (D. S. Farner and J. R. King, eds.), Vol. 5, pp. 77–128. Academic Press, New York.

Bilinski, E. (1963). *Can. J. Biochem. Physiol.* **41,** 107–112.

Bilinski, E. (1969). *In* "Fish in Research" (O. W. Neuhaus and J. E. Halver, eds.), pp. 135–149. Academic Press, New York.

Blem, C. R. (1973). *Ornithol. Monogr.* **14,** 96–121.

Blem, C. R. (1975). *Wilson Bull.* **87,** 543–549.

Blem, C. R. (1976). *Am. Zool.* **16,** 671–684.

Blokpoel, H., and Burton, J. (1975). *Bird-Banding* **46,** 311–328.

Blyumental, T. I. (1962). *Vses. Ornitol. Konf.* pp. 133–136 (in Russian).

Blyumental, T. I. (1963). *Prib. Ornitol. Konf.* pp. 17–19 (in Russian).

Blyumental, T. I. (1968). *Commun. Baltic Comm. Study Bird Migration* **5,** 146–154 (in Russian).

Blyumental, T. I., and Dolnik, V. R. (1970). *Zool. Zh.* **49,** 1069–1072 (in Russian).

Bower, E. B. and Helms, C. W. (1968). *Physiol. Zool.* **41,** 157–168.

Bradley, W. G., and O'Farrell, M. J. (1969). *In* "Physiological Systems in Semiarid Environ-

ment" (C. C. Hoff and M. L. Riedesel, eds.), pp. 85–96. Univ. of New Mexico Press, Albuquerque.
Brett, J. R. (1964). *J. Fish. Res. Board Can.* **21**, 1183–1226.
Brett, J. R. (1965a). *J. Fish. Res. Board Can.* **22**, 1491–1501.
Brett, J. R. (1965b). *Sci. Am.* **213** (2), 80–85.
Brett, J. R. (1967). *J. Fish. Res. Board Can.* **24**, 1731–1741.
Brett, J. R. (1971). *Am. Zool.* **11**, 99–113.
Brett, J. R. (1972). *Respir. Physiol.* **14**, 151–170.
Brett, J. R., and Glass, N. R. (1973). *J. Fish. Res. Board Can.* **30**, 379–387.
Brett, J. R., and Sutherland, D. B. (1965). *J. Fish. Res. Board Can.* **22**, 405–409.
Brisbin, I. L., Jr. (1968). *Ecology* **49**, 792–794.
Brisbin, I. L., Jr. (1969). *Oriole* **34**, 17–24.
Brodie, P. F. (1975). *Ecology* **56**, 152–161.
Brosemer, R. W., Vogell, W., and Bücher, T. (1963). *Biochem. J.* **338**, 854–910.
Brown, J. J., and Chippendale, G. M. (1974). *J. Insect Physiol.* **20**, 1117–1130.
Brown, R. H. J. (1961). *Symp. Zool. Soc. London* **5**, 95–99.
Bruderer, B. (1971). *Ornithol. Beob.* **68**, 89–158.
Bücher, T. and Klingenberg, M. (1958). *Angew. Chem.* **70**, 552–570.
Bursell, E. (1964). *In* "The Physiology of Insecta" (M. Rockstein, ed.), Vol. 1, pp. 323–361. Academic Press, New York.
Calder, W. A. and King, J. R. (1974) *In* "Avian Biology" (D. S. Farner and J. R. King, eds.), Vol. 4, pp. 259–413. Academic Press, New York.
Caldwell, L. D. (1973). *Comp. Biochem. Physiol. B* **44**, 493–497.
Caldwell, L. D., Odum, E. P., and Marshall, S. G. (1963). *Wilson Bull.* **75**, 428–434.
Carey, F. G., and Teal, J. M. (1966). *Proc. Natl. Acad. Sci. U.S.A.* **56**, 1464–1469.
Carey, F. G., and Teal, J. M. (1969a). *Comp. Biochem. Physiol.* **28**, 199–204.
Carey, F. G., and Teal, J. M. (1969b). *Comp. Biochem. Physiol.* **28**, 205–213.
Carey, F. G., Teal, J. M., and Kanwisher, J. W. (1971). *Am. Zool.* **11**, 135–143.
Carpenter, R. E. (1968). *Comp. Biochem. Physiol.* **24**, 951–961.
Carpenter, R. E. (1969). *Physiol. Zool.* **42**, 288–302.
Carr, A. (1964). *BioScience* **14**, 49–52.
Carr, A., and Goodman, D. (1970). *Copeia* pp. 783–786.
Casey, T. M. (1976a). *J. Exp. Biol.* **64**, 529–543.
Casey, T. M. (1976b). *J. Exp. Biol.* **64**, 545–560.
Cenedella, R. J. (1971). *Insect Biochem.* **1**, 244–247.
Chapman, F. M. (1940). *Bull. Am. Mus. Nat. Hist.* **77**, 381–438.
Chefurka, W. (1965a). *In* "The Physiology of Insecta" (M. Rockstein, ed.), Vol. 2, pp. 581–667. Academic Press, New York.
Chefurka, W. (1965b). *In* "The Physiology of Insecta" (M. Rockstein, ed.), Vol. 2, pp. 669–768.
Child, G. I. (1969), *Auk* **86**, 327–338.
Child, G. I., and Marshall, S. G. (1970). *Condor* **72**, 116–119.
Church, N. S. (1960a). *J. Exp. Biol.* **37**, 171–185.
Church, N. S. (1960b). *J. Exp. Biol.* **37**, 186–212.
Clegg, J. S., and Evans, D. R. (1961). *J. Exp. Biol.* **38**, 771–792.
Cloudsley-Thompson, J. L. (1970). *In* "Comparative Physiology of Thermoregulation" (G. C. Whittow, ed.), Vol. 1, pp. 15–77. Academic Press, New York.
Cloupeau, M., Devillers, J. F., and Devezeaux, D. (1979). *J. Exp. Biol.* **80**, 1–15.
Cockbain, A. J. (1961a). *J. Exp. Biol.* **38**, 163–174.
Cockbain, A. J. (1961b). *J. Exp. Biol.* **38**, 175–180.
Cockrum, E. L. (1969). *Misc. Publ., Mus. Nat. Hist., Univ. Kans.* **51**, 303–336.

Cohen, D. (1967). *Am. Nat.* **101**, 5–17.

Connell, C. E., Odum, E. P., and Kale, H. W., II (1960). *Auk* **77**, 1–9.

Crabtree, B., and Newsholme, E. A. (1972). *Biochem. J.* **130**, 697–705.

Crabtree, B., and Newsholme, E. A. (1975). *In* "Insect Muscle" (P. N. R. Usherwood, ed.), pp. 405–500. Academic Press, New York.

Davis, R. (1966). *Ecol. Monogr.* **36**, 201–237.

Davis, W. H., and Hitchcock, H. B. (1966). *J. Mammal.* **46**, 296–313.

Dawson, W. R., and Hudson, J. W. (1970). *In* "Comparative Physiology of Thermoregulation" (G. C. Whittow, ed.), Vol. 1, pp. 223–310. Academic Press, New York.

Diamond, A. W., Lack, P., and Smith, R. W. (1977). *Wilson Bull.* **89**, 456–465.

Dingle, H. (1972). *Science* **175**, 1327–1335.

Dodgen, G. L., and Blood, F. R. (1956). *Am. J. Physiol.* **187**, 151–154.

Dole, V. P. (1962). *In* "Fat as a Tissue" (K. Rodahl and B. Issekutz, Jr., eds.), pp. 250–259. McGraw-Hill, New York.

Dolnik, V. R. (1963). *Prib. Ornitol. Konf.* pp. 61–63 (in Russian).

Dolnik, V. R. (1968). *Int. Stud. Sparrows* **2**, 89–95.

Dolnik, V. R. (1969a). *Zh. Obshch. Biol.* **30**, 273–291.

Dolnik, V. R. (1969b). *In* "Ornithology in USSR," Vol. II, pp. 198–202. Ashkhabad (in Russian).

Dolnik, V. R. (1970). *Colloq. Int. C. N. R. S.* **172**, 351–364.

Dolnik, V. R. (1971a). *Ekologiya* **2**, 88–89 (in Russian).

Dolnik, V. R. (1971b). *Ekologiya* **2**, 89–91.

Dolnik, V. R. (1971c). *In* "Zoology of Vertebrates" (L. P. Poznanin, ed.), pp. 52–81. Moscow (in Russian).

Dolnik, V. R. (1972). *Dokl. Acad. Nauk SSSR* **206**, 247–249.

Dolnik, V. R. (1973). *In* "Productivity, Population Dynamics and Systematics of Granivorous Birds" (S. C. Kendeigh and J. Pinowski, eds.), pp. 103–109. The Hague.

Dolnik, V. R. (1974). *Int. Stud. Sparrows* **7**, 11–20.

Dolnik, V. R., and Blyumental, T. I. (1964). *Usp. Sovrem. Biol.* **58**, 280–301 (in Russian).

Dolnik, V. R., and Blyumental, T. I. (1967). *Condor* **69**, 435–468.

Dolnik, V. R., and Gavrilov, V. (1971). *In* "Ecological and Physiological Aspects of Bird Migration" (R. Potapov, ed.), pp. 1–244. Leningrad.

Dolnik, V. R., and Gavrilov, V. (1973). *In* "Bird Migrations: Ecological and Physiological Factors" (B. E. Bykhouskii, ed.), pp. 273–287. Halstead Press, New York.

Dolnik, V. R., Blyumental, T. I., Dobrynina, I. N., Orlova, G. M., and Keskpaik, Y. E. (1963a). *Prib. Ornitol. Konf.* pp. 63–65 (in Russian).

Dolnik, V. R., Gavrilov, V. M., and Ezerskas, L. J. (1963b). *Prib. Ornitol. Konf.* pp. 65–67. (in Russian).

Dorsett, D. A. (1962). *J. Exp. Biol.* **39**, 579–588.

Dorst, J. (1962). "The Migration of Birds." Houghton, Boston, Massachusetts.

Downer, R. G. H., and Matthews, J. R. (1976). *Am. Zool.* **16**, 733–745.

Drummond, G. I., and Black, E. C. (1960). *Annu. Rev. Physiol.* **22**, 169–190.

Duvall, A. J. (1966). *In* "Birds in Our Lives" (A. Stefferud, ed.), pp. 232–245. U.S. Fish Wildl. Serv., Washington, D. C.

Elder, H. Y. (1975). *In* "Insect Muscle" (P. N. R. Usherwood, ed.), pp. 1–74. Academic Press, New York.

Ellington, C. P. (1978). *In* "Comparative Physiology—Water, Ions and Fluid Mechanics" (K. Schmidt-Nielsen, L. Bolis, and S. H. P. Maddrell, eds.), pp. 327–345. Cambridge Univ. Press, London and New York.

Engles, W. L. (1962). *Biol. Bull. (Woods Hole, Mass.)* **123**, 542–554.

Evans, P. R. (1968). *Br. Birds* **61**, 281–303.

Evans, P. R. (1969). *Condor* **71**, 316–330.

Evans, P. R. (1970). *Nature (London)* **228**, 1121.

Ewing, W. G., Studier, E. H., and O'Farrell, M. J. (1970). *Comp. Biochem. Physiol.* **36**, 119–129.

Farner, D. S. (1960). *Proc. Int. Ornithol. Congr., 12th, 1958* pp. 196–208.

Farner, D. S. (1970). *Fed. Proc., Fed. Am. Soc. Exp. Biol.* **29**, 1649–1663.

Farner, D. S., Mewaldt, L. R., and King, J. R. (1954). *J. Comp. Physiol. Psychol.* **47**, 148–153.

Farner, D. S., Oksche, A., Kamemoto, F. I., King, J. R., and Cheyney, H. E. (1961). *Comp. Biochem. Physiol.* **2**, 125–142.

Farner, D. S., King, J. R., and Stetson, M. H. (1969). *Progr. Endocrinol., Proc. Int. Congr. Endocrinol., 1968* Int. Congr. Ser. No. 184, pp. 152–157.

Fast, P. G. (1964). *Mem. Entomol. Soc. Can.* **37**, 1–50.

Findley, J. S., and Jones, C. (1964). *J. Mammal.* **45**, 461–470.

Findley, J. S., Studier, E. H., and Wilson, D. E. (1972). *J. Mammal.* **53**, 429–444.

Fisher, H. I., and Bartlett, L. M. (1957). *Condor* **59**, 364–372.

Flynn, R. K., and Gessaman, J. A. (1979). *Comp. Biochem. Physiol. A* **63**, 511–514.

Fogden, M. P. L. (1972). *Ibis.* **114**, 548–552.

Freadman, M. Z. (1979). *J. Exp. Biol.* **83**, 217–230.

Fry, C. H. (1967). *Ibis* **109**, 118–120.

Fry, C. H., Ash, J. S., and Ferguson-Lees, I. J. (1970). *Ibis* **112**, 58–82.

Fry, F. E. J. (1971). *In* "Fish Physiology" (W. S. Hoar and D. J. Randall, eds.), Vol. 6, pp. 1–98. Academic Press, New York.

Fry, F. E. J., and Hochachka, P. W. (1970). *In* "Comparative Physiology of Thermoregulation" (G. C. Whittow, ed.), Vol. 1, pp. 79–134. Academic Press, New York.

Fulton, R. A., and Romney, V. E. (1940). *J. Agric. Res.* **61**, 737–743.

Gauthreaux, S. A., Jr. (1971). *Auk.* **88**, 343–365.

Gauthreaux, S. A., Jr. (1972). *Wilson Bull.* **84**, 136–148.

Gauthreaux, S. A., Jr., and Able, K. P. (1970). *Nature (London)* **228**, 476–477.

Gelineo, S. (1964). *In* "Handbook of Physiology Adaptation" (D. B. Dill, ed.), Sect. 4, pp. 259–282. Am. Physiol. Soc., Washington, D. C.

George, J. C., and Berger, A. J. (1966). "Avian Myology." Academic Press, New York.

George, J. C., and Jyoti, D. (1957). *J. Anim. Morphol. Physiol.* **5**, 57–60.

George, J. C., and Naik, D. V. (1962). *Q. J. Microsc. Sci.* **104**, 393–399.

George, J. C., and Stevens, E. D. (1978). *Environ. Biol. Fishes* **3**, 185–191.

George, J. C., and Talesara, C. L. (1961). *Q. J. Microsc. Sci.* **102**, 131–141.

George, J. C., and Vallyathan, N. V. (1964). *J. Appl. Physiol.* **19**, 619–622.

George, J. C., Vallyathan, N. V., and Scaria, K. S. (1958). *Experientia* **14**, 250–251.

Gifford, C. E., and Odum, E. P. (1965). *Condor* **67**, 383–403.

Gilbert, L. I. (1967). *Adv. Insect Physiol.* **4**, 69–211.

Gilby, A. R., Wyatt, S. S., and Wyatt, G. R. (1967). *Acta Biochim. Pol.* **14**, 83–100.

Gold, A. (1973). *Science* **181**, 275–276.

Goldspink, G. (1965). *Am. J. Physiol.* **209**, 100–104.

Goodridge, A. G. (1964). *Comp. Biochem. Physiol.* **13**, 1–26.

Goodridge, A. G., and Ball, E. G. (1967a). *Am. J. Physiol.* **213**, 245–249.

Goodridge, A. G., and Ball, E. G. (1967b). *Biochemistry* **6**, 1676–1682.

Goodridge, A. G., and Ball, E. G. (1967c). *Biochemistry* **6**, 2335–2343.

Gordon, M. S. (1968). *Science* **159**, 87–90.

Gould, L. L., and Heppner, F. (1974). *Auk* **91**, 494–506.

Greenewalt, C. H. (1975). *Trans. Am. Philos. Soc.* **65**, 1–67.

Griffin, D. R. (1970). *In* "Biology of Bats" (W. A. Wimsatt, ed.), pp. 233–264. Academic Press, New York.

Guppy, M., Hulbert, W. C., and Hochachka, P. W. (1979). *J. Exp. Biol.* **82**, 303–320.

Hails, C. J. (1979). *Comp. Biochem. Physiol. A* **63**, 581–585.

Hainsworth, F. R., and Wolf, L. L. (1969). *Am. Zool.* **9**, 1100–1101.

Hainsworth, F. R., and Wolf, L. L. (1972). *Am. Nat.* **106**, 589–596.

Harden-Jones, F. R. (1968). "Fish Migration." Arnold, London.

Hart, J. S., and Berger, M. (1972). *Proc. Int. Ornithol. Congr., 15th, 1970* pp. 189–199.

Hart, J. S., and Heroux, O. (1955). *Can. J. Biochem. Physiol.* **33**, 428–435.

Hart, J. S., and Roy, O. Z. (1966). *Physiol. Zool.* **39**, 291–305.

Hart, J. S., and Roy, O. Z. (1967). *Am. J. Physiol.* **213**, 1311–1316.

Hartman, F. A., and Brownell, K. A. (1961). *Condor* **63**, 403–409.

Hazelwood, R. L. (1972). *In* "Avian Biology" (D. S. Farner and J. R. Kings, eds.), Vol. 2, pp. 471–526. Academic Press, New York.

Hazelwood, R. L. (1976). *In* "Avian Physiology" (P. D. Sturkie, ed.), pp. 210–232. Springer-Verlag, Berlin and New York.

Heath, J. E., and Adams, P. A. (1967). *J. Exp. Biol.* **47**, 21–33.

Heath, J. E., Nanegan, J. L., Wilkin, P. J., and Heath, M. S. (1971). *Am. Zool.* **11**, 147–158.

Heinrich, B. (1974). *Science* **185**, 747–756.

Helms, C. W. (1963). *Auk* **80**, 318–334.

Helms, C. W. (1968). *Am. Zool.* **8**, 151–167.

Helms, C. W., and Smythe, R. B. (1969). *Wilson Bull.* **81**, 280–292.

Helms, C. W., Aussiker, W. H., Bower, E. B., and Fretwell, S. D. (1967). *Condor* **69**, 560–578.

Herreid, C. F., II (1963a). *J. Mammal.* **44**, 431–433.

Herreid, C. F., II (1963b). *Science* **142**, 1573–1574.

Hicks, D. L. (1967). *Condor* **69**, 387–399.

Hilditch, T. P., and Williams, P. N. (1964). "Chemical Composition of the Natural Fats," 4th ed. Wiley, New York.

Hill, L., Luntz, A. J., and Steele, P. A. (1968). *J. Insect Physiol.* **14**, 1–20.

Hintz, J. V., and Dyer, M. I. (1970). *J. Wildl. Manage.* **34**, 789–799.

Hochachka, P. W. (1969). *In* "Fish Physiology" (W. S. Hoar and R. J. Randall, eds.), Vol. 1, pp. 351–389. Academic Press, New York.

Hochachka, P. W., Hulbert, W. C., and Guppy, M. (1978). *In* "The Physiological Ecology of Tuna" (G. D. Sharp and A. E. Dizon, eds.), pp. 153–174. Academic Press, New York.

Hocking, B. (1953). *Trans. R. Entomol. Soc. London* **104**, 223–345.

Holmes, R. T. (1976). *Am. Midl. Nat.* **96**, 281–290.

Hulbert, W. C., Guppy, M., Murphy, B., and Hochachka, P. W. (1979). *J. Exp. Biol.* **82**, 289–301.

Humphrey, S. R., and Cope, J. B. (1968). *J. Mammal.* **49**, 329.

Hussell, D. J. T. (1969). *Auk* **86**, 75–83.

Idler, D. R., and Bitners, I. (1958). *Can. J. Biochem. Physiol.* **36**, 793–798.

Idler, D. R., and Bitners, I. (1959). *J. Fish. Res. Board Can.* **16**, 235–241.

Idler, D. R., and Bitners, I. (1960). *J. Fish. Res. Board. Can.* **17**, 113–122.

Idler, D. R., and Clemens, W. A. (1959). *Int. Pac. Salmon Fish. Comm. Prog. Rep.* pp. 1–80.

Idler, D. R., and Tsuyuki, H. (1958). *Can. J. Biochem. Physiol.* **36**, 783–791.

Irving, L. (1964). *In* "Handbook of Physiology" (W. O. Fenn and H. Rahn, eds.), Sect. 3, Vol. I, pp. 177–212. Am.Physiol. Soc., Washington, D. C.

Irving, L. (1972). "Arctic Life of Birds and Mammals." Springer-Verlag, Berlin and New York.

Issekutz, B., Jr., Paul, P., and Miller, H. I. (1967). *Am. J. Physiol.* **213**, 857–862.
Jensen, M. (1956). *Philos. Trans. R. Soc. London, Ser. B* **239**, 511–552.
Johnson, C. G. (1963). *Nature (London)* **198**, 423–427.
Johnson, C. G. (1965). In "The Physiology of Insecta" (M. Rockstein, ed.), Vol. 2, pp. 188–226. Academic Press, New York.
Johnson, C. G. (1969). "Migration and Dispersal of Insects by Flight." Methuen, London.
Johnston, D. W. (1964). *Ecology* **45**, 848–852.
Johnston, D. W. (1965). *Q. J. Fla. Acad. Sci.* **28**, 199–211.
Johnston, D. W. (1966). *Bird-Banding* **37**, 172–183.
Johnston, D. W. (1968). *Auk* **85**, 13–18.
Johnston, D. W. (1970). *Comp. Biochem. Physiol.* **34**, 827–832.
Johnston, D. W. (1973). *Condor* **75**, 108–113.
Johnston, D. W., and McFarlane, R. W. (1967). *Condor* **69**, 156–168.
Jutsum, A. R., and Goldsworthy, G. J. (1976). *J. Insect Physiol.* **22**, 243–249.
Kallapur, V. L., and George, C. J. (1973). *J. Insect Physiol.* **19**, 1035–1041.
Kammer, A. E. (1970). *J. Vergl. Physiol.* **68**, 334–344.
Karuhize, G. R. (1972). *Comp. Biochem. Physiol.* **43**, 563–570.
Keister, M., and Buck, J. (1964). In "The Physiology of Insecta" (M. Rockstein, ed.). Vol. 3, pp. 618–658. Academic Press, New York.
Kendeigh, S. C., West, G. C., and Cox, G. W. (1960). *Anim. Behav.* **8**, 180–185.
Kendeigh, S. C., Kontogiannis, J. E., Mazac, A., and Roth, R. (1969). *Comp. Biochem. Physiol.* **31**, 941–957.
Keskapik, J. (1968). *Izv. Akad. Nauk. Est. SSR, Biol.* **17**, 179–191.
Ketterson, E. D., and King, J. R. (1977). *Physiol. Zool.* **50**, 115–129.
Ketterson, E. D., and Nolan, V., Jr. (1976). *Ecology* **57**, 679–693.
King, J. R. (1961). *Condor* **63**, 128–142.
King, J. R. (1963). *Proc. Int. Ornithol. Congr., 13th, 1962* pp. 940–949.
King, J. R. (1972). *Proc. Int. Ornithol. Congr., 15th, 1970* pp. 200–217.
King, J. R., and Farner, D. S. (1956). *Proc. Soc. Exp. Biol. Med.* **93**, 354–359.
King, J. R., and Farner, D. S. (1959). *Condor* **61**, 315–324.
King, J. R., and Farner, D. S. (1963). *Condor* **65**, 200–223.
King, J. R., and Farner, D. S. (1965). *Ann. N. Y. Acad. Sci.* **131**, 422–440.
King, J. R., and Wales, E. E. (1965). *Physiol. Zool.* **38**, 49–68.
King, J. R., and Barker, S., and Farner, D. S. (1963). *Ecology* **44**, 513–521.
King, J. R., and Farner, D. S., and Morton, M. L. (1965). *Auk* **82**, 236–252.
Koerwitz, F. L., and Pruess, K. P. (1964). *J. Kans. Entomol. Soc.* **37**, 234–239.
Kokshaysky, N. V. (1970). *Zh. Obshch. Biol.* **31**, 527–549 (in Russian).
Kokshaysky, N. V. (1977). In "Scale Effects in Animal Locomotion" (T. J. Pedley, ed.), pp. 421–435. Academic Press, New York.
Kontogiannis, J. E. (1968). *Physiol. Zool.* **41**, 54–64.
Krogh, A., and Weis-Fogh, T. (1951). *J. Exp. Biol.* **28**, 344–357.
Krogh, A., and Zeuthen, E. (1941). *J. Exp. Biol.* **18**, 1–10.
Krueger, H. M., Saddler, J. B., Chapman, G. A., Tinsley, I. J., and Lowry, R. R. (1968). *Am. Zool.* **8**, 119–129.
Krulin, G. S., and Sealander, J. A. (1972). *Comp. Biochem. Physiol. A* **42**, 537–549.
Krzanowski, A. (1961). *Acta Theriol.* **4**, 249–264.
Kulzer, E., Nelson, J. E., McKeon, J. L., and Möhres, F. P. (1970). *Z. Vergl. Physiol.* **69**, 426–451.
Kuroda, N. (1964). *Misc. Rep. Yamashina Inst. Ornithol. Zool.* **4**, 91–105.
Kutty, M. N., and Saunders, R. L. (1973). *J. Fish. Res. Board Can.* **30**, 223–227.

Lack, D. (1960). *Br. Birds* **53**, 5–10.

Lagler, K. F., Bardach, J. E., Miller, R. R., and Passino, D. R. M. (1977). "Ichthyology," 2nd ed. Wiley, New York.

Lasiewski, R. C. (1962). *Condor* **64**, 324.

Lasiewski, R. C. (1963). *Physiol. Zool.* **36**, 112–140.

Lasiewski, R. C. (1972). *In* "Avian Biology" (D. S. Farner and J. R. King, eds.), Vol. 2, pp. 287–342. Academic Press, New York.

Lasiewski, R. C., Bernstein, M. H., and Ohmart, R. D. (1971). *Condor* **73**, 470–472.

LeFebvre, E. A. (1964). *Auk* **81**, 403–416.

Leveille, G. A., O'Hea, E. K., and Chakrobarty, K. (1968). *Proc. Soc. Exp. Biol. Med.* **128**, 398–401.

Lifson, N., Gordon, G. B., and McClintock, R. (1955). *J. Appl. Physiol.* **7**, 704–710.

Lighthill, J. (1977). *In* "Scale Effects in Animal Locomotion" (T. J. Pedley, ed.), pp. 365–404. Academic Press, New York.

Linderoth, L. S. (1975). *J. Appl. Physiol.* **39**, 501–502.

Lissaman, P. B. S., and Shollenberger, C. A. (1970). *Science* **168**, 1003–1005.

Lofts, B., Marshall, A. J., and Wolfson, A. (1963). *Ibis* **105**, 99–105.

Lord, R. D., Jr., Bellrose, F. C., and Cochran, W. W. (1962). *Science* **137**, 39–40.

Ludlow, A. (1966). *Ibis* **108**, 129–132.

McGreal, R. D., and Farner, D. S. (1956). *Northwest Sci.* **30**, 12–23.

McNeil, R. (1970). *Oiseau Rev. Fr. Ornithol.* **40**, 185–302.

McNeil, R. (1971). *Condor* **73**, 472–475.

McNeil, R., and de Itriago, M. C. (1968). *Can. J. Zool.* **46**, 123–128.

Mather, F. J., III (1962). *J. Cons. Cons. Int. Explor. Mer* **27**, 325–327.

May, M. L. (1979). *J. Exp. Biol.* **83**, 79–94.

Mayer, R. J., and Candy, D. J. (1969). *J. Insect Physiol.* **15**, 611–620.

Meinertzhagen, R. (1951). *Proc. Zool. Soc. London* **121**, 81–132.

Meinertzhagen, R. (1955). *Ibis* **97**, 81–117.

Merkel, F. W. (1958). *Z. Vergl. Physiol.* **41**, 154–178.

Moreau, R. E. (1961). *Ibis* **103a**, 373–427, 580–618.

Moreau, R. E., and Dolp, R. M. (1970). *Ibis* **112**, 209–228.

Morton, M. L. (1967). *Condor* **69**, 491–512.

Morton, M. L., and Liebmann, H. A. (1974). *Comp. Biochem. Physiol.* **48**, 329–335.

Mueller, H. C., and Berger, D. D. (1966). *Bird-Banding* **37**, 83–112.

Mullen, R. K. (1973). *In* "Ecological Energetics of Homeotherms" (J. A. Gessaman, ed.), Monogr. Ser. 20, pp. 32–43. Utah State Univ. Press, Logan.

Murray, B. G., Jr. (1965). *Wilson Bull.* **77**, 122–133.

Nachtigall, W. (1976). *Symp. R. Entomol. Soc. London* **7**, 31–47.

Nakamura, T. (1963). *Mem. Fac. Lib. Arts Educ., Yamanashi Univ.* **14**, 141–149.

Nakamura, T. (1964). *Mem. Fac. Lib. Arts Educ., Yamanashi Univ.* **15**, 148–153.

Neveille, A. C., and Weis-Fogh, T. (1963). *J. Exp. Biol.* **40**, 111–121.

Newton, I. (1968). *Condor* **70**, 323–332.

Newton, I. (1969). *Physiol. Zool.* **42**, 96–107.

Nielsen, E. T. (1961). *Biol. Medd. Dan. Vid. Selsk.* **23**, 1–81.

Nisbet, I. C. T. (1963a). *Bird-Banding* **34**, 57–67.

Nisbet, I. C. T. (1963b). *Bird-Banding* **34**, 139–159.

Nisbet, I. C. T. (1967). *Bird-Banding* **38**, 306–308.

Nisbet, I. C. T. (1970). *Bird-Banding* **41**, 207–240.

Nisbet, I. C. T. (1971). *Bird-Banding* **42**, 134.

Nisbet, I. C. T., Drury, W. H., Jr., and Baird, J. (1963). *Bird-Banding* **34**, 107–138.

Odum, E. P. (1960a). *Proc. Int. Ornithol. Congr., 12th, 1958* pp. 563–576.

Odum, E. P. (1960b). *Am. J. Clin. Nutr.* 8, 621–629.
Odum, E. P. (1965). *In* "Handbook of Physiology" (A. E. Renold and G. F. Cahill, Jr., eds.), Sect. 5, pp. 37–43. Am. Physiol. Soc., Washington, D. C.
Odum, E. P., and Connell, C. E. (1956). *Science* 123, 892–894.
Odum, E. P., and Major, J. C. (1956). *Condor* 58, 222–228.
Odum, E. P., and Perkinson, J. D., Jr. (1951). *Physiol. Zool.* 24, 216–230.
Odum, E. P., Connell, C. E., and Stoddard, H. L. (1961). *Auk* 78, 515–527.
Odum, E. P., Rogers, D. T., and Hicks, D. L. (1964). *Science* 143, 1037–1039.
Odum, E. P., Marshall, S. G., and Marples, T. G. (1965). *Ecology* 46, 901–904.
Oehme, H. (1977). *In* "Scale Effects in Animal Locomotion" (T. J. Pedley, ed.), pp. 479–494. Academic Press, New York.
Oehme, H., Dathe, H. H., and Kitzler, U. (1977). *Fortschr. Zool.* 24, 257–273.
O'Farrell, M. J., and Bradley, W. G. (1977). *Comp. Biochem. Physiol. A* 58, 223–227.
O'Farrell, M. J., and Studier, E. H. (1970). *Comp. Biochem. Physiol.* 35, 697–703.
Olson, J. B. (1965). Ph.D. Dissertation, University of Illinois, Urbana.
Orr, R. T. (1970). "Animals in Migration." Macmillian, New York.
Osman, M. F. H., and Schmidt, G. H. (1961). *Biochem. Z.* 334, 441–450.
Page, G., and Middleton, A. L. A. (1972). *Bird-Banding* 43, 85–96.
Pagels, J. F. (1975). *Comp. Biochem. Physiol. A.* 50, 237–246.
Pagels, J. F., and Blem, C. R. (1973). *Comp. Biochem. Physiol. A* 45, 497–501.
Pearson, D. J. (1971). *Ibis* 113, 173–184.
Pearson, D. J., and Backhurst, G. C. (1976). *Ibis* 118, 78–105.
Pearson, D. J., Phillips, J. H., and Backhurst, G. C. (1970). *Ibis* 112, 199–208.
Pearson, O. P. (1950). *Condor* 52, 145–152.
Pearson, O. P. (1964). *Condor* 66, 182–185.
Pennycuick, C. J. (1968). *J. Exp. Biol.* 49, 527–555.
Pennycuick, C. J. (1969). *Ibis* 111, 525–556.
Pennycuick, C. J. (1972). "Animal Flight." Arnold, London.
Pennycuick, C. J. (1975). *In* "Avian Biology" (D. S. Farner and J. R. King, eds.), Vol. 5, pp. 1–75. Academic Press, New York.
Pennycuick, C. J. (1978). *Oikos* 30, 165–176.
Phelps, W. H. (1961). *Auk* 78, 93–94.
Phillips, A. M., Jr. (1969). *In* "Fish Physiology" (W. S. Hoar and D. J. Randall, eds.), Vol. 1, pp. 391–432. Academic Press, New York.
Pohl, H. (1971). *Ibis* 113, 185–193.
Prange, H. D. (1976). *J. Exp. Biol.* 64, 1–12.
Pringle, J. W. S. (1957). "Insect Flight." Cambridge Univ. Press, London and New York.
Pringle, J. W. S. (1965). *In* "The Physiology of Insecta" (M. Rockstein, ed.), Vol. 2, pp. 283–329. Academic Press, New York.
Rainey, R. C. (1958). *Q. J. Roy. Meteorol. Soc.* 84, 334–354.
Rainey, R. C. (1976). *In* "Insect Flight" (R. C. Rainey. ed.), pp. 75–112. Wiley, New York.
Rao, G. M. M. (1968). *Can. J. Zool.* 46, 781–786.
Rao, R. A. and Dutt, S. (1965). *Curr. Sci.* 34, 695.
Raveling, D. G. and LeFebvre, E. A. (1967). *Bird-Banding* 38, 97–113.
Rayner, J. M. V. (1977). *In* "Scale Effects in Animal Locomotion" (T. J. Pedley, ed.), pp. 437–443. Academic Press, New York.
Rayner, J. M. V. (1979). *J. Exp. Biol.* 80, 17–54.
Reeder, W. G., and Cowles, R. B. (1951). *J. Mammal.* 32, 389–403.
Richardson, W. J. (1972). *Am. Birds* 26, 10–17.
Richardson, W. J. (1974). *Ibis* 116, 172–193.
Richardson, W. J. (1976). *Ibis* 118, 309–332.

Rogers, D. T. (1965). *Bird-Banding* **36**, 115–116.

Rogers, D. T., and Odum, E. P. (1964). *Auk* **81**, 505–513.

Rogers, D. T., and Odum, E. P. (1966). *Wilson Bull.* **78**, 415–433.

Sacktor, B. (1965). In "The Physiology of Insecta" (M. Rockstein, ed.), pp. 483–580. Academic Press, New York.

Sacktor, B. (1970). *Adv. Insect Physiol.* **7**, 267–347.

Sacktor, B. (1974). In "The Physiology of Insecta" (M. Rocksein, ed.), 2nd ed., Vol. 4, pp. 272–353. Academic Press, New York.

Sacktor, B. (1975). In "Insect Biochemistry and Function" (D. J. Candy and B. A. Filby, eds.), pp. 1–58. Chapman & Hall, London.

Salomonsen, F. (1969). "Vogelzug." Bayerischer Landwirtsch., Munich.

Savage, S. B., Newman, B. G., and Wong, D. T.-M. (1979). *J. Exp. Biol.* **83**, 59–77.

Schaefer, G. W. (1968). *Ibis* **110**, 413–414.

Schaefer, G. W. (1976). *Symp. R. Entomol. Soc.* **7**, 157–197.

Scheid, P., and Piiper, J. (1970). *Respir. Physiol.* **11**, 308–314.

Schmidt-Nielsen, K. (1972a). *Science* **177**, 222–228.

Schmidt-Nielsen, K. (1972b). "How Animals Work." Cambridge Univ. Press, London and New York.

Schmidt-Nielsen, K., Kanwisher, J., Lasiewski, R. C., Cohn, J. E., and Bretz, W. L. (1969). *Condor* **71**, 341–352.

Schmidt-Nielsen, K., Hainsworth, F. R., and Murrish, D. E. (1970). *Respir. Physiol.* **9**, 263–276.

Searle, B. (1956). *Ibis* **98**, 307–311.

Seibert, H. C. (1949). *Auk* **66**, 128–153.

Shilov, I. A. (1973). "Heat Regulation in Birds." Moscow Univ. Press, Moscow (translated from Russian by Bureau Sport Fish. Wildlife).

Sinnhuber, R. O. (1969). In "Fish in Research" (O. W. Neuhaus and J. E. Halver, eds.), pp. 245–259. Academic Press, New York.

Smith, R. H., and Suthers, R. (1969). *Physiologist* **12**, 358.

Smith, V. H. (1966). *Ibis* **108**, 492–512.

Sotavalta, O. (1954). *Ann. Zool. Soc. Zool. Bot. Fenn. Vanamo* **16**, 1–22.

Steidinger, P. (1972). *Ornithol. Beob.* **69**, 20–39.

Stevenson, E. (1968). *J. Insect Physiol.* **14**, 179–198.

Struhsaker, T. (1961). *J. Mammal.* **42**, 152–159.

Studier, E. H., and Howell, D. J. (1969). *J. Mammal* **50**, 842–845.

Studier, E. H., and O'Farrell, M. J. (1972). *Comp. Biochem. Physiol. A* **41**, 567–595.

Sun, G. Y., and Brookes, V. J. (1968). *Comp. Biochem. Physiol.* **24**, 177–185.

Suthers, R. A., Thomas, S. P., and Suthers, B. J. (1972). *J. Exp. Biol.* **56**, 37–48.

Tashima, L. and Cahill, G. F. (1965). In "Handbook of Physiology" (A. E. Renold and G. F. Cahill, Jr., eds.), Sect. 5, pp. 55–58. Am. Physiol. Soc., Washington, D. C.

Taylor, L. R. (1958). *Proc. Linn. Soc. London* **169**, 67.

Teal, J. M. (1969). *Zoologica (N. Y.)* **54**, 17–23.

Thomas, D. G., and Dartnall, A. J. (1970). *Emu* **70**, 87.

Thomas, K. K. (1974). *J. Insect Physiol.* **20**, 845–858.

Thomas, S. P. (1975). *J. Exp. Biol.* **63**, 273–293.

Thomas, S. P., and Suthers, R. A. (1970). *Fed. Proc., Fed. Am. Soc. Exp. Biol.* **29**, 265.

Thomas, S. P., and Suthers, R. A. (1972). *J. Exp. Biol.* **57**, 317–335.

Thompson, S. H., and Bennett, R. B. (1971). *J. Insect Physiol.* **17**, 1555–1563.

Tietz, A. (1961). *J. Lipid Res.* **3**, 421–426.

Tietz, A. (1965). In "Handbook of Physiology" (A. E. Renold and G. F. Cahill, Jr., eds.), Sect. 5, pp. 45–54. Am. Physiol. Soc., Washington, D. C.

Tietz, A. (1967). *Eur. J. Biochem.* **2**, 236–242.
Tomlinson, J. T., and McKinnon, R. S. (1957). *Condor* **59**, 401.
Torre-Bueno, J. R. (1976). *J. Exp. Biol.* **65**, 471–482.
Torre-Bueno, J. R. (1978). *J. Exp. Biol.* **75**, 231–236.
Torre-Bueno, J. R., and Larochelle, J. (1978). *J. Exp. Biol.* **75**, 223–229.
Tucker, V. A. (1966). *Science* **154**, 150–151.
Tucker, V. A. (1968a). *J. Exp. Biol.* **48**, 55–66.
Tucker, V. A. (1968b). *J. Exp. Biol.* **48**, 67–87.
Tucker, V. A. (1969). *Sci. Am.* **220** (5), 70–78.
Tucker, V. A. (1970). *Comp. Biochem. Physiol.* **34**, 841–846.
Tucker, V. A. (1971). *Am. Zool.* **11**, 115–124.
Tucker, V. A. (1972a). *Am. J. Physiol.* **222**, 237–245.
Tucker, V. A. (1972b). *Respir. Physiol.* **14**, 75–82.
Tucker, V. A. (1973a). *J. Exp. Biol.* **58**, 689–709.
Tucker, V. A. (1973b). *In* "Comparative Physiology" (L. Bolis, K. Schmidt-Nielsen, and S. H. P. Maddrell, eds.), pp. 63–75. North-Holland Publ., Co., Amsterdam.
Tucker, V. A. (1974). *Nuttall Ornithol. Club* **15**, 298–328.
Tucker, V. A. (1975a). *Am. Sci.* **63**, 413–419.
Tucker, V. A. (1975b). *Symp. Zool. Soc. London* **35**, 49–63.
Tucker, V. A. (1975c). *In* "Swimming and Flying in Nature" (T. Y. Wu, C. J. Brokaw, and C. Brennen, eds.), pp. 845–867. Plenum, New York.
Tucker, V. A. (1977). *In* "Scale Effects in Animal Locomotion" (T. J. Pedley, ed.), pp. 497–509. Academic Press, New York.
Tucker, V. A., and Schmidt-Koenig, K. (1971). *Auk* **88**, 97–107.
Turĉek, F. J. (1960). *Proc. Int. Ornithol. Congr., 12th, 1958* Vol. 2, pp. 724–729.
Twente, J. W., Jr. (1956). *J. Mammal.* **37**, 42–47.
Urquhart, F. A. (1960). "The Monarch Butterfly." Univ. of Toronto Press, Toronto.
Utter, J. M., and LeFebvre, E. A. (1970). *Comp. Biochem. Physiol.* **35**, 713–719.
Vague, J., and Fenasse, R. (1965). *In* "Handbook of Physiology" (A. E. Renold and G. F. Cahill, Jr. eds.), Sect. 5, pp. 25–36. Am. Physiol. Soc., Washington, D. C.
Vallythan, N. V., and George, J. C. (1964). *Pavo* **2**, 55–60.
Van Handel, E., and Nayar, J. K. (1972). *Insect Biochem.* **2**, 203–208.
Villa, R. B., and Cockrum, E. L. (1962). *J. Mammal.* **43**, 43–64.
Walker, A. T. (1964). *Physiol. Zool.* **37**, 57–64.
Walker, P. R., and Bailey, E. (1970). *J. Insect Physiol.* **16**, 499–509.
Walker, P. R., Hill, L., and Bailey, E. (1970). *J. Insect Physiol.* **16**, 1001–1016.
Wallgren, H. (1954). *Acta Zool. Fenn.* **84**, 1–110.
Ward, P. (1963). *Ibis* **105**, 109–111.
Ward, P. (1964a). *Ibis* **106**, 256–257.
Ward, P. (1964b). *Ibis* **106**, 370–375.
Ward, P. (1969). *Physiol. Zool.* **42**, 85–95.
Wardle, C. S. (1977). *In* "Scale Effects in Animal Locomotion" (T. J. Pedley, ed.), pp. 299–313. Academic Press, New York.
Webb, P. W. (1971a). *J. Exp. Biol.* **55**, 489–520.
Webb, P. W. (1971b). *J. Exp. Biol.* **55**, 521–540.
Webb, P. W. (1975). *Bull. Fish. Res. Board Can.* **190**, 1–158.
Weber, N. S., and Findley, J. S. (1970). *J. Mammal.* **51**. 160–162.
Weise, C. M. (1956). *Ecology* **37**, 275–287.
Weise, C. M. (1963). *Proc. Int. Ornithol. Congr., 13th, 1962* pp. 983–993.
Weis-Fogh, T. (1952a). *Trans. Int. Congr. Entomol., 9th, 195* 341–347.
Weis-Fogh, T. (1952b). *Philos. Trans. R. Soc. London, Ser. B* **237**, 1–36.

Weis-Fogh, T. (1956). *Philos. Trans. R. Soc. London Ser. B* **238**, 459–510.

Weis-Fogh, T. (1961). *In* "The Cell and the Organism " (J. A. Ramsey and V. B. Wigglesworth, eds), pp. 283–300. Cambridge Univ. Press, London and New York.

Weis-Fogh, T. (1964a). *J. Exp. Biol.* **41**, 229–256.

Weis-Fogh, T. (1964b). *J. Exp. Biol.* **41**, 257–271.

Weis-Fogh, T. (1967a). *J. Exp. Biol.* **47**, 561–587.

Weis-Fogh, T. (1967b). *In* "Insect and Physiology" (J. W. Beament and J. E. Treherne, eds.), pp. 143–159. Oliver & Boyd, Edinburgh.

Weis-Fogh, T. (1972). *J. Exp. Biol.* **56**, 79–104.

Weis-Fogh, T. (1976). *Symp. R. Entomol. Soc. London* **7**, 48–74.

Weis-Fogh, T. (1977). *In* "Scale Effects in Animal Locomotion" (T. J. Pedley, ed.), pp. 405–420. Academic Press, New York.

Weis-Fogh, T., and Jensen, M. (1956). *Philos. Trans. R. Soc. London, Ser. B* **239**, 415–458.

West, G. C. (1960). *Auk* **77**, 306–329.

West, G. C., and Meng. M. S. (1968). *Comp. Biochem. Physiol.* **25**, 535–540.

Whittow, G. C. (1976a). *In* "Avian Physiology" (P. D. Sturkie, ed.), 3rd ed., pp. 146–173. Springer-Verlag, Berlin and New York.

Whittow, G. C. (1976b). *In* "Avian Physiology" (P. D. Sturkie, ed.), 3rd ed., pp. 174–184. Springer-Verlag, Berlin and New York.

Williams, C. B. (1958). "Insect Migration." Macmillan, New York.

Williams, T. C., Berkeley, P., and Harris, V. (1977). *Bird-Banding* **48**, 1–10.

Wyatt, G. R. (1967). *Adv. Insect Physiol.* **4**, 287–360.

Yapp, W. B. (1956). *Wilson Bull.* **68**, 312–319.

Yapp, W. B. (1962). *Ibis* **104**, 86–89.

Yarbrough, C. G. (1970). *Auk* **87**, 100–110.

Yarbrough, C. G., and Johnston, D. W. (1965). *Wilson Bull.* **77**, 175–191.

Zebe, E. (1953). *Naturwissenschaften* **40**, 298.

Zebe, E. (1954). *Z. Vergl. Physiol.* **36**, 290–317.

Zebe, E., and McShan, W. H. (1959). *Biochim. Biophys. Acta* **31**, 513–518.

Zimmerman, J. L. (1965a). *Auk* **82**, 278–279.

Zimmerman, J. L. (1965b). *Wilson Bull.* **77**, 55–70.

4 *Physiology of Migration*

ALBERT H. MEIER AND ALBERT J. FIVIZZANI

I. General Introduction

Migration involves elaborate organization within an individual animal and between an animal and its environment so that all aspects are properly integrated in time and place. We have chosen among a plethora of opportunities to provide an overall view of migration physiology and to concentrate on several areas that are significant because of their central roles,

Animal Migration, Orientation, and Navigation

historic interest, and probable importance for further investigation. Of the animal taxa studied, most attention has been accorded to fish and birds; and we have reviewed these investigations in more detail. We have emphasized research dealing with accumulation and mobilization of energy stores, timing and synchronization of migration with other events, and with special problems associated with migration (e.g., osmoregulatory adjustments of fish and crustaceans moving to waters of different salinities).

Special concern has been given to neuroendocrine mechanisms that direct preparation for and adjustment to migration, and to circadian and circannual systems that provide the bases for temporal organization and even for the neuroendocrine control of physiological and behavioral conditions. Temperature and photoperiodic stimulations of migratory readiness as well as environmental cues for orientation are considered as they involve circadian systems. Some aspects of migration physiology are treated superficially or omitted because they are treated in depth elsewhere in this volume (see Blem, Chapter 3).

II. Bird Migration

A. INTRODUCTION

Physiological studies of avian migration stem from a background of intense interest in avian behavior and natural history. Ornithologists became physiologists in order to better understand the whole organism. This origin for avian physiology produced an orderly approach in which individual mechanisms were studied in order to relate them to the total biology of a species. The success of this approach is evidenced in part by the feedback this knowledge has had in generating new insight in avian natural history.

Several areas of special interest with respect to avian migration have drawn much attention. These interests were often stated in the form of a question with a note of wonderment or even incredulity: How is it possible for a small bird to fly across the Gulf of Mexico from the Louisiana coast to the Yucatan Peninsula? How do birds recognize the seasons so that their migrations occur at about the same time year after year? How is it possible for a bird to fly vast distances to ancestral wintering areas that it has never seen and return to specific breeding territories? To be sure none of these questions has been fully resolved but the quest for answers has been joined with great enthusiasm.

Physiological investigations of avian migration have been helped immeasurably by findings that indices of migratory behavior could be studied in birds under controlled environmental conditions. Caged birds disposed

to migrate during the migratory seasons display this urge by restless fluttering (Palmgren, 1949). Provided that the birds have access to appropriate
environmental cues the fluttering is directed during the appropriate seasons
to that side of the cage which faces toward the breeding or wintering territories (Kramer, 1951; Emlen and Emlen, 1966). The high metabolic rate of
birds also makes it feasible to monitor changes in body weight and
amounts of energy stores in experiments of short duration.

B. METABOLISM

Fat is the principal energy store in avian migrants. Whereas nonfat body
components vary little (McGreal and Farner, 1956; Connel *et al.*, 1960),
premigratory fat deposits, especially in the body cavity and subcutaneous
tissue, may range from as little as 3–5% during the intermigratory periods
to as much as 20–50% of the nonfat body weight during migration.
Premigratory fattening has been described in more than a hundred
passerine species (Wolfson, 1945, 1954; Farner, 1950, 1960; Wallgren,
1954; Odum, 1960a; Blyumental, 1973; King, 1972; Blem, 1976). In typical
migrants with body weights of 20–30 gm, fat deposition begins shortly
before the onset of migration and proceeds in captivity at the rate of 0.5–2
gm/day (Wallgren, 1954; Helms and Drury, 1960; King, 1961). Fat accumulation in wild birds occurs at about half the rate of that observed in
captive birds (King and Farner, 1956; Odum and Major, 1956; King, 1961).
Because more fat can be metabolized during a single flight than is produced
in a day, pauses of several days for replenishment of fat stores is common
among migrants (Wolfson, 1954; Szulc-Olech, 1965).

Increase in fat reserves in preparation for migration is accompanied by
decrease in carbohydrate (glycogen) reserves (Farner *et al.*, 1961).
Glycogen can supply energy for white muscle which responds well during
brief periods of vigorous physical activity but fatigues rapidly. Higher fat
content in migrants is accomplished by a greater proportion of red muscle
which is capable of aerobic oxidation of fats and is practically in-
defatiguable in the presence of sufficient fat stores (George and Berger,
1966; Baggott, 1975). Not only does a given weight of fat supply more than
twice the amount of ATP derivable from protein and carbohydrate but fat
also occupies less space and requires less water because it is not hydrated
(King and Farner, 1965; Hicks, 1967; Helms *et al.*, 1967; Johnston, 1973).
This savings in weight and space has obvious advantages for an airborne
migrant.

Cytological investigations of fat depots in indigo buntings, *Passerina
cyanea*, indicate that premigratory fattening is accomplished by an increase
in lipid content per adipocyte rather than by an increase in adipocytes

(Johnston, 1973). Hicks (1967) came to a similar conclusion on the basis of studies with thrushes. Premigratory fat is largely composed of 16- and 18-carbon fatty acids (McGreal and Farner, 1956; Walker, 1964; Bower and Helms, 1968; Nakamura, 1969). Variations in the composition of fatty acids during the year result primarily from changes in diet (Nakamura, 1962; West and Peyton, 1972; Morton and Liebman, 1974; Thomas and George, 1975) and probably have little influence on migratory metabolism. The fatty acid composition in the migratory Hindian sparrow, *Passer domesticus bactrianus*, was virtually identical to that of the closely related sedentary house sparrow, *Passer domesticus domesticus* (Dolñik and Gavrilov, 1975).

Lipogenesis is largely restricted to the liver; very little occurs within adipose tissue of birds (Goodridge and Ball, 1966, 1967; O'Hea and Leveille, 1969; Wheeland *et al.*, 1976). Lipogenesis is characterized by increased activity of several key enzymes, malic, citrate cleavage, and malate dehydrogenase, which can supply the NADPH for reduction of acetyl-CoA to fatty acids (O'Hea and Leveille, 1968; Wheeland *et al.*, 1976). However, it was concluded on the basis of lipogenic studies in the white-throated sparrow, *Zonotrichia albicollis*, that the principal difference between fat birds during the vernal migratory season and lean birds during the summer is a greater capacity for transport of triglycerides in fat birds (Meier, 1977). After formation in the liver, lipids are transported in the blood to adipose tissue as low density lipoproteins (O'Hea and Leveille, 1969).

Insulin is a potent lipogenic hormone in mammals but has little or no apparent direct regulatory role in premigratory fattening. Cytological studies of the pancreatic islets of many small birds indicate that β-cell activity does not correlate directly with time of fattening (Epple, 1963). Insulin was also ineffective in promoting incorporation of [^{14}C]glucose and [^{14}C]acetate into hepatic lipid *in vitro* (Goodridge, 1964).

Interest in lipogenic activities of prolactin originated in studies of pigeons, *Columba livia* (Schooley *et al.*, 1941; Bates *et al.*, 1962). Prolactin injections prevented weight and body fat losses in hypophysectomized birds. Growth hormone was also partially effective and thyroxine and corticosteroid hormone augmented the prolactin and growth hormone effects. Prolactin did not stimulate lipogenesis when added to hepatic tissue cultures of pigeons (Goodridge, 1964); however, prolactin injections *in vivo* for at least 3 days did promote incorporation of [^{14}C]glucose, [^{14}C]pyruvate, and [^{14}C]acetate into lipids of liver slices tested *in vitro* (Goodridge and Ball, 1967). Lipogenesis was not significantly increased by prolactin when tested 1 day following the hormone injection. Prolactin was similarly ineffective when given at the beginning of a 4-hr incubation period in live white-throated sparrows, *Zonotrichia albicollis* (Meier,

1977). In Japanese quail, *Coturnix c. japonica*, an inhibitor of prolactin synthesis (CB 154) reduced hepatic lipogenesis when given for 2 weeks prior to testing (Wheeland *et al.*, 1976). These findings indicate that the lipogenic activity of prolactin is not an immediate direct effect on the lipogenic pathway. The lag in lipogenic response suggests that induction of enzyme or other protein synthesis is involved. Some possibilities are hepatic enzymes involved in lipogenesis, membrane receptors for other hormones or for the plasma lipoprotein carriers of triglycerides, and plasma lipoprotein carriers. It is also conceivable that prolactin's lipogenic effect is mediated by the release of another hormone.

Prolactin might also increase lipogenesis indirectly by increasing food consumption. Hyperphagia accompanies fattening in migrants (King and Farner, 1956, 1965; Odum, 1960b), and fat deposition, though not migratory restlessness, can be prevented by limiting food availability (King and Farner, 1963; Lofts *et al.*, 1963). Fasting reduces activities of hepatic enzymes involved in lipogenesis and feeding restores their activities (Goodridge, 1964). Prolactin injections also stimulate hyperphagia that accompanies fattening in *Zonotrichia* species (Meier and Farner, 1964; Meier, 1969). Although increased food consumption may well be expected to contribute toward fattening by supplying necessary substrates, the limiting effects of fasting on fat deposition may result from reduced production of proteins (enzymes, receptors, carriers) that are parts of the lipogenic pathway. Inasmuch as prolactin can stimulate fattening in starved as well as fed fish (Meier, 1969), it seems probable that prolactin may increase both food consumption and hepatic lipogenesis by separate means.

Temporal variations in lipogenesis and in response to prolactin should be considered in the design of experiments. Lipogenesis is largely restricted to specific times of day (Meier, 1977) and prolactin can stimulate fattening at certain times of day only (Meier, 1976).

Many hormones have fat mobilizing activities and may be expected to have roles in migration. The lipolytic activities are mediated by cyclic AMP in adipose tissue and involve activation of a hormone sensitive lipase (Boyd *et al.*, 1975; Goodridge, 1975; Meier and Burns, 1976). Free fatty acid (FFA) is carried in the blood in combination with albumen and made available for β oxidation in red muscle. Three important lipolytic hormones in mammals, ACTH, epinephrine, and norepinephrine, appear to have little effect in birds (Carlson *et al.*, 1964; Goodridge and Ball, 1965; Langslow and Hales, 1969).

The pancreatic hormone, glucagon, has marked lipolytic activities both *in vivo* and *in vitro* (Goodridge and Ball, 1965; Grande, 1968; Langslow and Hales, 1969; John *et al.*, 1973; Lefebvre, 1975). Interestingly, insulin enhanced glucagon's effect *in vitro* (Langslow and Hales, 1969), an activity that is the reverse of that in mammals. Glucagon maintains high levels of blood glucose and FFA in birds (Farina et al., 1975).

Growth hormone and corticosterone appear to be necessary to permit the expression of glucagon activities (Assenmacher, 1973). Growth hormone injections stimulate an increase in plasma FFA (John et al., 1973) and short-term (10 min) exercise increases the levels of plasma growth hormone in pigeons, Columba livia (McKeown et al., 1974). However, prolonged exercise more characteristic of migration causes a decrease in plasma growth hormone concentration, perhaps by a negative feedback of plasma FFA (McKeown et al., 1974).

Vasotocin and oxytocin have also shown to stimulate increases in blood concentration of FFA (Kook et al., 1964; Farner et al., 1967; John et al., 1973). Because vasotocin is thought to be an antidiuretic agent (Farner et al., 1967; Follett and Farner, 1966), stimulation of its production during flight by water loss might serve to supply both energy and metabolic water. Oxidation of fat can provide sufficient metabolic water for migrations across the Gulf of Mexico (Odum et al., 1964; Johnston, 1968; Child, 1969), but dehydration does occur in some migrants that cross the Mediterranean (Fogden, 1972). Loss of water usually exceeds production of metabolic water only at higher temperatures (Berger and Hart, 1974).

Thyroid hormones have permissive effects for both lipogenesis and fat mobilization and increase metabolic rate in mammals. For these reasons and others, considerable attention was given the thyroid by early investigators of avian migration. The results have been inconsistent. Histological evidence of increased thyroid activity correlate with migration in some species (Putzig, 1937, 1938; George and Naik, 1964) and not in others (Putzig, 1937, 1938; Wilson and Farner, 1960, Robiller, 1975). Similarly thyroxine injections have variable effects on fat stores and locomotor activity (Wagner, 1930; Merkel, 1938, 1960; Putzig, 1938; Schildmacher and Rautenberg, 1952). The evidence seems to indicate that the thyroid is important in the metabolic support of migration but that it is only one of a complex of control mechanisms which interact synergistically.

Induction of autumnal fattening in migrants of the temperate zone differ from that of the vernal phase. Vernal premigratory fattening is induced by an increasing daylength whereas autumnal fattening is an indirect consequence of the vernal increase in daylength. Autumnal fattening may take longer to develop and be less than vernal fattening. However, there is no evidence that the energetics of lipogenesis and fat mobilization in autumnal migration differs significantly from that in vernal migration (King et al., 1963).

It has been suggested that fat mobilization may not only supply FFA for the support of muscular activity but may also provide a means for stimulating migratory restlessness (Dolńik, 1967, 1970; Yablonkevich, 1975). This suggestion is attractive in that it could explain why migratory activity is generally more intense in birds with heavier fat stores (Dolńik and Blyumental, 1964; Dolńik, 1968). The sensitivity of adipose tissue to fat mobilizing hormones in chickens, *Gallus domesticus*, increases with greater fat content and releases more FFA into the blood (Freeman and Manning, 1977).

C. Timing

Perhaps in no other area of biology has the question "when?" been posed with more persistence than in avian physiology. Because migration involves precise internal organization as well as external integration of the total animal in its social and physical environment, accurate timing is literally a life and death proposition. Migration must relate properly with reproduction and molt, metabolism must support behavior, behavior must provide for metabolism, and the total animal must find its way to the ancestral breeding and wintering areas at the appropriate times.

In order to migrate at seasonally appropriate times, birds respond to environmental cues that change predictably during the year. In tropical regions both rainfall (Baker, 1938; Serventy and Marshall, 1959) and daylength (Wolfson, 1959; Thapliyal et al., 1963) appear to be important environmental timers. In temperate regions daylength is the most important environmental cue. Rowan (1926) was first to demonstrate in the dark-eyed junco, *Junco hyemalis*, that migratory as well as reproductive indices could be induced in midwinter by artificially lengthening the daily photoperiod. Similar findings have been reported since for many other migrants; they indicate that vernal migration and subsequent reproductive behavior are timed by increasing daylength in spring (Farner, 1960; Follett and Farner, 1966).

After migrants reach the breeding territory, migration ceases and migratory indices are suppressed. It has been suggested that each species has a specific capacity and time-course for migratory activity and when that potential has been achieved migration ends (von Lucanus, 1923; Gwinner, 1968, 1977; Berthold, 1977). During the breeding season gonadal hormones might suppress migratory behavior in that high doses of gonadal hormones can inhibit premigratory fattening and vernal migratory restlessness (Wagner, 1956, 1961; Wagner and Thomas, 1957). However gonadal hormones do not prevent photoperiodic stimulation of premigratory fattening and migratory restlessness (Yokoyama, 1977). Following breeding many species become reproductively refactory to long daylengths and the gonads regress. Reproductive photorefractoriness is accompanied by metabolic refractoriness and usually by a postnuptial molt. This period also marks the time when migratory orientation changes from the vernal to the autumnal direction. The increasing daylength in spring is thought to initiate a series of endogenous events that eventually lead to the autumnal migratory condition (King, 1963, 1968; Sansum and King, 1976). Although not an obligatory stimulus, decreasing daylength in the autumn may accelerate autumnal migration (Gifford and Odum, 1965). Following fall migration, some birds once more become metabolically refractory and do not fatten in response to artificially supplied long daily photoperiods until early winter, several weeks after reestablishment of reproductive photosensitivity (Shank, 1959; King et al., 1960). Reestablishment of reproductive and metabolic sensitivity can be induced prematurely by imposing an interval (6 weeks) of short daily photoperiods (Wolfson, 1954).

A need for endogenous timing in addition to an ability to recognize and respond to seasonal changes in environmental cues has long been recognized (Bissonnette, 1937; Marhsall, 1961). Some migrants that winter in equatorial areas are insensitive to the slight changes in daylength. Other potential environmental cues fluctuate irregularly (Van Oordt, 1959; Marshall, 1961). In passerine species that migrate between temperate regions, the summer intermigratory and fall migratory conditions follow sequentially in birds kept on long daily photoperiods following establishment of vernal conditions (King, 1968; Meier and Fivizzani, 1975).

An endogenous timing mechanism for seasonal events is more clearly defined by experiments in which birds were maintained on constant daily photoperiods for a year or longer. Appropriate seasonal changes in migratory and reproductive indices have been observed in several species maintained on constant daylengths longer as well as shorter than 12 hr (Merkel, 1963; Zimmerman, 1966; Gwinner, 1967, 1968; King, 1968; Berthold et al., 1971; for reviews, see Farner, 1970; Gwinner, 1972, 1975, 1977; Sansum and King, 1976). Circannual periods range from 7 to 15 months. Fat deposition and migratory restlessness coincide in most instances and alternate with molt and gonadal development as expected. Occurrence of this rhythmicity for as long as three cycles (Berthold et al., 1972) justifies the designation, endogenous circannual rhythm. For photoperiodic species it seems clear that the annual cycle of daylength is the principal environmental entrainer of circannual rhythms. The increasing daylength in spring is probably the most significant exogenous cue for many birds that overwinter in temperature areas. Timing of migration is more precise in spring than in autumn (Preston, 1966). A circadian basis for the circannual mechanism is discussed elsewhere (Meier, 1976; see Section II,D).

Although it may last for several weeks or even several months, avian migration is essentially a daily process. Among nocturnal migrants, nocturnal locomotor activity is initiated during the migratory period and is the index of migratory restlessness. Feeding activity is retained during the day and is necessary, of course, to support migration (Morton, 1967). In diurnal migrants as well, migratory activity occurs at a particular interval (early during the day, usually) and feeding activity occurs at another time. Only in species that feed during flight (swifts, swallows) are migratory and feeding activities related temporally. In birds, as in other animals, daily periods of locomotor activity are expressions of circadian organization. The daily rhythms of nocturnal migratory activity as well as diurnal feeding activity persist (free-run) in white-throated sparrows (*Zonotrichia albicollis*) under conditions of continuous dim light (DD) (Wagner, 1930; McMillan *et al.*, 1970). Nocturnal migratory restlessness, but not feeding activity, dampens out in white-crowned sparrows, *Zonotrichia leucophrys gambelii*, maintained in LL (Morton, 1967). Entrainment of the locomotor activity rhythms is by the daily light–dark cycle.

Several endocrine systems have been implicated in the maintenance and control of locomotor activity rhythms. In the migratory *Zonotrichia albicollis* (McMillan, 1972) as well as the sedentary *Passer domesticus* (Menaker, 1968; Gaston and Menaker, 1968), pinealectomy removes the rhythm in birds that normally retain a free-running rhythms of locomotor activity in DD. Inasmuch as pinealectomy has no apparent influence on the locomotor activity of birds retained on normal photoperiodic schedules, a satisfactory explanation concerning the physiological or behavioral significance of this observation has not been offered.

The thyroid also influences circadian rhythms. Thyroxine shortens the period of the free-running rhythm of locomotor activity in *Passer domesticus* kept in DD and thiouracil (an inhibitor of thyroxine production) lengthens the period (unpublished data). Free-running rhythms of metabolic activity that dampen out after several days in pigeons, *Colomba livia*, kept in LL are retained in birds treated with thyroid hormone (John *et al.*, 1972). It was suggested that thyroid hormones permit the expression of some circadian rhythms in vertebrates (Meier, 1975).

Corticosterone and prolactin have been implicated in the induction of premigratory fattening and migratory restlessness in the white-throated sparrow, *Zonotrichia albicollis*. Because these hormones organize many other behavioral and metabolic events the discussion of their effects will be dealt with more appropriately in Section II,D.

Because migratory activity occurs during specific daily intervals, it follows that metabolism should also be apportioned so that it relates appropriately. Measurements of body weight and of body and hepatic lipid content at various times of day support a predictable conclusion: fat deposition occurs during and perhaps for several hours after feeding activity (Oakeson, 1953; Fischer and Bartlett, 1957; Farner et al., 1961; Dolńik, 1966). In a nocturnal migrant, lipogenesis peaks during the afternoon and first few hours of darkness (Meier, 1977). Plasma level of free fatty acid, an index of lipolytic activity, peaks during the period of migratory activity (Yablonkevich, 1975). The roles of circadian hormone rhythms in regulating metabolic rhythms and conditions are considered in Section II,D.

A role for circadian rhythms in avian orientation was first demonstrated in a diurnal migrant, the European starling, Sturnus vulgaris (Kramer, 1953). During its migratory period this bird exhibited restless fluttering (migratory restlessness) on that side of the cage facing the appropriate destination. Because of disorientation when the sun was not visible and as a result of experiments using mirrors to change the apparent position of the sun, it was concluded that the birds were able to orient by using as a cue the sun's position. A circadian mechanism was proposed to reconcile constant orientation with a progressively changing cue that moves approximately 180° across the horizon during the course of a day The circadian mechanism in sun-compass orientation is entrained by the daily photoperiod. Delay or advance of dawn influenced orientation in predictable ways (Kramer, 1953; Matthews, 1955).

Seasonal changes in orientation are the result of an endogenous circannual mechanism. Rowan (1931) provided the first indication that the seasonal change in orientation in a diurnal migrant, the common crow, Corvus brachyrhynchos, was a consequence of change in interpretation of the environmental cues rather than a change in the cues themselves. Rowan produced the vernal migratory conditions in birds by photoperiodic adjustment and found that some of the birds which he released in the fall were recovered north of the release point near Edmonton, Canada. Similar conclusions were reached by Emlen (1969) on the basis of his investigations of the indigo bunting, (Passerina cyanea). This nocturnal migrant was shown to exhibit northward orientation (toward the breeding grounds) in the autumn when placed in the vernal migratory condition by photoperiodic manipulation and southward orientation in spring when placed in the autumnal migratory condition. Vernal orientation in the autumn by birds in the vernal migratory condition has also been demonstrated in the white-throated sparrow, Zonotrichia albicollis (Miller and Weise, 1978).

Given the prodigeous feats of some migratory species there has been a tendency to search for unique mechanisms among migrants not present in sedentary birds. This approach has largely succeeded in promoting a counterconclusion that sedentary and migratory birds differ little with respect to basic physiological mechanisms. In many instances it appears that the capacities for both behaviors are present within a species and that one life style may be blocked and the other released by either genetic or environmental factors. An illustration of a simple genetic switch is found in a study of song sparrows, *Melospiza melodia*, wherein both sedentary and migratory behaviors were demonstrated among siblings of a single nesting (Nice, 1937). Even among the predominantly sendentary house sparrow *Passer domesticus domesticus*, there is a migratory race, *P. d. bactrianus*. Although the sedentary house sparrow is poorly equipped for the storage of fat it nevertheless exhibits directed movements that seem more appropriate for migrants. Banded individuals released in the autumn had a tendency to disperse southward whereas those released in the spring dispersed northward (Preiser, 1957). Efforts to demonstrate orientation in caged house sparrows, however, were unsuccessful (Dolñik and Gavrilov, 1975).

Migratory restlessness appears to be an exaggeration of one of two discrete periods of daily activity. The presence of two daily activity peaks has been described as a general characteristic of many vertebrates including birds (Aschoff, 1967). These activities appear to be the expressions of two separate circadian oscillators which may free run in changing relations with one another (McMillan *et al.*, 1970; Gwinner, 1974; Hoffmann, 1971). Although speculative, seasonal changes in migratory orientation may be adaptations of the orientations characteristic for each of the two daily activity periods. In roosting birds, or in any animal that returns each day to a specific area, there are two basic orientations: one away from the roosting area and the other toward it. Perhaps one of these daily activities and the accompanying orientations are exaggerated to form the migratory restlessness of spring and the other daily activity becomes the autumnal migratory behavior.

D. REGULATORY MODEL FOR AN AVIAN MIGRANT

It can be advantageous to focus a discussion of migration on a single species in order to present a model for comparison and further research with other species. The white-throated sparrow, *Zonotrichia albicollis*, is well suited for this purpose because its life history and annual cycle have been extensively investigated (Odum and Perkinson, 1951; Eyster, 1954; Weise, 1956; Wolfson, 1958; Helms, 1968; Meier *et al.*, 1969). *Zonotrichia albicollis* is a nocturnal migrant that winters largely in the southeastern United States and breeds in the northeastern United States and much of Canada. Northward migration in spring is stimulated by increasing daylength; premigratory fattening, nocturnal migratory restlessness, as well as reproductive development, can be induced during the winter by artificially lengthening the daily photoperiod. Migration and the accompanying heavy fat stores are terminated when the birds reach the breeding grounds although migratory restlessness may be demonstrated in caged birds until the summer molt. Following breeding there is a complete postnuptial molt and the reproductive system regresses. Body fat stores are extremely low in both caged and feral birds, and nocturnal activity is absent in caged birds. Because these regressive changes occur during midsummer when daylengths are equivalent to those that are stimulatory in the spring, it has been concluded that *Z. albicollis* undergoes seasonal changes in photosensitivity and photorefractoriness.

Reestablishment of metabolic sensitivity to long daylengths occurs in December when daylengths are short, several weeks after reestablishment of reproductive photosensitivity (Shank, 1959). Metabolic and reproductive photosensitivity (vernal reproductive and migratory responsiveness) can be induced prematurely in *Z. albicollis* by artificially providing short daylengths for 6 weeks (Wolfson, 1954). Castration experiments performed in early winter demonstrate that some reproductive recrudescence is necessary for the fattening and migratory responses to increasing daylength of spring (Weise, 1967; see also Gwinner *et al.*, 1971; Stetson and Erickson, 1972). Injections of testosterone in castrated *Zonotrichia* sp. allow for the vernal responses (Stetson and Erickson, 1972). However, castration carried out later in winter or early spring does not prevent photoperiodic stimulation of premigratory fattening or migratory restlessness (Miller, 1960; see also Morton and Mewaldt, 1962). Mammalian studies indicate that gonadal hormones may stimulate both the production of prolactin and the responsiveness of target tissues. Both effects could be expected to change metabolic responsiveness to long daylengths (Meier, 1973). Apparently the effect of gonadal hormone persists for an extended period.

A circadian basis for reproductive photoperiodism is well established in birds including *Zonotrichia* species (review, Meier and MacGregor, 1972; Meier, 1976; Farner, 1976). Circadian systems in migratory responses to changes in daylength have received less study and much of its restricted to *Z. albicollis*. A daily rhythm of photoresponsiveness was first demonstrated by interrupted night experiments (Jenner and Engels, 1952). In birds kept on short nonstimulatory daily photoperiods, 2-hour intervals of light were stimulatory for fattening and reproductive development when provided between 14 and 18 hr after the onset of the daily photoperiods but not when the interruptions were given at other times of day. The circadian basis of the photoinducible phase for fattening was demonstrated in *Z. albicollis* kept on a nonstimulatory 72-hr "day" (LD 6:66) (Alexander, reported in Meier, 1976). Six-hour light interruptions at 36–40 hr after the onset of light caused large increases in body fat stores and gonadal growth whereas 6-hr interruptions at 24–30 hr were ineffective. Stemming from the early interpretation of photoperiodism (Bünning, 1960) a current explanation of these results is that a circadian rhythm of photosensitivity or of a photoinducible phase is entrained by the daily photoperiod. If light is present during this phase it induces fattening, perhaps by introducing another circadian system that interacts with the entraining rhythm.

Photoperiodic entrainment of the photoinducible phases for fattening and migratory restlessness apparently involves corticosteroid hormones. In photosensitive Z. *albicollis* kept on a nonstimulatory 6-hr daily photoperiod, daily injections of corticosterone at 18 hr before the onset of light stimulated fattening and nocturnal migratory restlessness whereas corticosterone injections at 6 and 12 hr before the onset of light were ineffective (Meier and Martin, 1971). Entrainment of a photoinducible phase for migratory indices was also apparent in Z. *albicollis* kept on a stimulatory daily photoperiod (LD 14:10) and disturbed daily for 21 days (Meier, 1976). Disturbances (expected to release endogenous corticosteroids) reduced lipid stores when given 14 hr before the 10-hr dark interval but not at 8 or 2 hr before dark. Thus disturbances were inhibitory when the photoinducible phase was set 18–24 hr later, in the dark.

Entrainment of the photoinducible phase for migratory indices by corticosterone injection or disturbance may be explained in two ways. Corticosterone may set the photoinducible phase by acting directly on neural pathways or receptors for photostimulation or it may set the phase of a neural oscillator that directs both the photoinducible phase as well as the rhythm of endogenous corticosteroid hormone. The latter effect is attractive because it could be accounted for by negative feedback effects of corticosterone in neural control mechanisms for the adrenal.

Assays of endogenous plasma corticosteroid hormone in Z. *albicollis* support the conclusions based on injection experiments. In birds caged outdoors during the vernal migratory period the daily rise of plasma corticosterone occurred near dawn (Dusseau and Meier, 1971). In birds maintained indoors during the early spring on either a nonstimulatory (LD 10:14) or stimulatory (LD 16:8) regimen the daily rise of plasma corticosterone occurred 12 hr after the onset of dark (Meier and Fivizzani, 1975). If one equates the rise of plasma corticosteroid with injection of corticosterone, a photoinducible phase would be expected 18–24 hr later. Thus the photoinducible phase would occur during the dark in birds kept on the short photoperiod and partly during the light in birds kept on the long photoperiod.

Seasonal changes in photosensitivity and photorefractoriness are thought to be the consequences of changing temporal relations between two circadian systems (Meier, 1973, 1976; see Fig. 1). One circadian system involves

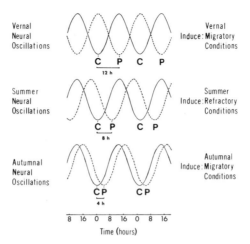

Fig. 1.   Circadian basis for a circannual sequence of physiological and behavioral conditions in the white-throated sparrow, *Zonotrichia albicollis*. According to this model the temporal relations of two circadian neural oscillators change seasonally. Interaction between these oscillators produce endocrine and neural complexes that stimulate development of reproductive and migratory conditions. The circadian rhythms of plasma corticosteroid and prolactin are expressions of the two oscillators and account directly for some of the seasonal changes. Daily injections of corticosterone and prolactin entrain the neural oscillators by way of negative feedback mechanisms and thereby set the phase of the neural circannual clock. C and P represent the times of peak plasma levels of corticosterone and prolactin, respectively.

a rhythm of prolactin release from the pituitary. Prolactin release occurs during the afternoon and early evening in *Z. albicollis* in vernal migratory condition and during the night in summer refractory birds (Meier *et al.*, 1969). Stimulation of these relations by daily injections of prolactin elicited the appropriate changes in fat stores (Meier and Davis, 1967) and nocturnal restlessness (Meier, 1969). Similar findings with respect to fattening responses have been reported for the chaffinch, *Fringilla coelebs* (D'yachenko, 1972).

The other circadian system involves the daily rhythms of fattening and locomotor responses to prolactin. The response rhythm is probably set by the daily photoperiod and involves the daily rhythm of plasma corticosteroid hormone. Daily injections of corticosterone entrain daily rhythms of responses to prolactin in *Z. albicollis* kept in LL (Meier *et al.*, 1971b). Prolactin injections at 12 hr following corticosterone injections (12-hr relation) stimulated large increases in body fat stores and gonad weights, an 8-hr relation inhibited fat stores and gonadal weights, and a 4-hr relation stimulated fattening but not gonadal growth. It was concluded that the 12-hr, 8-hr, and 4-hr relations account for spring, summer, and autumn conditions, respectively.

Seasonal regulation of migration by way of changing circadian relations was further indicated by studies of nocturnal restlessness and orientation in Z. albicollis kept in LL and treated with corticosterone and prolactin (Martin and Meier, 1973). Both the 4- and 12-hr relations of corticosterone and prolactin injections stimulated fattening and migratory restlessness in birds tested in cages under the open night sky whereas the 8-hr relation caused loss of fat stores and did not induce nocturnal restlessness. In addition, the 4-hr relation stimulated activity oriented southward whereas the 12-hr relation promoted northward orientation (toward the breeding area). These studies were carried out at different times of year indicating that seasonal orientation is set by changing circadian systems. Establishment of vernal conditions in autumn and autumn conditions in spring by photoperiodic manipulation also confirms in Z. albicollis that seasonal changes in orientation result from endogenous changes in the interpretation of environmental cues rather than in changes of the cues themselves (Miller and Weise, 1978). Our laboratory (B. R. Ferrell and A. H. Meier, unpublished) has recently observed that once vernal migratory and reproductive conditions are established in Z. albicollis during autumn by a 12-hr relation of corticosterone and prolactin injections, summer and autumn conditions develop sequentially during the next several months in birds kept on a constant regimen (LD 14L:10D). This finding supports a conclusion that the circannual mechanism can be set by injections of corticosterone and prolactin in specific temporal relations.

The seasonal alteration in the manner in which the corticosteroid and prolactin rhythms relate with one another seems to be a part of a circannual mechanism. In Z. albicollis maintained on a constant daily regimen (LD 16:8) beginning in winter, an annual sequence of changes in body fat stores, nocturnal restlessness, molt, and reproductive indices were associated with appropriate changes in the circadian rhythms of plasma corticosteroid (Meier and Fivizzani, 1975). Of further interest was the finding that following establishment of summer reproductive and metabolic refractoriness the corticosteroid rhythm dampened out completely for a time. The corticosteroid rhythm was reestablished when the birds exhibited autumnal migratory condition and behavior.

Thyroid hormones have been implicated in a permissive way for corticosteroid rhythms in vertebrates (Meier, 1976). Low levels of thyroid hormones may account for the dampening of the corticosteroid rhythm in birds during summer photorefractoriness. Thyroid influence of the circannual sequence during the autumn has recently been obtained in our laboratory (B. R. Ferrell, unpublished). Corticosteroid and prolactin injections in the 12-hr relation that promote vernal fattening and migratory behavior when given at other times of year were ineffective when attempted in October. Addition of thyroxine at the time of prolactin injection but not at the time of corticosterone injection promoted fattening and nocturnal activity. These findings suggest that thyroxine is necessary to permit the expression of prolactin activities and may be a limiting factor during fall migration. Thyroxine injections stimulate migratory activity in autumn but not in spring (Wagner, 1930; Putzig, 1938).

E. SUMMARY REMARKS

Much of the attention of avian physiologists interested in migration has focused on the metabolic support and timing of migratory behavior. Circadian mechanisms appear to have crucial roles in many aspects of metabolism and behavior, including premigratory fattening, migratory restlessness, and migratory orientation. Although daylength is an important environmental synchronizer for migration, the principle timer for the annual cycle of migratory and reproductive conditions is an endogenous circannual mechanism. The circannual mechanism appears to involve seasonal changes in the temporal relations of two circadian oscillators. Photoperiodism also depends on circadian rhythms and probably utilizes in part the circannual oscillators. Research is needed to determine whether other environmental synchronizers (i.e., rainfall) of annual cycles are able to set the circannual clock.

The circadian rhythms of corticosteroid hormone and prolactin appear to be manifestations of the two neuroendocrine oscillators and have important roles in directing metabolism and behavior during the annual cycle. Simulation of the daily hormone rhythms present in the white-throated sparrow, *Zonotrichia albicollis,* during the migratory seasons by daily injections of corticosterone and prolactin can produce, out of season, metabolic and behavioral indices appropriate for either vernal or autumnal migration (Section II,D). Many other hormones, as well, probably contribute in directing mobilization of fat stores and responsiveness of the nervous system to environmental cues. Although not confirmed in a significant manner, the possibility of direct neural stimulation of migratory restlessness without endocrine mediation should be considered. The stimulation of hormone production may be supportive of the migratory urge and orientation without being obligatory. Exogenous hormones (corticosterone and prolactin) may establish by way of feedback mechanisms temporal relations of neural oscillators that in turn direct the entire repertoire of endogenous neural and endocrine mechanisms that regulate migration.

III. Fish Migration

A. INTRODUCTION

Although regulation of fish migration is treated as a singular phenomenon for simplicity of presentation, it should be recognized that considerable diversity exists with respect to the taxa studied, the habitats of the fish, and the kinds of migration. Many of the strong migrants (salmon and eels) are members of the modern (teleost) bony fishes, Osteichthyes, but there are also migrants that belong to the Chondrichthyes (sharks) and Agnatha (lampreys). Differences in evolutionary development may well have produced variations in regulatory mechanisms involved in migration.

The habitats of migratory fish differ considerably with respect to salinity. Some species exist in and are adapted for seawater whereas others are found exclusively in freshwater. Migrators such as salmon and eel which move between fresh- and seawater must be able to adapt to both extremes of salinity (that is, they are amphihaline). In addition, these adaptations to salinity changes must be integrated with other important events during the life cycle. The spawning migration of salmon occurs primarily in fresh water whereas that of eels occurs primarily in seawater. This relation with salinity is reversed in the trophic migrations of the immature fish.

Because fish are ectothermic their metabolism and behavior are greatly influenced by variations in water temperature. Seasonal fluctuations in temperature may impose annual cycles in some fish and set the times for migration. Migration itself may bring the fish into waters of different temperatures. Because temperature also changes with water depth, vertical movements of fish may produce further metabolic alterations. For example, a fish may seek its prey in deep colder water during the day and migrate at night in warmer surface water where it might better use celestial cues for orientation (Bitinkov, 1959).

Regulatory mechanisms have multiple functions which may be divided into preparative and adaptive activities. These activities might best be visualized by citing a specific example. The European eel, *Anguilla anguilla*, matures in freshwater ponds and river tributaries where it is known as the yellow eel. Before migration the yellow eel accumulates a store of body fat and undergoes metamorphosis into the silver eel. The silver eel differs dramatically from the yellow eel in that its osmoregulatory mechanism is prepared for transition to seawater whereas the yellow eel cannot survive a direct transfer to seawater. The heavy fat stores and the osmoregulatory adjustment are distinctly preparatory in that they occur before the time when they are needed. Adaptive changes may include the mobilization of fat and protein energy stores for the metabolic support of migration to the spawning grounds in the Sargasso Sea. Further adaptive changes also occur in the osmoregulatory mechanisms in response to increased salinity as the silver eel enters seawater.

B. STIMULI AND TIMING

Preparation for migration and the final proximate stimulus depend on an interaction of endogenous and exogenous factors. Preparation includes development of systems responsive to the proximate stimuli. Responsiveness may be periodic so that a fully matured animal may be influenced by proximate environmental cues only at certain times of year. Because an annual cycle of reproductive and migratory indices persists in birds held at constant conditions of temperature and daylength, it seems likely that seasonal changes in migratory readiness in fish may also be part of an endogenous circannual mechanism. A circannual rhythm of reproductive changes has been reported in a teleost (Sundararaj et al., 1973), but the possibility of a circannual mechanism regulating migratory indices has not been examined in fish. Changes in environmental cues could be expected to set the phase of such a circannual rhythm. Seasonal changes in daylength (Baggerman, 1957, 1960; McInerney, 1964; Spieler et al., 1976a) and especially temperature (review Shulman, 1974) are involved in regulating annual physiological and behavioral changes.

Change in water temperature is an important environmental stimulus for the initiation of migration in some species. A decrease in body temperature initiates migration in the Azov anchovy, Engraulis encrasicholus (Shulman, 1960; Taranenko, 1966). The fattest fish leave first; more drastic decreases in temperature are necessary to induce leaner fish to migrate. Similar relations between temperature decrease and amount of fat stores in stimulation of migration have been reported in several other fishes (Shulman, 1974). Other environmental stimuli that initiate migration in responsive fish include flooding (European eel, Anguilla anguilla: M. Fontaine, 1975) and weather fronts (Atlantic salmon, Salmo sp.; Ellis, 1962).

Studies of daily and seasonal patterns of locomotor activity have provided valuable insight for understanding bird migration and similar studies of lesser scope have been reported in fish. Juvenile herring, Clupea harengus, are most active at sunset (Stickney, 1972) whereas the brown trout, Salmo trutta (Swift, 1964), and several other species (Gibson, 1970;

Olla and Studholme, 1972) are most active at sunrise. Whether this variation in daily activity constitutes evidence for species differences is uncertain
because the phase of the activity rhythm may also differ within a species
depending on the time of year tested and on the developmental stage of the
fish. The fry of sockeye salmon, *Oncorhynchus nerka* (Byrne, 1968), and
pink salmon, *O. gorbuscha* (Hoar et al., 1957), are nocturnal initially, but
become diurnally active after several weeks to several months. Mature
migrating salmonids, however, are largely nocturnal (Hasler, 1971). During the winter most Atlantic salmon, *Salmo salar,* are nocturnally active
and in spring become more active at sunrise and sunset (Richardson and
McCleave, 1974).

During the anadromous migration of the alewife, *Alosa pseudoharengus,*
a bimodal rhythm of migratory activity with peaks at sunrise and sunset
was usually observed (Richkus, 1974). However, activity was spread
throughout the day when it was cloudy and single peaks occurred near
sunset early in the migratory period (April) and near sunrise at the end of
migration (June). Although one of the daily activity periods may dominate,
bimodal patterns of activity have also been described for many salmonids
(Ellis, 1962; Banks, 1969; Groot, 1972), herring (Stickney, 1972) and other
migrants.

Sculpins, *Cottus gobio* and *C. poecilopus* (Andreasson, 1973), and the
Arctic burbot, *Lota lota* (Müller, 1973), are diurnally active during the
winter and nocturnally active in summer. During the phase shift in spring
and fall there is an interval when *C. poecilopus* is active at sunrise and
sunset. Andreasson suggested that activity is regulated by two separate
oscillators, one day active and the other night active. Müller demonstrated
experimentally that the Arctic burbot became nocturnal when daylength
was long and diurnal when the daylength was short. The phase changed
when the daily photoperiod approximated 12 hr. This finding may explain
why there was little synchrony among a population of Atlantic salmon,
Salmo salar, parr maintained on an LD schedule of 12:12 (Richardson and
McCleave, 1974). Thirty-five fish were diurnally active, 20 were nocturnal,
and 32 were active at the light changes.

A fascinating interaction of temperature and photoperiod in the regulation of locomotor activity was observed in juvenile sockeye salmon, *Oncorhynchus nerka* (Byrne, 1968). At each photoperiodic schedule tested
there was a temperature at which, when exceeded, the fish changed from
diurnally active to nocturnally active. Higher temperatures coupled with
shorter daily photoperiods promoted nocturnal activity. Inasmuch as daily
rhythms of locomotor activity are synchronized by the daily photoperiod
in *O. nerka* (Byrne, 1968) and other salmonids (Ali, 1964; Swift, 1964;
Varanelli and McCleave, 1974), it follows that there is a critical threshold

temperature that determines the phase angle of the activity rhythm with respect to the entraining light–dark (LD) schedule. A similar effect of temperature in changing the phase of daily hormonal rhythm (plasma prolactin concentration) was observed in the gulf killifish, *Fundulus grandis* (Spieler *et al.*, 1978). The significance of the temperature–photoperiodic interaction in the integration of metabolic and behavioral activities is discussed later (see Section III,F).

In addition to entrainment of rhythms, the LD cycle is also necessary in some species to maintain the expression of a daily rhythm. Under constant conditions of illumination (LL or DD), circadian rhythms of locomotor activity persisted for only several (2–5) days in juvenile Atlantic herring, *Clupea harengus* (Stickney, 1972), sockeye salmon, *Oncorhynchus nerka* (Byrne, 1968), Atlantic salmon, *Salmo salar* (Ali, 1964), adult European minnow, *Leucaspius dilineatus* (Seigmund and Wolff, 1973), and sea lamprey, *Petromyzon marinus* (Kleerekoper *et al.*, 1961). In blennies, *Blennius pholis*, a circadian rhythm persists in DD and the fish are very resistive to LD entrainment. However, they readily reacted to changes in hydrostatic pressure resulting from local tidal cycles and exhibited activity rhythms with a period of 12 hr. In sea lampreys kept in LL, a single scent entrained an activity rhythm that persisted for at least 5 days (Kleerekoper *et al.*, 1961).

Nonoptic photoreceptors for photoperiodic entrainment have been reported. Photoperiodic entrainment of the circadian activity rhythm was disrupted in *Anguilla anguilla* by placing an aluminum foil shield over the head (Van Veen *et al.*, 1976). However, blinding and pinealectomy had no effect. In a European minnow, *Phoxinus phoxinus*, the photoreceptor may be in the ependymal area of the third ventricle (Oksche and Hartwig, 1975).

Studies of neuroendocrine regulation of fish migration have centered on the thyroid and corticosteroid hormones. An active thyroid appears to be characteristic of many fish during the preparatory period. There is an active hyperplasia of thyroid (Hoar, 1939) and of pituitary TSH cells (Olivereau, 1954) during smoltification of salmonids. Secretion of thyroid hormone also increased at the beginning of smoltification (Leloup and Fontaine, 1960), and injections of thyroxin into parr induce some of the features of the smolt such as silvering (Landgrebe, 1941; Robertson, 1948; Fontaine and Baraduc, 1954) and increased porphyropsin, a visual pigment better adapted for vision at sea (Beatty, 1972). Both thyroxine and prolactin stimulated an increase in porphyropsin in *Salmo gairdneri* (Cristy, 1974).

There is a histological evidence of increased thyroidal activity in the European eel, *Anguilla anguilla*, prior to the catadromous migration of the adult (Callamand and Fontaine, 1942; Bernardi, 1948) and the anadromous migration (upstream) of the elver (Sklower, 1930). These periods correspond with transformation of the yellow eel to the silver eel and with metamorphosis of the leptocephali larva to the elver, respectively.

Pituitary ACTH and interrenal activity also increase in salmonid smolt (Fontaine and Olivereau, 1949; Milne et al., 1971), and cortisol participates with thyroxine to cause silvering (M. Fontaine, 1975). Similarly plasma cortisol concentration increases during the transformation of the yellow eel to the silver eel (Leloup-Hatey, 1964). In fact cortisol injections for 7–10 days caused yellow eels to change color to that of silver eels (Epstein et al., 1971).

Among mammals both thyroid and corticosteroid hormones increase excitability of the nervous system (Timiras and Woodbury, 1956). Thyroxine may also influence neural excitability in teleosts as indicated by increased locomotor activity. Thyroxine or TSH injections increased swimming in salmon fry (Hoar et al., 1955) and in many other species (Woodhead, 1970). Both corticosteroids and thyroid hormones potentiate catecholamine activities in mammals, and may be expected to stimulate greater noradrenergic activities. Noradrenergic neurons are more active during the active waking periods of the day and influence many functions associated with locomotor activity. For example, in salmon, epinephrine increases during migration and stimulates blood pressure (Randall and Stevens, 1967), dilation of gill vessels (Keys and Bateman, 1932), and increase in muscle extracellular space (Gras et al., 1971).

C. Orientation Cues

Fish use a multitude of environmental cues for orientation during migration. Because these are discussed by Abel (Chapter 5), no attempt will be made here to enumerate and evaluate them as such. Responses to cues will be considered as they indicate roles for regulating mechanisms.

Some fish orient by the sun and have a time-compensated mechanism to take into account the change in the sun's position during the day (Hasler et al., 1958; Johnson and Groot, 1963; Groot, 1965; Goodyear and Ferguson, 1969; Goodyear, 1970). The sun may provide useful navigational aids during the seaward migrations of juvenile salmon Oncorhynchus nerka (Groot, 1965; Healy, 1967); however, Groot and many other researchers doubt that the sun is an essential navigational cue for fish migration. Although cutthroat trout, Salmo clarki, employed the sun as a reference in laboratory experiments (Jahn, 1969), field tracking studies indicated that blinded fish find the home river very well (McCleave, 1967; McCleave and LaBar, 1972). In addition many migrations occur at night and at great depths where nocturnal celestial cues would have little use. Nevertheless the presence of a time-compensated sun compass is indicative of a circadian neural mechanism that can influence behavior.

Because the expression of rheotaxis depends on the recognition of visual cues or surface contact as reference for realization of current (Gregory and Fields, 1962; Ryapolova, 1964; Tesch, 1967), rheotaxis is most important for migration in relatively clear and shallow flowing streams. Other environmental cues may trigger the occurrence or change the sign of rheotaxis. Rainbow trout fry, *Salmo gairdneri*, are negatively rheotactic at low temperatures but exhibit positive rheotaxis with a sharp rise in temperature (Tesch, 1975). Temperature is probably sensed in the brain where warm temperature has been shown to promote the activities of a different set of neurons than cold temperature in brook trout, *Salvelinus fontinalis* (Greer and Gardner, 1970).

Odor is an important cue in the anadromous migrations of salmonids (Hasler, 1966) and eels (Creutzberg, 1959). The elver eel reacts to odors, perhaps of conspecifics, from upstream sources never before frequented by the elver whereas the adult salmon reacts to odors imprinted on them in their freshwater habitat before they undertook the catadromous migration. M. Fontaine (1975) reported that the secretion of melanophore-stimulating hormone (MSH) was greater in *Salmo salar* smolt than in parr and suggested that MSH may have a role in learning the home odor. MSH increased sensitivity to external stimuli (Segawa *et al.*, 1973), attention (De Wied *et al.*, 1968; Kastin *et al.*, 1973), and retention of learned tasks (De Wied and Bohus, 1966; Sandman *et al.*, 1969) in rodents. Given the chemical similarity of MSH and ACTH, it is not surprising that ACTH acts similarly in rodents by increasing retention of a learned response (Miller and Ogawa, 1962). Pituitary ACTH cells appear very active in *Salmo salar* smolt (Milne *et al.*, 1971). These results suggest that the release of large amounts of MSH and ACTH in smolts may have the effect of making memory permanent (imprinting). This possibility deserves to be examined in greater detail.

Because salinity gradients are usually very small, the detection of differences probably has limited usefulness in migratory orientation. Notable exceptions might include the layering of seawater by freshwater streams combined with tidal fluctuations. However some fish can perceive differences as little as 0.0005 M NaCl, and McInerney (1964) postulated that salmon may use this sense by trial and error searches to move through estuaries.

Preference to salinity has been investigated under a variety of experimental treatments in artificial salinity gradients. Salinity preference varied seasonally and as a result of daylength in several species of immature salmon (Baggerman, 1959; McInerney, 1964; Otto and McInerney, 1970), the gulf killifish, *Fundulus grandis* (Spieler *et al.*, 1976a), and the three-spined stickleback, *Gasterosteus aculeatus* (Braggerman, 1959). The movement of many estuarine fish into less saline waters during spring and summer has been attributed to increasing temperature (Gunter, 1967). Although it ought not be applied too rigorously, the available reports appear to indicate that long daylengths and warm temperatures promote movement into and preference for less saline water in many fish.

Because of the proposed involvement of the thyroid in migration (Hoar, 1939), Braggerman (1960, 1963) made a thorough investigation of the effects of thyroid hormones on salinity preference in salmonids. She found that thyroid hormones stimulated a preference for more saline waters, a finding that is consonant with increased thyroidal activity prior to movement toward the sea. On the other hand thyroid hormones promoted a preference for less saline waters in the three-spined stickleback, *Gasterosteus aculeatus* (Baggerman, 1957). The differences in the thyroidal activities suggest that thyroxine may not act directly but rather may indirectly influence salinity preference perhaps by augmenting the effects of other regulatory agents. Thyroxine is known to have permissive effects for the activities of other hormones and for the expression of circadian rhythms (Meier, 1975).

Our laboratory has demonstrated a role for the circadian rhythms of corticosteroids and prolactin in regulating the seasonal migratory behavior and metabolism in a sparrow (see Section II). This system involves an annual cycle of changes in the temporal relations of the hormone rhythms. Recently we (Fivizzani and Meier, 1978) applied this system to test for possible control of salinity preference in the gulf killifish, *Fundulus grandis.* Daily injections of prolactin were given at several different times relative to daily injections of cortisol. After several days the preferences of treated fish were tested in a salinity gradient. Most of the experimental groups preferred salinities similar to that preferred by controls, but the group that received simultaneous injections of cortisol and prolactin preferred waters that were distinctly more saline (see Section III,F).

Recent researches have demonstrated that electric and magnetic fields are used as cues for orientation of migration in several species (Branover *et al.*, 1971; Gleiser and Khodorkovskii, 1971; Rommel and McCleave, 1971, 1973; Martin and Lindauer, 1973; Tesch, 1974). Although a possible neuroendocrine control over the manner in which these cues are used have not been investigated, directional changes during migration without apparent change in electromagnetic field, as well as developmental changes in orientation, suggest that this possibility should be studied. Particularly relevant is a study by Tesch (1974) of the European eel, *Anguilla anguilla,* believed to rely strongly for orientation on geomagnetic forces (Branover *et al.*, 1971; Vasiliev and Gleiser, 1973). The directions taken by silver eels and homing yellow eels were opposite when placed in the ocean.

D. OSMOREGULATION

Fish in fresh water have osmoregulatory problems that are the opposites in several ways from those experienced by fish in seawater. Movements of some of the great migrators (salmon, eels) from one medium to another during migration demand that the fish be able to osmoregulate in both. The osmoregulatory adjustments, in turn, must be integrated appropriately not only with the physiological requirements of migration itself but also in many instances with reproductive maturation and with differences in water temperature.

Regulation in fresh water involves the removal of excess water and the retention and uptake of ions. To this end fish in fresh water produce a copious dilute urine; they obtain minerals by active transport of ions in the gill and from feeding; and the integument is relatively impermeable to water. In seawater the extracellular environment must be maintained at ionic concentrations that are lower than those in seawater so that the principal problems involve the retention and replenishment of water and the removal of excess ions. Teleosts in seawater conserve water by reducing urine flow and obtain water by swallowing. The swallowing of seawater satisfies the requirement for uptake of water but further exacerbates the problem of excess ions. Some of the ions, especially the divalent cations, are actively removed by way of the kidney and gut. Most of the sodium and chloride is actively removed at the gill filaments. A high level of calcium ion appears responsible for maintaining a relatively impermeable integument in seawater (Potts and Fleming, 1970, 1971; Ogawa, 1974, 1975). There are vaiations among the taxa concerning this generalized scheme of osmoregulation (Johnson, 1973).

In fresh water adapted fish both sodium and chloride ions are obtained from the medium by active transport in the gill chloride cells. The cation pump responsible for a net movement of Na^+ from the medium into the cell has been described as a Na^+-ATPase pump (Pfeiler and Kirschner, 1972). In contrast to the ubiquitous Na^+K^+-ATPase pumps present in most cell membranes, the Na^+-ATPase pump does not depend on K^+ exchange but probably utilizes NH_4^+ and H^+ as counterions (Hochachka and Somero, 1973). Both NH_4^+ and H^+ stimulate the cation pump. Chloride ion is also taken up at the gill in exchange for bicarbonate ions by an anion pump that is stimulated by both Cl^- and HCO_3^- (Kerstetter and Kirschner, 1974; de Rensis and Bornancin, 1977). Inasmuch as both the cation counterions (NH_4^+ and H^+) as well as the bicarbonate counterions are products of metabolism, their stimulatory effects on the sodium and chloride pumps could assist in the hydromineral balance of fish migrating in fresh water. The increased osmoregulatory efficiency may be of particular value during the anadromous migration of salmon when vigorous locomotor activity and high gluconeogenic activities of corticosteroids assure high levels of all the counterions.

In seawater-adapted fish, as in freshwater-adapted fish, the most important effector organ in sodium and chloride movements is the branchial gill apparatus. Sodium ion is actively transported to the exterior medium by a Na^+,K^+-ATPase pump dependent on external potassium as the counterion (Hochachka and Somero, 1973). There is also an anion pump in *Fundulus heteroclitus* that transports chloride ion to the exterior (Karnaky *et al.*, 1977). The bicarbonate and hydrogen ions which accumulate in the blood as products of cellular respiration are thought to diffuse into branchial cells where they are converted by the action of carbonic anhydrase into carbonic acid and subsequently to water and carbon dioxide which diffuse into the external medium (Hochachka and Somero, 1973).

Because adaptation to salinity changes occurs after the fact, preparatory adjustments that precede migration of fish into water of different salinity are more interesting from a regulatory point of view. Smoltification in salmon is accompanied by an increase in chloride cells (Hoar, 1951) and in Na^+,K^+-ATPase activity which functions to eliminate NA^+ (Giles, 1969; Zaugg and McLain, 1972). Smolts transferred into a hypertonic medium behave with respect to osmotic concentration of blood and urine volume as though they were marine and not freshwater fish (Zaks and Sokolova, 1961). Interestingly smolting and an increase in Na^+,K^+-ATPase activity in the gills of steelhead trout, *Salmo gairdneri*, do not occur when temperatures rise above 15°C (Adams *et al.*, 1973). Osmoregulatory adjustments occur as well during the anadromous migration before salmon (*Oncorhynchus masu* and *O. nerka*) reach fresh water (Kubo, 1960; Kashiwagi, 1971).

Transformation of the yellow eel to the silver eel in fresh water is accompanied by osmoregulatory preparation for seawater. Silver eels, *Anguilla anguilla*, have more gill chloride cells (Colombo and Cecchini, 1959; Thomson and Sargent, 1977) and greater Na^+,K^+-ATPase activity (Thomas and Sargent, 1977) than yellow eels. The number of cells and amount of enzyme activity in the silver eels kept in fresh water were equivalent to those found in yellow eels following adaptation to seawater (Thomas and Sargent, 1977).

The principal osmoregulatory hormones in teleosts appear to be corticosteroids and prolactin. Corticosteroids induce preparatory changes for seawater and prolactin is more active in preparation for fresh water. Eels (*Anguilla anguilla* and *A. japonica*) have been extensively studied in this

regard. Pituitary prolactin cells are more active in elvers during anadromous migration (Knowles and Vollrath, 1966; Vollrath, 1966) and
plasma cortisol concentrations increase during the transformation of the
yellow eel to the silver eel (Leloup-Hatey, 1964).

Cortisol injections for 7–10 days caused silvering in the yellow eel and
increased both the number of chloride cells and the activity of
Na$^+$,K$^+$-ATPase (Epstein et al., 1971; Doyle and Epstein, 1972; see also
Maetz, 1969; Pickford et al., 1970b). Cortisol injections also enabled inter-
renalectomized eels to survive in seawater (Mayer et al., 1967) and increased the absorption of water at the intestine (Utida et al., 1972) and
urinary bladder (Hirano et al., 1973).

Prolactin injections, on the other hand, have opposing activities and tend
to prepare fish for fresh water. Prolactin reduced gill permeability (Ogawa,
1975, 1977), Na$^+$ efflux (Lam, 1969; Lahlou and Giordan, 1970), and activity of Na$^+$,K$^+$-ATPase involves in the outward transport of Na$^+$ in the gill
(Pickford et al., 1970a; Kamiya, 1972). The pituitary prolactin cells are
also more active in elvers moving into fresh water than in adult silver eels
moving to sea (Knowles and Vollrath, 1966; Vollrath, 1966). Evidences of
prolactin involvement in anadromous migrations have also been reported
for the salmonids, Oncorhynchus keta (Nagahama and Yamamoto, 1970)
and O. nerka (McKeown and van Overbeeke, 1972), the three-spined
stickleback, Gasterosteus aculeatus (Lam, 1972), the ayu, Plecoglossus
altivelis (Honma and Yoshie, 1974), and for the mullet, Mugil cephalus and
M. captio (Blanc-Livni and Abraham, 1970).

Although prolactin levels in the pituitary and plasma are usually greater
in fresh water than seawater, prolactin production does not cease in
seawater. There is considerable prolactin cell activity in the pituitary of
adult salmon, Oncorhynchus nerka, at sea (McKeown and van Overbeeke,
1972) and there is even increased prolactin cell activity in the mullet, Mugil
cephalus, during the migration from fresh water to seawater (Abraham,
1974). In the flounder, Platichthys stellatus, prolactin promotes
osmoregulation in seawater by increasing water reabsorption from the
urine (Hirano et al., 1971). These results and others that ascribe a multitude
of activities for prolactin in fish (Bern and Nicoll, 1968) suggest that a simple interpretation of direct prolactin support for osmoregulation in fresh
water is inadequate.

E. Metabolism

Some species support metabolic demands by increased feeding during
migration. Many others migrate in waters that do not consistently support
adequate foraging and may cease feeding entirely during migration; the

digestive tract may even degenerate extensively (salmon, eel) (Woodhead, 1975). Large energy stores are necessary for the nonforaging great migrators.

Of the three types of energy stores (proteins, fats, and carbohydrates), the carbohydrate store (glycogen) is least important in migration (Fontaine and Hatey, 1954; Swift, 1955; Black, 1958). Fat stores are most important but protein may also be metabolized extensively, especially when fat is deficient. Nitrogen excretion in teleosts is about $3\frac{1}{2}$ times that in mammals (Butler, 1968). Fat accumulation occurs during specific intervals that are related to stage of development and time of year in a particular species. However, there is considerable variation in timing among species. Body growth (protein) and fattening are often linked.

Fat stores range from about 10 to 50% of the total weight of teleosts at the beginning of migration (Kleimenov, 1962; Nikolskii, 1963). The amount varies according to species, but there is also a direct relation between the amount of fat stores and the distance traveled by various populations within a species (Kiselevich, 1924; Clark and Clough, 1926; Pentegov et al., 1928; Levanidov, 1932; Nikolskii, 1963). Lampreys also accumulate large fat stores (30-50% body weight) (Nikolskii, 1963). Fish store primarily triglycerides (75% of all reserve fat) (Katada et al., 1960; Ackman, 1964; Linko, 1964). Most of the remaining body lipid is structural.

Fat depots are widely scattered among subcutaneous connective tissue (eels and herring), skeletal muscle (anchovies, salmon, mackerels), abdominal mesentery (especially on nonmigrants), liver (most fish), and interosseus tissue (salmon) (Morawa, 1955; Kleimenov, 1962; Shulman, 1974). Shulman pointed out that the various deposits have special functions. For example, mesenteric fat may accumulate more fat soluble vitamins and contribute toward development and maturation of eggs.

Most of the neutral fat stores as well as significant amounts of protein are used up during migration (Robertson and Wexler, 1960; Lobanov-Rostovsky, 1969). Glycogen levels may well be higher at the end of migration than at the beginning and blood glucose levels are maintained high throughout migration (Robertson, 1955). Both glycogen and glucose levels are maintained at the expense of lipids and proteins. During the spawning migration of the sockeye salmon, Oncorhynchus nerka, 10% of the energy is used for gonad growth in females and 0.5% in males (Idler and Tsuyuki, 1958). Females used 1.1 cal/kg/km of travel whereas males used 0.8 cal/kg/km. The oxidation of fatty acids occurs in the liver in O. nerka (Brown and Tappel, 1959).

Relatively few hormones are likely to stimulate an increase in lipogenesis and fat storage. Insulin would be so indicated according to the mammalian literature. Insulin is produced by the teleostean pancreas (Epple, 1969; Lewis and Epple, 1972) and has some of the activities that it has in mammals (de Vlaming and Pardo, 1975). It decreased serum free fatty acids (prevents mobilization) in cyclostomes and teleosts (Leibson *et al.*, 1968; Plisetskaya and Mazina, 1969; Minick and Chavin, 1972) although not in the eel (Larsson and Lewander, 1972). It also stimulates the formation of proteins (Seshadri 1959; Tashima and Cahill, 1968) and decreases the plasma concentrations of both amino acids and glucose (Epple, 1969; Thorpe and Ince, 1974).

The most potent lipogenic hormone in teleosts appears to be prolactin. Our laboratory has shown that daily injections of ovine prolactin can more than double body fat stores within 1 week in *Fundulus chrysotus* (Lee and Meier, 1967) and *F. grandis* (Joseph and Meier, 1971). The fattening effect, however, is limited to a specific interval during the day (about 8 hr after the onset of light). At other times prolactin is inhibitory to body fat stores. The daily variation in fattening response to prolactin has also been demonstrated in several other teleosts (Mehrle and Fleming, 1970; de Vlaming and Sage, 1972). In *Fundulus grandis* the phase of the response rhythm is set by the onset of the daily photoperiod (Joseph and Meier, 1971) apparently by way of a circadian rhythm of plasma corticosteroid concentration (Meier *et al.*, 1971c). Both plasma cortisol (Garcia and Meier, 1973; Srivastava and Meier, 1972) and prolactin (Spieler *et al.*, 1978) concentrations vary during the day in *Fundulus grandis*. Both daylength (Spieler, 1975) and temperature (Spieler *et al.*, 1978) influence the phase of the prolactin rhythm. Corticosteroid (Boehlke *et al.*, 1966; Singley and Chavin, 1971) and prolactin (Leatherland and McKeown, 1973; Spieler and Meier, 1975; Leatherland *et al.*, 1974; Spieler *et al.*, 1976b) rhythms have been reported in other teleosts as well.

On the basis of an *in vitro* study of *Notemigonus cryosoleuces*, Pardo and de Vlaming (1976) concluded that prolactin may directly stimulate hepatic lipogenesis. Using radiolabeled acetate (a precursor in fatty acid synthesis) as a marker, our laboratory has not been able to demonstrate a direct lipogenic effect of prolactin in *Fundulus grandis*. The mode of prolactin's lipogenic activity needs further research.

According to mammalian studies, many hormones may be expected to have a role in the mobilization of fat. Some of the principal hormones include the catecholamines, glucagon, growth hormone, and ACTH. Thyroid hormones and the corticosteroids also are important fat mobilizers in that they augment the activities of the other hormones.

Considerable attention has been given the corticosteroids partly because they are produced in large amounts during the spawning migrations of Pacific salmon. However, high activity of corticosteroid hormone is not an obligate requirement during migration. In the Atlantic salmon, *Salmo salar*, maximal activity of the interrenal gland occurred in the adults on the feeding grounds and the return to fresh water coincided with a reduction in interrenal activity (Olivereau, 1975). Evidences of hyperadrenocorticism were found during the prespawning and spawning periods in both the migratory steelhead and nonmigratory rainbow trout, *Salmo gairdneri* (Robertson *et al.*, 1961) Little or no circulating corticosteroids were found in the serum of migrating lampreys (Weisbart and Idler, 1970; Buus and Larsen, 1975).

Although there is not a direct causative relation between corticosteroid activity and migration, the functions of corticosteroids in fish would seem to be important adjuncts in maintaining metabolic support. The corticosteroids (cortisol in most teleosts) direct a change of proteins and fats to carbohydrates (gluconeogenesis). This activity increases the excretion of nitrogen and may increase the levels of blood glucose and liver glycogen (Butler, 1968; Tashima and Cahill, 1968; Patent, 1970; De Roos and De Roos, 1972). ACTH appears to stimulate the conversion of glycogen to glucose (Ball *et al.*, 1966; Swallow and Fleming, 1966). Epinephrine also raises blood glucose levels in cyclostomes (Bentley and Follett, 1965; Plisetskaya and Mazina, 1969), chondrichthyes (Patent, 1970; De Roos and De Roos, 1972), and teleosts (Perrier, 1971; Larsson, 1973; Mazeaud, 1973).

The fat-mobilizing activity of catecholamines would be more important than the hyperglycemic activity in migrants. Increases in the free fatty acid (FFA) level in the blood after catecholamine treatment have been reported (Leibson *et al.*, 1968; Larsson, 1973; Mazeaud, 1973). However, *in vitro* studies of adipose tissue of eel (Farkas, 1969) and trout (Y. A. Fontaine, 1975) indicated an absence of fat-mobilizing activity by catecholamines. In fact glucagon, ACTH, and MSH were also ineffective in causing increases in formation of FFA

Lack of fat-mobilizing activities by hormones recognized for these activities in other vertebrates seems incongruous in fish that depend so much on fat stores to maintain locomotor activity. Perhaps the most judicious attitude at this juncture would be to consider the results in a manner consistent with the well-documented lipolytic activities of these hormones among many vertebrate taxa. Along these lines, then, Mazeaud (1973) suggested that newly synthesized prostaglandins might block lipolytic activities of catecholamines. This suggestion has merit on the basis of mammalian studies (Ramwell and Shaw, 1970). Because corticosteroid hormone inhibits prostaglandin synthesis, this hormone may be required to allow the expression of lipolytic activities of other hormones. Addition of corticosteroid hormone may be essential for *in vitro* studies testing lipolytic effects of other hormones.

Another possible explanation for a lack of lipolytic activity of several hormones in *in vitro* studies (Farkas, 1969; Y. A. Fontaine, 1975) is that specific adipose tissues may have other functions than the support of locomotor activity and should therefore lack the responsive system (for example, cell membrane receptors) in those tissues. The abdominal adipose tissue, a principal tissue used in *in vitro* studies, is thought to be involved primarily in the support of reproduction (Shulman, 1974). In an *in vitro* study utilizing a fat-containing tissue (lateral line muscle) more likely to have a supportive role in locomotor activity, Bilinski and Lau (1969) demonstrated a 20% increase in lipolytic activity after 2 hr incubation with epinephrine.

The catecholamines have many other activities that enhance or support locomotor activity. For example, they modify gill ventilation (Peyraud-Waltzenegger, 1972; Steen and Kruysse, 1964) and blood pressure (Fange, 1962; Reite, 1969; Chester-Jones et al., 1969). However, regulation of these changes by way of increasing plasma catecholamine levels may well be limited to specific stresses (Mazeaud, 1964, 1972; Nakono and Tomlinson, 1967). It seems more likely that the long-term utilization of catecholamine function may better be served by increasing the effectiveness of the catecholamines. Both the thyroid hormones and the corticosteroids greatly potentiate catecholamine activities in mammals by inhibiting the synthesis of enzymes (monoamine oxidase and catecholamine-*O*-methyltransferase) that inactivate the catecholamines. Perhaps some of the conflicting results concerning a role for the thyroid in the regulation of oxygen consumption in fish (Gorbman, 1969; Woodhead, 1975) is resolvable in terms of catecholamine participation. A catecholamine–thyroid hormone synergism could also account for glycogenolytic and lipolytic effects of injected thyroxine (Fontaine and Hatey, 1953; Baraduc, 1954) and may be expected to influence activity of noradrenergic neurones in the central nervous system.

Glucagon (pancreatic hormone), vasotocin (neurohypophyseal hormone), and growth hormone (adenohypophyseal hormone) (McKeown and van Overbeeke, 1972) have also been implicated in migration by virtue of their fat-mobilizing activities. Growth hormone is of special significance in that mammalian studies have demonstrated that exercise and hypoglycemia may stimulate its release and that growth hormone tends to conserve proteins while making fatty acids available for metabolism.

F. Regulatory Model for a Minimigrant

The gulf killifish, *Fundulus grandis*, is a euryhaline teleost of the estuarine zone along the southeastern United States shores of the Gulf of Mexico. Like many other estuarine species, including the closely related

Fundulus heteroclitus (Childester, 1920; Lotrich, 1975), *F. grandis* is found in water of higher salinity concentration during the fall and winter than in spring and summer (Gunter, 1954; Perret *et al.*, 1971). Seasonal migrations are thought to be stimulated by changes in both temperature (Gunter, 1967) and daylength (Spieler *et al.*, 1976a). Heavy fat stores are found during the winter when the reproductive system matures (Fivizzani, 1977). Spawning occurs primarily early in the spring.

Because prolactin has an important role in the migration and annual cycle of the white-throated sparrow, *Zonotrichia albicollis* (Meier, 1973), our laboratory also explored the possiblity that prolactin may be involved in regulating the annual cycle of migration and reproduction of *F. grandis.* Daily injections of prolactin (ovine) were found to influence amounts of fat stores, gonad weights, and plasma chloride concentrations. However, the time of day of the injections was critically important. Injections at 8 hr after the onset of the daily photoperiod (LD cycles: 8:16, 12:12, and 16:8) produced large gains in body fat stores within 1 week of injections, but injections at other times were either ineffective or even reduced the level of body fat (Joseph and Meier, 1971). In addition to increasing body fat stores, injections at 8 hr after the onset of light (LD 14:10) were stimulatory for gonad weights (Fivizzani, 1977).

In view of the well-documented stimulatory effect of prolactin on plasma electrolyte concentrations in hypophysectomized fish, the temporal effects of prolactin on plasma chloride concentrations in *F. grandis* are especially significant. In fish maintained on LD 11:13, prolactin injection early (2–5 hr after onset of light) in the day caused an increase in plasma chloride concentration (about 25 mEq/liter more than the mean control value, 120 mEq/liter) determined at 4 hr postinjection whereas injections at 8 hr after the onset of light stimulated a sharp decrease in plasma chloride (about 25 mEq/liter less than the mean control value, 125 mEq/liter). Both the stimulatory and inhibitory effects of prolactin are apparently mediated at the cellular level by a prostaglandin. PGE_1 injections at the two times of day had virtually the same stimulatory and inhibitory effects as prolactin on plasma chloride concentrations. Inhibition of prostaglandin synthesis by simultaneous injections of either aspirin or indomethecin with prolactin blocked the chloride responses to prolactin (Horseman and Meier, 1978).

Photoperiodic entrainment of the daily rhythms of responses to prolactin also involves the interrenal gland. In fish maintained in continuous light to prevent photoperiodic entrainment, responses to daily injections of prolactin varied depending on the temporal relation between the prolactin injections and daily injections of cortisol. Injections of both hormones at the same time of day (0-hr relation) stimulated fattening (Meier *et al.*, 1971c) increase in gonad weights, and a preference for water of higher salinity (Fivizzani, 1977). Other temporal relations of hormone treatment were ineffective or decreased the parameters tested. It should be noted, then, that the 0-hr relation of cortisol and prolactin treatment in fish kept on LL produced results comparable to those obtained by prolactin injections at 8 hr after the onset of a daily photoperiod.

Assays of plasma concentrations of corticosteroid and prolactin further indicate that a temporal synergism of these hormones is involved in the regulation of seasonal conditions in *F. grandis*. The peak of the daily rhythm of plasma cortisoteroid concentration occurred at 8 hr after the onset of light in both summer and winter and in water temperatures of about 20°–22°C to 30°–32°C (Garcia and Meier, 1973; Srivastava and Meier, 1972).

The prolactin rhythm in *F. grandis* is more influenced by temperature (Spieler *et al.*, 1978). In fish kept in water at 20°C the daily peak occurred 8 hr after the onset of light (LD 12:12) whereas at 28°C the peak occurred at "dawn." The results indicate that the phase of the circadian rhythm of plasma prolactin with respect to the light–dark cycle is influenced by temperature.

A comparison of the daily rhythms of plasma corticosteroid and prolactin rhythms in fish acclimated to late winter–early spring (20°–22°C) and summer (28°–32°C) temperatures reveals that both hormones peak about 8 hr after the onset of light at the low temperature and that the prolactin peak occurs about 16 hr after (8 hr before) the corticosteroid peak at the high temperature. The 0-hr relation was shown by injection to stimulate conditions present during the winter and early spring (fattening, gonadal growth, preference for water of higher salinity, and increased capacity to reduce plasma chloride concentration) whereas a 16-hr relation produced conditions appropriate for summer (loss of fat, gonadal regression, preference for lower salinity, and a reduction in plasma chloride concentration) (see Fig. 2).

The effects of the temporal hormonal synergism are probably both direct at the tissue level as well as indirect by way of the nervous system. Influences on the gonads may best be explained as an influence on the mechanisms that control the release of gonadotropic hormone. Influences on chloride concentrations may be results of direct interaction of corticosteroid and prolactin at the tissue level. Inasmuch as inhibitors of prostaglandin synthesis block the stimulatory action of prolactin on plasma chloride concentration it follows that corticosteroid, which also inhibits prostaglandin synthesis (Tashjian *et al.*, 1975), should have a similar effect. Our results are consonant with this conclusion in that prolactin injections stimulate an increase in plasma chloride concentrations when plasma corticosteroid levels are high and a decrease in plasma chloride when the corticosteroid concentration is low. This relation appears to hold also for the

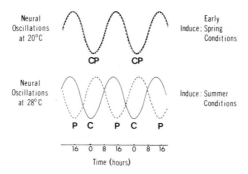

Fig. 2.   Circadian basis for temperature control of physiological and behavioral conditions in the gulf killifish, *Fundulus grandis*. According to this model the temporal relations of two circadian neural oscillators vary according to ambient temperature and account for many of the effects of temperature on metabolic and behavioral conditions. The circadian rhythms of plasma corticosteroid and prolactin are expressions of the two oscillations and account directly for some of the temperature effects. Daily injections of cortisol and prolactin in time relations that stimulate peak levels of the endogenous hormones in the plasma induce the appropriate physiological and behavioral changes. Thus simultaneous injections of cortisol and prolactin (to stimulate hormone rhythms at 20°C, a characteristic ambient temperature of early spring habitat) produce conditions present in early spring (increases in gonad weights, body fat stores, and capacity to reduce blood chloride concentrations and a preference for water of greater salinity). Daily injections of prolactin at 16 hours after cortisol injections (to simulate hormone rhythms at 28°C, a characteristic temperature of water in summer) induce conditions appropriate for summer (decreases in gonad weights and body fat stores, increase in capacity to raise blood chloride concentration, and preference for water of lower salinity). In addition to direct effects at the peripheral level, cortisol and prolactin injections are thought to set the phases of the neural oscillations and thus indirectly to produce changes in metabolism and behavior. C and P represent the times of peak plasma levels of corticosteroid and prolactin, respectively.

normal circadian rhythm of plasma chloride concentration in *F. grandis* maintained in water at 25°C. The plasma levels increase during the early part of the day, reach a peak at about 8 hr after the onset of light, and then fall sharply during the next several hours (Spence *et al.*, 1977). That is, endogenous prolactin appears to account for both the early rise in plasma chloride when corticosteroid levels are low and also for the reduction in chloride concentrations when plasma corticosteroid levels are high.

G. Summary Remarks

A very extensive literature exists concerning physiological mechanisms involved in fish migration. Osmoregulation and metabolic support of locomotor activity in particular have received much attention. Adrenocorticosteroid (interrenal) hormone and prolactin have central roles.

Corticosteroid is associated with adjustment to saline water and with mobilization of energy stores. Prolactin, on the other hand, enhances survival in fresh water and stimulates increases of energy stores. However, an interpretation of the hormonal roles must also take into account evidences that hormonal complexes, involving both synergistic and antagonistic relations, regulate migratory conditions and behavior. For example, marked corticosteroid activity for the metabolic support of vigorous activity is found in Pacific salmon during the anadromous migration when the fish are in fresh water. Similarly there are evidences of prolactin production in fish in seawater. Observations of the effects of temporal relations of the circadian rhythms of hormones as well as relative amounts of hormone productions offer an explanation for these apparent contradictions.

Preparation for migration and adjustments during migration involve an intricate interaction of endogenous systems with exogenous stimuli. Responses to cues vary depending on the physiological readiness of the animal. A report of opposite directions taken by silver and yellow eels released at sea illustrates this important principle (Tesch, 1974). Physiological readiness of the metabolic and behavioral systems involved in migration depends on circadian mechanisms that include the daily rhythms of corticosteroid and prolactin. Because of the central role of temperature in environmental timing of fish physiology and behavior, it is especially noteworthy that temperature can alter phases of circadian rhythms and thereby direct physiological and behavioral conditions (Section III,F). Further research dealing with the roles of circadian, lunar, and circannual rhythms in the organization of fish biology is clearly projected by the evidence.

IV. Migration in Other Vertebrates

A. AMPHIBIANS

Many frogs, toads, and salamanders migrate short distances to water breeding areas and some amphibians migrate to winter hibernating locations (Oliver, 1955). Although their migrations are comparatively short, amphibian migrants share mechanisms that have important roles in the great migrators. Circadian mechanisms have been demonstrated in a variety of behaviors important in migration. Daily rhythms of locomotor activity persist under continuous light conditions and are entrained by 24-hr LD schedules (Higginbotham, 1939; Cloudsley-Thompson, 1967; Adler, 1970). Both the eyes and extraoptic photoreceptors perceive light for photoentrainment of rhythms. In the southern cricket frog, *Acris gryllus*, removal of the eyes as well as occlusion of light from the brain were required to prevent adjustment of a sun-compass mechanism to a shift in an LD regimen (Taylor and Ferguson, 1970). Interestingly an extraoptic photoreceptor as well as the eyes were used to orient correctly on a compass course. The frontal organ has been implicated as an extraoptic photoreceptor for entrainment of the locomotor activity rhythm in the green frog, *Rana clamitans* (Adler, 1971).

As in teleosts and birds, prolactin has an important role in migration of some amphibians. Movement of the terrestrial stage of the newt, *Notophthalmus viridescens*, to water for breeding and the accompanying adaptations for aquatic survival are triggered by prolactin (Chadwick, 1940; Grant and Grant, 1958). The water drive response occurs within about 1 week following prolactin injections. However, sensitivity of the response varies with respect to a daily photoperiod and with time of corticosterone injection in newts maintained in LL (Meier *et al.*, 1971a).

B. Reptiles

Little can be added in this section that has not been reviewed elsewhere in this volume. Although fat stores have a role in supporting migration of sea turtles, the regulation of fattening and fat mobilization has not been explored. Studies of a strictly sedentary lizard, *Anolis carolinensis*, indicate that lipogenesis may be induced by daily injections of corticosteroid hormone and prolactin in specific temporal relations (Meier *et al.*, 1971c; Trobec, 1974). These findings are similar to those reported for fish and bird migrants, and they suggest that circadian mechanisms have important roles in the metabolic preparation of reptilian migrants as well. Circadian rhythms of locomotor activity that are entrained by daily cycles of either light or temperature fluctuations and that persist under constant environmental conditions have also been noted in sedentary species (Taylor, 1962; Cloudsley-Thompson, 1965). Temporal interactions of daylength and temperature have been shown to be important determinants of fattening in a lizard (Noeske and Meier, 1977), findings that might also be important in understanding environmental regulation of migratory indices.

C. Mammals

The longest migration by a mammal is that performed by the gray whale, *Eschrichtius robustus*, which travels 9000–12,000 km from feeding grounds in Arctic waters of the Bering Sea to calving grounds in bays of northeastern Mexico (Rice and Wolman, 1971). Enormous amounts of energy are stored in blubber coats when they leave the feeding grounds in autumn. This fat supply must last for about 6 months because little food is available for this species during migration and calving. The metabolic demands of pregnancy and lactation are superimposed on that of migration in the female. Apparently migration to warmer water for calving is necessary for reasons of thermoregulation. Subcutaneous fat or blubber is an effective insulation in cold water which can transfer heat away from a submerged mammal about 25 times as fast as air (Bryden, 1964; Ridgway, 1972). *Eschrichtius robustus* calves are born with little blubber. The percentages of blubber to body weight in adult cetaceans are reported from 21 to 45 (Slijper, 1962).

The level of energy store and changes in daylength are thought to be important determinants of migration in marine mammals (Dawbin, 1966; Kinne, 1975). Definitive research of neuroendocrine regulation is apparently lacking. Curiously the migration routes of male and female seals, *Zalophus californianus*, differ radically. Females stay near the breeding grounds or move southward whereas males migrate northward (Fry, 1939). Although orientational and metabolic capabilities are highly evolved in marine mammals, both resident and migratory populations are often found in the same species.

Owing to their considerable dependence on flying insects, bats of the temperate regions have employed both hibernation and migration to cope with the winter loss of their food supply. Fat is accumulated prior to either hibernation or migration (Orr, 1954; Twente, 1955). In some species there are both hibernating and migrating populations. Some individuals hibernate as well as migrate. These findings reemphasize a conclusion that physiological and behavioral mechanisms for migration are present in diverse taxa but may be blocked in favor of another life style. Circadian rhythms entrained by a daily LD schedule have been reported (Menaker, 1961; DeCoursey and DeCoursey, 1964). Regulatory mechanisms for migration have received scant attention in this interesting group.

A review of physiological mechanisms that might be expected to contribute toward the support of mammalian migration would be too exhaustive for the purposes of this treatise. If it were attempted, a major conclusion of the review would be that migration requires the total physiological attention of the animal. The student of animal migration can ill afford to set aside areas of investigation as irrelevant. Perhaps special attention should be given to research dealing with the physiology and biochemistry of the brain as they relate to circadian and neuroendocrine regulation. This topic is discussed in Section VI.

V. Invertebrate Migrations

A. MARINE INVERTEBRATES

Extensive movements of invertebrate species occur to and from estuaries. In some instances spawning must occur in concentrated seawater for survival of the eggs and larvae, but the young mature in fresh water. In other cases most of the life cycle is spent in seawater but spawning occurs in more dilute media. Lobsters (*Panuliris, Homarus,* and *Jasus* species) and crabs (*Callinectes species*) may migrate 8–50 km (Bainbridge, 1961). Extensive open sea migrations are also carried out by some species. A squid, *Loligo vulgaris,* migrates in spring and summer from the Eastern Atlantic to the North Sea where they spawn and return to the Atlantic in autumn and winter (Tinbergen and Verwey, 1945). The migratory route of another squid, *Todarodes pacificus,* is over 1000 km (Tashiro *et al.,* 1972).

Movements to and from estuaries require remarkable adaptations for metabolic support and osmoregulation. The blue crab, *Callinectes sapidus,* illustrates these adaptations particularly well (Mangum and Towle, 1977). In the autumn after it copulates with a male, the female leaves the tributaries and migrates to more saline water where it burrows into the sand for a period of winter dormancy. When the temperature rises above 10°C in the spring, the crabs emerge and spawn. After development and metamorphosis to the adult form, the young ascend the estuaries to feed and grow in dilute waters. The upstream migration may be as much as 200 km (Fiedler, 1930) and requires the development of locomotor muscles that fortunately are large enough for humans to eat. Blood osmolality in *C. sapidus* conforms to an osmotic change in the medium until the water reaches a salinity of about 27%. At lower salinities the crab changes to hyperosmotic regulation of the blood although homeostasis is not maintained. During migration from water of 35 to 1% salinity, the reduction in total blood salt is about 37%. The intracellular fluid establishes osmotic equilibrium with the dilute blood less by losing salt than by decreasing the size of a mobile pool of osmotically active free amino acids which involves an increase in deamination (Gerard and Gilles, 1972). The resultant increase in blood NH_4^+ is thought to serve as a counterion for Na^+ and stimulate a greater intake of Na^+ from the medium by sodium pumps in the gills (Towle and Taylor, 1976; Mangum and Towle, 1977). The economy of relating a byproduct of energy demands (NH_4^+) with osmoregulatory requirements in crabs moving upstream is similar to that observed in the anadromous migration of salmon (Section III,D). Much of the body fat in crabs is stored in the hepatopancreas or digestive gland (Giese, 1966).

Many invertebrate forms utilize the tides for horizontal migration. The postlarval form of the pink shrimp, *Panaeus duorarum*, is positively rheotactic during flood tide off the Everglades of Florida when the water is more saline and moves downstream during ebb tide when the water is less saline (Tabb *et al.*, 1962; Hughes, 1972). Similar examples have been cited which allow animals to move in either direction with respect to tides. They may stay on the bottom during ebb tide and let the flood tide carry them shoreward (Creutzberg, 1975). Because directed movements occur at specific stages of development and times of year, regulatory mechanisms are implicit in these findings.

Regulation of locomotor activity is complicated in estuarine species in that both circadian and tidal rhythms may be present. The daily photoperiod is the most conspicuous entrainer of circadian rhythms. In the crayfish, *Procambarus clarkii*, for example, there is a bimodal rhythm of locomotor activity under a LD 12:12 light cycle (Page and Larimer, 1972). The activity peak associated with "lights on" was purely exogenous and disappeared following retinal ablation or when the crayfish were kept in continuous dark (DD). However, the peak associated with "lights off" persisted in the DD and also in LD following retinal ablation. Thus retinal photoreception does not entrain the circadian rhythm of locomotion in *P. clarkii*; extraocular receptors for photoentrainment are probable for decapod crustacians as well as other invertebrate taxa (Arechiga and Naylor, 1976).

Temperature (Williams and Naylor, 1969), mechanical stimulation (Enright, 1963, 1965), pressure (Naylor and Atkinson, 1972), and light (Rodriques and Naylor, 1972; Palmer, 1973) changes have all been shown to entrain or influence tidal rhythms of locomotion in crustaceans (period: 24.8 hr). Whether separate mechanisms are involved in regulating circadian and circatidal rhythms has not been satisfactorily resolved. The eyestalk has been implicated in the regulation of both circadian and circatidal rhythms (Palmer, 1973). The eyestalk generally exerts an inhibitory influence on locomotion (Roberts, 1941; Brown, 1961). The X-organ/sinus gland complex of the eyestalk is thought to be responsible for the periodic elaboration and release of a neurohormone; eyestalk extracts have been shown to have tidal (Naylor *et al.*, 1973) and daily (Arechiga, 1974) rhythms of hormonal potencies in crayfish. However, the locomotor activity rhythm may not be maintained solely by a cyclical hormone release. Motoneurons may be activated directly by neuronal oscillators. Temperature above 8°C is required for the expression of a tidal rhythm of locomotor activity in a crab, *Carcinus maenas* (Creutzberg, 1975).

Complex interactions of light, temperature, pressure, and sound in causing timed and directed locomotor activity is of considerable interest and might form the basis for physiological research. This subject has been considered elsewhere (Papi, 1955; Enright, 1962; Palmer, 1973). A common feature of many of these studies involving environmental stimuli and directional cues is the involvement of circadian or circatidal rhythms. For example time-compensated sun-compass orientation has been described for amphipod crustaceans (Pardi and Papi, 1952).

Interaction of circadian and circatidal rhythms offer numerous possibilities for elaboration of lunar rhythms. Because there are tidal fluctuations of temperature and salinity in many areas, circadian rhythms of sensitivity or responsiveness to these environmental stimuli may be expected to form lunar rhythms of temperature and salinity effects. Numerous lunar cycles involving population movements have been described for both invertebrate and vertebrate marine species (Palmer, 1973).

B. INSECTS

Migratory displacements of insects serve to transport individuals from suboptimal to optimal habitats for overwintering, feeding, or reproduction. The displacements range from a few meters (mosquitoes) to thousands of kilometers (some grasshoppers and butterflies). Migratory flights usually occur early in the life of the individual before reproduction and are timed by temperature and daylength (Johnson, 1969; Dingle, 1974).

The migratory syndrome involves properly coordinated behavioral, morphological, and metabolic states. Fat is the primary energy source for sustained flight; glycogen serves only for short-term flight (Weis-Fogh, 1952). Lipid is advantageous in that it requires less storage weight per caloric equivalent than carbohydrate or protein (Beenakkers, 1965). Sufficient fat is deposited in the fat bodies prior to migration to sustain flight for 35 hr in the desert locust, *Schistocerca gregaris*, and for 117 hr in the monarch butterfly, *Danaus plexippus* (Johnson, 1974).

Polymorphism of the flight apparatus is a common characteristic among species with migratory individuals. Migratory forms develop wings and functional wing muscles (alatery) whereas nonmigratory forms often lack flight structures (aptery) or have poorly developed flight structures (brachyptery). The flight apparatus degenerates following migration coincident with reproductive development (Johnson, 1969; Dingle, 1972).

Prenatal or larval exposure of true bugs (Heteroptera) and aphids (Aphididae) to specific daylengths (Lees, 1967; Sutherland and Mittler, 1969; Vepsalainen, 1971; Rankin, 1974) and temperature (Brinkhurst, 1959; Southwood, 1961; Schaefers and Judge, 1971) can determine whether migratory or nonmigratory forms develop. Crowding also induces development of migratory forms and behavior in aphids (Lees, 1961; Toba et al., 1967) and locusts, Schistocerca gregaria and Locusta migratoria migratoriodes (Carlisle and Ellis, 1959; Johnson, 1969). Some of the effects of crowding, though not all, result from changes in food quality (Johnson and Birks, 1960; Schaefers and Judge, 1971). Deficiencies of some amino acids in the larval diet stimulate development of migratory forms (Mittler and Dadd, 1966).

The prothoracic gland and corpora allata have been implicated in the regulation of migratory behavior. The prothoracic gland has an inhibitory influence. Homogenates and implants reduced locomotor activity of gregarious (migratory form) immature (Carlisle and Ellis, 1963) and mature (Michel, 1972) locusts. Ecdysone, a major hormone of the prothoracic gland, reduced the spontaneous motoneuron activity from the mata-thoracic ganglia of the locust, Schistocerca gregaria (Haskell and Moorhouse, 1963). The prothoracic gland is smaller in the immature migratory locust than in the nonmigratory form; it eventually disappears in the migrant but remains in the nonmigrant (Carlisle and Ellis, 1959).

The corpora allata (CA) conversely have stimulatory influences on migratory indices in a variety of species. Removal of the CA severely decreased locomotor activity in Schistocerca gregaria (Odhiambo, 1965, 1966; Strong, 1968). Although Strong did not observe reduced activity in Locusta migratoria following allatectomy, Wajc and Pener (1971) reported that allatectomy reduced flight intensity and Cassier (1963, 1964) found that CA implants increased speed of movement in gregarious L. migratoria. Implantation of CA or application of juvenile hormone, a principle secretion of CA, induced long sustained flights in the adult milkweed bug, Oncopeltus fasciatus. Similar treatment was effective in the ovariectomized but not in the intact female (Caldwell and Rankin, 1972; Rankin, 1974). Elimination of migratory activity by starvation was accompanied by a decline of juvenile hormone; reestablishment of migratory activity was accomplished by treatment with synthetic juvenile hormone (Rankin and Riddiford, 1977). These results support the general hypothesis of Johnson (1969) who proposed that migration is stimulated by low levels of juvenile hormone in the absence of ecdysone.

A role for corpora allata in determination of direction of migration has been proposed for the female cockchafer, Melolontha melolontha (Stengel, 1974). During the reproductive season female cockchafers undergo two or three ovarian cycles each of which involves oriented migration from feeding areas to egg laying sites (preovulatory flight) followed by reserve migration (prefeeding flight). Implantation into prefeeding recipients of CA obtained from females during the preovulatory flight caused a reversal of migration. Similar findings were obtained in the Colorado beetle, Lep-

tinotarsa decemlineata, which migrates from host plant to soil preparatory for diapause in response to short daylengths. This migration could be induced out of season in animals exposed to long daylengths by removal of the CA. Reimplantation of CA reversed the migration (de Wilde and de Boer, 1961, 1969).

Studies of locusts indicate that adipokinetic hormone from the corpora cardiaca mobilize lipid from the fat bodies for support of sustained flight. Carbohydrates provide the energy source for the first 30 min of flight (Weis-Fogh, 1952). It is thought that the flight process itself stimulates release of adipokinetic hormone which promotes lipid utilization by flight muscles (Goldsworthy *et al.,* 1975; Robinson and Goldsworthy, 1977). Hemolymph levels of adipokinetic hormone are elevated in locusts that have flown for more than 30 min (Mayer and Candy, 1969; Houben and Beenakkers, 1973).

Neuronal regulation of locomotor activity appears to involve serotonergic neurons. Serotonin injections increased both the intensity and duration of night flights in moths, *Noctua pronumba* and *Agrotis ipsilon* (Hinks, 1967), whereas inhibition of serotonin activity with reserpine decreased locomotor activity in crickets, *Acheta domesticus* (Cymborowki, 1973). Daily rhythms of brain neurosecretory material and serotonin content are inversely related to locomotor activity rhythms in crickets (Cymborowski, 1970) and fruit flies (*Drosophila melanogaster*) (Fowler *et al.,* 1972), respectively. Decreasing levels of neurosecretory material and serotonin suggest greater neuronal activity. Whether neuronal regulation of locomotor activity rhythms is direct or mediated hormonally has not been ascertained (see Brady, 1974; Saunders, 1976). Circadian organization in the neuroendocrine control of migratory behavior and metabolism in insects is poorly understood.

VI. Conclusions and Projections

In view of scientific reductionism prevailing during the previous two decades, it may not be superfluous to emphasize the obvious: Migration involves the total organism. Because a migrant is also involved in other important processes, such as growth and reproduction, migratory behavior and metabolism must be properly integrated within the entire biology of the species. Temporal organization is crucial so that metabolism supports behavior and that one event, such as migration, leads to another, such as reproduction, within an appropriate temporal framework. Temporal organization is manifested by the existence of developmental changes and of daily, lunar, and annual cycles. Persistence of circadian and circannual rhythms under constant daily and annual environmental conditions point to the presence of endogenous neuroendocrine oscillators that direct the rhythms.

Circadian rhythms are the basic units of temporal organization. Circadian neural and endocrine rhythms account for daily variations in behavior and metabolism and they interact with one another to produce circannual cycles. Alterations of the phase relations of two or more circadian neuroendocrine oscillators apparently induce seasonal changes in metabolic and behavioral conditions that constitute a circannual cycle. The daily rhythms of corticosteroid hormone and prolactin appear to be the expressions of two circadian oscillators. Daily injections of these hormones in various temporal relations are thought to set the phase angles of neural oscillators so that all of the hormones and other regulatory systems appropriate for specific seasons of the year are produced. Daylength is the most important environmental synchronizer of circannual mechanisms in photoperiodic animals.

Probably because temperature rather than daylength is the most important environmental regulator of metabolism and reproduction in most ectothermic vertebrates, relatively little attention has been given to circadian mechanisms among this group. However, our studies of the gulf killifish, *Fundulus grandis*, seem to implicate circadian mechanisms in temperature control of osmoregulation, lipid metabolism, reproduction, and behavior (see Section III,F). Temperature was shown to control the phase angle of the circadian rhythm of prolactin with respect to the daily photoperiod (Spieler *et al.*, 1978) and thus to set the temporal relation of the circadian rhythms of cortisol and prolactin. Apparently temperature influences a system in fish analogous to a circannual mechanism in birds. Whether a temperature effect in *F. grandis* is imposed on a circannual mechanism has not as yet been ascertained.

The possibility that circadian mechanisms mediate temperature control of metabolism and behavior in ectothermic animals is attractive from several aspects. Some of the abrupt changes in behavior and metabolism that occur at specific temperatures are much more readily understood in terms of phase changes in circadian systems than in terms of graded thermal influences on biochemical activities. Diurnally active salmon, for example, become nocturnal when specific temperatures are exceeded (Byrne, 1968) and trout change from negative to positive rheotaxis at a specific temperature (Tesch, 1975). Cirannual mechanisms, photoperiodism, and thermoregulation may all utilize the same basic circadian mechanisms. A role of circadian systems in temperature regulation of metabolism and behavior may be of signal importance, and merits further investigation.

Orchestration of the hormonal rhythms is an obvious function of the brain. However, almost nothing has been done as yet to delineate these neural mechanisms for the specific purpose of understanding migration. Recent studies of neurotransmitters, which convey nerve impulses from one neuron to another, indicate that this methodology may prove extremely useful for investigations of neural oscillators. Mammalian work has demonstrated that serotonergic, noradrenergic, cholinergic, and dopaminergic neurons are parts of the pathways that regulate corticosteroid and prolactin release. The daily rhythm of plasma corticosteroid concentration is abolished by p-chlorophenylalanine (PCPA), an inhibitor of serotonin synthesis (Krieger and Rizzo, 1969). A precursor of serotonin, 5-hydroxytryptophan, as well as a potentiator of serotonergic activity, fluoxetine, stimulate large increases of plasma corticosteroid concentration (Fuller et al., 1976). Cholinergic activity is apparently necessary for serotonergic stimulation (Krieger and Krieger, 1967).

Prolactin production by the mammalian adenohypophysis is under inhibitory control by dopaminergic neurons in the hypothalamus (Krulich et al., 1975). A precursor of dopamine, l-dopa, decreases plasma prolactin concentrations. Serotonergic activity in turn is thought to inhibit the dopaminergic inhibition of prolactin release (Kamberi et al., 1970; Clemens et al., 1977). The short-loop negative feedback of prolactin is thought to occur by prolactin stimulation of dopamine synthesis (Kamberi et al., 1970). Inhibitors and stimulators of neurotransmitter synthesis and activity have also been utilized to delineate types of neurons involved in the productions of other hormones.

The sleep–wakefulness cycle, as well, has been studied in mammals with respect to neurotransmitters. Serotonergic neuronal activity dominates during sleep whereas noradrenergic and dopaminergic activities are more important during wakefulness. Dopamine and norepinephrine stimulate locomotor activity in rodents (Geyer et al., 1972; Benkurt et al., 1973; Ranje and Ungerstedt, 1977) whereas a catecholamine inhibitor decreases activity (Gorden and Shellenburger, 1974). On the other hand, inhibitors of serotonin synthesis increase spontaneous locomotor activity (Fibiger and Campbell, 1971; Jacobs et al., 1975). In the whole brain and in most specific regions of the brain, assays have demonstrated a gradual rise of serotonin content during the waking period and a subsequent fall during sleep (Albrecht et al., 1956; Quay, 1965; Hery et al., 1972). A reversal of this pattern is found for brain catecholamine content (Gorden and Shellenburger, 1974).

Studies of neurotransmitter roles in the control of locomotor activity of nonmammalian organisms have just begun. In the gulf killifish, Fundulus grandis, injections of serotonin, which presumably crosses the blood–brain barrier, decreased swimming activity (Fingerman, 1976). Brain serotonin levels tended to increase during the daily locomotor activity period and to decrease during the period of quiescence. The daily decline of brain serotonin was attributed to activity of serotonergic neurons that inhibit locomotor activity. Interestingly in insects serotonergic activity seems to be stimulatory for locomotor activity (Hinks, 1967; Cymborowski, 1973).

Preliminary investigations in our laboratory using neurotransmitter inhibitors and stimulators to study avian migration have produced promising results (L. J. Miller, unpublished). Vernal migratory and reproductive indices were induced in autumn in the white-throated sparrow, *Zonotrichia albicollis*, by injections of several drugs that included an inhibitor (PCPA) of serotonin synthesis, a stimulator or serotonergic activity (fluoxetine), and a stimulator of noradrenergic activity (L-dopa). Daily injections of the drugs in specific temporal relations for 10 days induced gonadal growth, fattening, nocturnal migratory restlessness, and northward (toward the breeding grounds) orientation. These results seem to indicate that it is possible to set the phase of a circannual clock and thereby produce a specific complex of seasonal physiological and behavioral conditions.

The suprachiasmatic nucleus (SCN) of the hypothalamus appears to be an important link in the pathway of one of the neuronal oscillators. Bilateral lesions of the SCN in rats obliterated circadian rhythms of locomotor activity (Stephan and Zucker, 1972; Rusak, 1977; Stephan and Nunez, 1977), sleep–wakefulness (Ibuka and Kawamura, 1975), and adrenal corticosterone (Moore and Eichler, 1972). These results were obtained in animals kept on a LD regimen. The SCN receives abundant serotonergic terminals from the raphe nucleus which is also an important center for the induction of sleep (Fuxe, 1965).

Pinealectomy has no influence on the locomotor activity rhythm of either the sedentary house sparrow, *Passer domesticus* (Menaker, 1968), or the migratory white-throated sparrow, *Zonotrichia albicollis* (McMillan, 1972), maintained on normal LD schedules. However, pinealectomy did dampen or remove the free-running rhythm in DD. Because these results are not found consistently in vertebrates or even in *Z. albicollis* (Meier, 1975), it seems probable that the pineal is not an obligatory coupler but rather a modifier of more basic neuronal oscillators. Many hormones have been shown to have modifying effects on circadian rhythms. Because melatonin, a pineal hormone, can stimulate an increase in brain serotonin content in rats (Anton-Tay *et al.*, 1968), the loss of circadian activity in sparrows kept on DD may be a consequence of a loss in a circadian rhythm of serotonergic inhibition.

The permissive and intensifying effects of heavy fat stores on migration may also be rationalized in a manner that involves serotonergic activity. Adipose tissue of fat animals is more sensitive to lipolytic hormones and releases more FFA upon stimulation than that of lean animals (Freeman and Manning, 1977). Free fatty acids, as well as tryptophan (no other amino acids) (Wurtman and Fernstrom, 1975), are carried by albumin. Competition between FFA and tryptophan, then, for albumin sites might promote hepatic and peripheral removal of tryptophan, reduce the supply of precursor for serotonin synthesis in the brain, and thereby reduce serotonergic inhibition of locomotor activity.

Migration of fish and birds has often been observed to occur in waves of greater intensities. Salmon move into the rivers with a passage of warm fronts (Ellis, 1962). Warm fronts are also stimulatory for bird migration in spring whereas cold fronts are stimulatory in autumn (Bagg, 1950; Bennett, 1952; Pettingill, 1972). The effects of weather fronts have generally been ascribed to changes in barometric pressure. Except for the seasonal alternation in birds, locomotor activity intensifies with increasing pressure in endothermic animals, including rabbits (Johnson and Hendrickson, 1958; Chapman and Tretheway, 1972) and rodents (Sprott, 1967; Truchan and Boyer, 1972). On the other hand, locomotor activity intensifies with decreasing pressure in ectothermic animals such as insects (Underhill, 1940; Haufe, 1946; Wellington, 1954; Edwards, 1961), marine invertebrates (Brown et al., 1955), fish (Petterson, 1972), frogs (Knepton, 1952; Oliver, 1955), and lizards (Bowker, 1977).

Correlations of locomotor activity with changes in atmospheric pressure may be misdirected. Olivereau (1971) pointed out that ion changes occur with atmospheric pressure changes and therefore also correlate with changes in locomotor activity. Air ions undergo bimodal daily fluctuations as well as annual changes that correlate well with the changes in atmospheric pressure. Positive ions tend to increase with decreasing pressure and negative ions increase with increasing pressure. When air ions were maintained constant in experiments with rats, the correlation between activity and atmospheric pressure disappeared (Olivereau, 1971). Other experiments with rats demonstrated that locomotor activity is generally enhanced by exposure to negatively ionized atmospheres (Herrington and Smith, 1935; Bachman et al., 1966) and depressed by exposure to increasing concentrations of positive ions (Bachman et al., 1966). However, in an ectothermic insect, Calliphora vicina, positive ions stimulated flight activity (Edwards, 1960). The effects of ions on locomotor activity may be mediated by way of ion effects on serotonin concentrations. Negatively charged air increased blood serotonin in rats (Krueger et al., 1968). Experimental studies are needed to determine whether air ions might have significant roles in timing migratory activity.

Whereas most behaviorists seem impressed with the diversity among animals, physiologists in general are more inspired by the conservativeness

of physiological mechanisms. Although physiological as well as many
behavioral innovations are present among migrants, the evidence en-
courages a conservative attitude in a evaluation of regulatory mechanisms.
There is is little reason to believe that the physiological mechanisms of
migrants differ basically from one another or even from those of sedentary
species. The expression of migratory or sedentary behavior is often determined by slight genetic variation or environmental influence. This dual
capacity in the genetic blueprints of many species of the major vertebrate
classes as well as of many invertebrate phyla suggest an ancient common
evolutionary background that involved periods of both lifestyles. An alternation of migratory and sedentary behaviors may well have been repeated
often in the divergent taxa. Convergent evolution of migratory behavior,
then, may occur at a relatively superficial level; basic physiological
mechanisms seem to have diverged little.

Acknowledgment

We thank Ms. Deborah M. McMullan for typing the manuscript.

References

Abraham, M. (1974). *Annee Biol.* **13**, 143–149.
Ackman, R. G. (1964). *J. Fish. Res. Board Can.* **21**, 247–254.
Adams. B. L., Zaugg, W. S., and McLain, L. R. (1973). *Comp. Biochem. Physiol. A* **44**, 1333–
    1339.
Adler, K. (1970). *J. Herpetol.* **4**, 99–112.
Adler, K. (1971). *In* "Biochronometry" (M. Menaker, ed.), pp. 351–361. Natl. Acad. Sci.,
    Washington, D. C.
Albrecht, P., Visscher, M. B., Bittner, J. J., and Halberg, F. (1956). *Proc. Soc. Exp. Biol. Med.*
    92, 703–710.
Ali, M. A., (1964). *Proc. Indian Acad. Sci., Sect. B* **60**, 249–263.
Andreasson, S., (1973). *Oikos* **24**, 16–23.
Anton-Tay, F. C., Chou, S., and Wurtman, R. J. (1968). *Science* **162**, 277–278.
Arechiga, H. (1974). *In* "The Neurosciences: Third Study Program" (F. O. Schmitt, ed.),
    pp. 517–523. MIT Press, Cambridge, Massachusetts.
Arechiga, H., and Naylor, E. (1976). *In* "Biological Rhythms in the Marine Environment"
    (P. J. DeCoursey, ed.), pp. 1–16. Univ. of South Carolina Press, Columbia.
Aschoff, J. (1967). *Proc. Int. Ornithol. Congr., 14th, 1966* pp. 81–105.
Assenmacher, I. (1973). *In* "Avian Biology" (D. S. Farner, J. R. King, and K. C. Parkes, eds.),
    Vol. 3, pp. 183–286. Academic Press, New York.
Bachman, C. H., McDonald, R. D., and Lorenz, P. J. (1966). *Int. J. Biometeorol.* **10**, 39–46.
Bagg, A. M. (1950). *Wilson Bull.* **62**, 5–19.
Baggerman, B. (1957). *Arch. Neerl. Zool.* **12**, 105–317.

Baggerman, B. (1959). *In* "Comparative Endocrinology" (A. Gorbman, ed.), pp. 24–37. Wiley, New York.

Baggerman, B. (1960). *J. Fish. Res. Board Can.* **17**, 295–322.

Baggerman, B. (1963). *Can. J. Zool.* **41**, 307–319.

Baggott, G. K. (1975). *J. Zool.* **175**, 299–314.

Bainbridge, R. (1961). *In* "The Physiology of the Crustacea" (T. H. Waterman, ed.), Vol. 2, pp. 431–463. Academic Press, New York.

Baker, J. R. (1938). *Proc. Zool. Soc. London* **108**, 557–582.

Ball, J. N., Giddings, M. R., and Hancock, M. P. (1966). *Am. Zool.* **6**, 388.

Banks, J. W. (1969). *J. Fish. Biol.* **1**, 85–136.

Baraduc, M. M. (1954). *C. R. Hebd. Seances Acad. Sci.* **283**, 728–730.

Bates, R. W., Miller, R. A., and Garrison, M. M. (1962). *Endocrinology* **71**, 610–614.

Beatty, D. D. (1972). *Vision Res.* **12**, 1947–1960.

Beenakkers, A. M. T. (1965). *J. Insect Physiol.* **11**, 879–888.

Benkurt, O., Gluba, H., and Matussek, N. (1973). *Neuropharmacology* **12**, 177–186.

Bennett, H. F. (1952). *Wilson Bull.* **64**, 197–220.

Bentley, P. J., and Follett, B. K. (1965). *J. Endocrinol.* **31**, 127–137.

Berger, M., and Hart, J. S. (1974). *In* "Avian Biology" (D. S. Farner and J. R. King, eds.), pp. 415–477. Academic Press, New York.

Bern, H. A., and Nicoll, C. S. (1968). *Recent Prog. Horm. Res.* **24**, 681–720.

Bernardi, C. (1948). *Riv. Biol.* **40**, 186–228.

Berthold, P. (1977). *Naturwissenschaften* **64**, 389.

Berthold, P., Gwinner, E., and Klein, H. (1971). *Experientia* **27**, 399.

Berthold, P., Gwinner, E., and Klein, H. (1972). *J. Ornithol.* **113**, 170–190.

Bilinski, E., and Lau, Y. C. (1969). *J. Fish. Res. Board Can.* **26**, 1857–1866.

Bissonnette, T. H. (1937). *Wilson Bull.* **49**, 249–270.

Bitinkov, E. P. (1959). *Dokl. Akad. Nauk SSSR* **128**, 179–182.

Black, E. C. (1958). *In* "The Investigation of Fish-power Problems" (P. A. Larkin and H. R. MacMillan, eds.), pp. 51–67. University of Columbia Institute of Fisheries, Vancouver.

Blanc-Livni, N., and Abraham, M. (1970). *Gen. Comp. Endocrinol.* **14**, 184–197.

Blem, C. R. (1976). *Am. Zool.* **16**, 671–684.

Blyumental, T. I. (1973). *In* "Bird Migrations, Ecological and Physiological Factors" (B. E. Bykhovskii, ed.), pp. 125–218. Wiley, New York.

Boehlke, K. W., Church, R. L., Tiemeier, O. W., and Elefthériou, B. E. (1966). *Gen. Comp. Endocrinol.* **7**, 18–21.

Bower, E. B., and Helms, C. W. (1968). *Physiol. Zool.* **41**, 157–168.

Bowker, R. W. (1977). Master's Thesis, Louisiana State University, Baton Rouge.

Boyd, J. A., Wiesser, P. B., and Fain, N. (1975). *Gen. Comp. Encocrinol.* **26**, 243–247.

Brady, J. (1974). *Adv. Insect Physiol.* **10**, 1–115.

Branover, G. G., Vasiliev, A. S., Gleiser, S. I., and Tsinober, A. B. (1971). *Vopr. Ikhtiol.* **11**, 720–727.

Brinkhurst, R. O. (1959). *J. Anim. Ecol.* **28**, 211–230.

Brown, F. A., Jr. (1961). *In* "The Physiology of Crustacea" (T. H. Waterman, ed.), Vol. 2, pp. 401–430. Academic Press, New York.

Brown, F. A. Jr., Bennett, M. F., Webb, H. M., and Ralph, C. L. (1955). *Anat. Rec.* **122**, 462–463.

Brown, W. D., and Tappel, A. L. (1959). *Arch. Biochem. Biophys.* **85**, 149–158.

Bryden, M. M. (1964). *Nature (London)* **203**, 1299–1300.

Bünning, E. (1960). *Cold Spring Harbor Symp. Quant. Biol.* **25**, 249–256.

Butler, D. G. (1968). *Gen. Comp. Endocrinol.* **10**, 85–91.

Buus, O, and Larsen, L. O. (1975). *Gen. Comp. Endocrinol.* **26**, 96–99.

Byrne, J. E. (1968). Ph.D. Thesis, University of British Columbia, Vancouver, Canada.

Caldwell, R. L., and Rankin, M. (1972). *Gen. Comp. Endocrinol.* **19**, 601–605.

Callamand, O., and Fontaine, M. (1942). *Arch. Zool. Exp. Gen.* **82**, 129–136.

Carlisle, D. B., and Ellis, P. E. (1959). *C. R. Hebd. Seances Acad. Sci.* **249**, 1059–1060.

Carlisle, D. B., and Ellis, P. E. (1963). *Nature (London)* **200**, 603–604.

Carlson, L. A., Liljedahl, S., Verdy, M., and Wirsen, C. (1964). *Metab., Clin. Exp.* **13**, 227–231.

Cassier, P. (1963). *C. R. Hebd. Seances Acad. Sci.* **257**, 4048–4049.

Cassier, P. (1964). *C. R. Hebd. Seances Acad. Sci.* **258**, 723–725.

Chadwick, C. S. (1940). *Proc. Soc. Exp. Biol. Med.* **45**, 335–337.

Chapman, J. A., and Trethewey, D. E. C. (1972). *J. Wildl. Manage.* **36**, 1221–1226.

Chester-Jones, I., Chan, D. K. O., and Rankin, J. C. (1969). *J. Endocrinol.* **43**, 21–31.

Child, G. I. (1969). *Auk* **86**, 327–338.

Childester, F. E. (1920). *Am. Nat.* **54**, 551–557.

Clark, E. D., and CLough, R. W. (1926). *Rep. U.S. Comm. Fish.* No. 501.

Clemens, J. A., Sawyer, B. D., and Cerimele, B. (1977). *Endocrinology* **100**, 692–698.

Cloudsley-Thompson, J. L. (1965). *J. Zool.* **146**, 55–69.

Cloudsley-Thompson, J. L. (1967). *J. Zool.* **152**, 43–54.

Colombo, G., and Cecchini, S. (1959). *Rend. Sedute Accad. Naz. Lincei* **27**, 136–137.

Connel, C. E., Odum, E. P., and Kale, H. (1960). *Auk* **77**, 1–9.

Creutzberg, F. (1959). *Nature (London)* **184**, 1961–1962.

Creutzberg, F. (1975). *In* "Marine Ecology" (O. Kinne, ed.), Vol. 2, pp. 555–655. Wiley, New York.

Cristy, M. (1974). *Gen. Comp. Endocrinol.* **23**, 58–62.

Cymborowski, B. (1970). *Zool. Pol.* **20**, 127–149.

Cymborowski, B. (1973). *J. Insect Physiol.* **19**, 1423–1440.

Dawbin, W. H. (1966). *In* "Whales, Dolphins, and Porpoises" (K. S. Norris, ed.), pp. 145–170. Univ. of California Press, Berkeley.

DeCoursey, G., and DeCoursey, P. J. (1964). *Biol. Bull. (Woods Hole, Mass.)* **126**, 14–27.

de Rensis, G., and Bornancin, M. (1977). *Biochim. Biophys. Acta* **467**, 192–207.

De Roos, R., and De Roos, C. C. (1972). *Gen. Comp. Endocrinol., Suppl.* **3**, 192–197.

de Vlaming, V. L., and Pardo, R. J. (1975). *Comp. Biochem. Physiol. B* **51**, 489–497.

de Vlaming, V. L., and Sage, M. (1972). *Contrib. Mar. Sci.* **16**, 59–63.

De Wied, D., and Bohus, B. (1966). *Nature (London)* **212**, 1484–1486.

De Wied, D., Bohus, B., and Greven, H. M. (1968). *In* "Endocrinology and Human Behaviour" (R. P. Michael, ed.), pp. 188–196. Oxford Univ. Press, London and New York.

de Wilde, J., and de Boer, J. A. (1961). *J. Insect Physiol.* **6**, 152–161.

de Wilde, J., and de Boer, J. A. (1969). *J. Insect Physiol.* **15**, 661–675.

Dingle, H. (1972). *Science* **175**, 1327–1335.

Dingle, H. (1974). *In* "Experimental Analysis of Insect Behaviour" (L. B. Browne, ed.), pp. 329–342. Springer-Verlag, Berlin and New York.

Dolńik, V. R. (1966). *Zool. Zh.* **45**, 827–909.

Dolńik, V. R. (1967). *Tr. Zool. Inst. Akad. Nauk SSSR* **40**, 96.

Dolńik, V. R. (1968). *Zool. Zh.* **47**, 1205.

Dolńik, V. R. (1970). *Colloq. Int. C.N.R.S.* **172**, 351–364.

Dolńik, V. R. (1974). *Zh. Obshch. Biol.* **35**, 543–552.

Dolńik, V. R., and Blyumental, T. I. (1964). *Usp. Sovrem. Biol.* **58**, 280–301.

Dolńik, V. R., and Gavilov, V. M. (1975). *Ekol. Pol.* **23**, 211–226.

Doyle, W. L., and Epstein, F. H. (1972). *Cytobiology* **6**, 58–73.

Dusseau, J. W., and Meier, A. H. (1971). *Gen. Comp. Endocrinol.* **16**, 399–408.

D'yachenko, V. P. (1972). *Dokl. Akad. Nauk SSSR* **206**, 211–212.

Edwards, D. K. (1960). *Can. J. Zool.* **38**, 1079–1091.

Edwards, D. K. (1961). *Can. J. Zool.* **39**, 623–635.

Ellis, D. V. (1962). *J. Fish. Res. Board Can.* **19**, 137–148.

Emlen, S. T. (1969). *Science* **165**, 716–718.

Emlen, S. T., and Emlen, J. T. (1966). *Auk* **83**, 361–367.

Enright, J. T. (1962). *Comp. Biochem. Physiol.* **7**, 131–145.

Enright, J. T. (1963). *Z. Vergl. Physiol.* **46**, 276–313.

Enright, J. T. (1965). *Science* **147**, 864–867.

Epple, A. (1963). *Proc. Int. Ornithol. Congr., 13th, 1962* pp. 974–982.

Epple, A. (1969). *In* "Fish Physiology" (W. S. Hoar and D. J. Randall, eds.), Vol. 2, pp. 275–319. Academic Press, New York.

Epstein, F. H., Cynamon, M., and McKay, W. (1971). *Gen. Comp. Endocrinol.* **16**, 323–328.

Eyster, M. B. (1954). *Ecol. Monogr.* **24**, 1–28.

Fange, R. (1962). *Pharmacol. Rev.* **14**, 281–316.

Farina, J., Pinto, J., Basabe, J. C., and Chieri, R. A. (1975). *Gen. Comp. Endocrinol.* **27**, 209–213.

Farkas, T. (1969). *Acta Biochim. Biophys. Acad. Sci. Hung.* **4**, 237–249.

Farner, D. S. (1950). *Condor* **52**, 104–122.

Farner, D. S. (1960). *Proc. Int. Ornithol. Congr., 12th, 1958* pp. 197–208.

Farner, D. S. (1970). *Environ. Res.* **3**, 119–131.

Farner, D. S. (1976). *Proc. Int. Ornithol. Congr., 16th 1974* pp. 369–382.

Farner, D. S., Oksche, A., Kamemoto, F. I., King, J. R., and Cheyney, H. E. (1961). *Comp. Biochem. Physiol.* **2**, 125–142.

Farner, D. S., Wilson, F. E., and Oksche, A. (1967). *Neuroendocrinology* **2**, 529–582.

Fibiger, H. C., and Campbell, B. A. (1971). *Neuropharmacology* **10**, 25–32.

Fiedler, R. H. (1930). *Fish. Gaz.* **47**, 18–21.

Fingerman, S. W. (1976). *Comp. Biochem. Physiol. A* **54**, 49–53.

Fischer, H. J., and Bartlett, L. M. (1957). *Condor* **59**, 364–372.

Fivizzani, A. J., Jr. (1977). Ph.D. Dissertation, Louisiana State University, Baton Rouge.

Fivizzani, A. J., Jr., and Meier, A. H. (1976). *Assoc. Southeast. Biol. Bull.* **23**, 58.

Fivizzani, A. J., Jr., and Meier, A. H. (1978). Submitted for publication.

Fogden, M. P. L. (1972). *Ibis* **114**, 548–552.

Follett, B. K., and Farner, D. S. (1966). *Gen. Comp. Endocrinol.* **7**, 111–124.

Fontaine, M. (1930). *C. R. Hebd. Seances Acad. Sci.* **191**, 680.

Fontaine, M. (1960). *Experientia* **16**, 433–438.

Fontaine, M. (1975). *Adv. Mar. Biol.* **13**, 231–355.

Fontaine, M., and Baraduc, M. M. (1954). *C. R. Hebd. Seances Acad. Sci.* **248**, 1942.

Fontaine, M., and Hatey, J. (1953). *Physiol. Comp. Oecol.* **3**, 37–52.

Fontaine, M., and Hatey, J. (1954). *C. R. Hebd. Seances Acad. Sci.* **239**, 319–321.

Fontaine, M., and Olivereau, M. (1949). *C. R. Hebd. Seances Acad. Sci.* **228**, 772–774.

Fontaine, Y. A. (1975). *Biochem. Biophysi. Perspect. Mar. Biol.* **2**, 139–212.

Fowler, D. J., and Goodnight, C. J. (1966). *Am. Zool.* **6**, 187–193.

Fowler, D. J., Goodnight, C. J., and Labrie, M. M. (1972). *Ann. Entomol. Soc. Am.* **65**, 138–141.

Freeman, B. M., and Manning, A. C. C. (1977). *Comp. Biochem. Physiol. A* **57**, 211–214.

Fry, D. H. (1939). *Calif. Fish Game* **25**, 245–250.

Fuller, R. W., Snoddy, H. D., and Molloy, B. B. (1976). *Life Sci.* **19**, 337–345.

Fuxe, K. (1965). *Acta Physiol. Scand.* **64**, 37–85.

Garcia, L. E., and Meier, A. H. (1973). *Biol. Bull. (Woods Hole, Mass.)* **144**, 471–479.

Gaston, S., and Menaker, M. (1968). *Science* **160**, 1125–1127.

George, J. C., and Berger, A. J. (1966). "Avian Myology." Academic Press, New York.
George, J. C., and Naik, D. V. (1964). *Pavo* **2**, 37–49.
Gerard, J. F., and Gilles, R. (1972). *J. Exp. Mar. Biol. Ecol.* **10**, 125–136.
Geyer, T. A., Segal, D. S., and Mandell, A. J. (1972). *Physiol. Behav.* **8**, 653–658.
Gibson, R. N., (1970). *Anim. Behav.* **18**, 539–543.
Giese, A. G. (1966). *Physiol. Rev.* **46**, 244–298.
Gifford, C. E., and Odum, E. P. (1965). *Condor* **67**, 383–403.
Giles, M. A. (1969). Ph.D. Dissertation, University of British Columbia, Vancouver, Canada.
Gleiser, S. I., and Khodorkovskii, V. A. (1971). *Dokl. Akad. Nauk SSSR* **201**, 964–967.
Goldsworthy, G. J., Jutsum, A. R., and Robinson, N. L. (1975). *J. Endocrinol.* **64**, 66P–67P.
Goodridge, A. G. (1964). *Comp. Biochem. Physiol.* **13**, 1–26.
Goodridge, A. G. (1975). *Fed. Proc., Fed. Am. Soc. Exp. Biol.* **34**, 117–123.
Goodridge, A. G., and Ball, E. G. (1965). *Comp. Biochem. Physiol.* **16**, 367–381.
Goodridge, A. G., and Ball, E. G. (1966). *Am. J. Physiol.* **211**, 803–808.
Goodridge, A. G., and Ball, E. G. (1967). *Biochemistry* **6**, 2335–2343.
Goodyear, C. P. (1970). *Science* **168**, 603–605.
Goodyear, C. P., and Ferguson, D. E. (1969). *Anim. Behav.* **17**, 636–640.
Gorbman, A. (1969). *In* "Fish Physiology" (W. S. Hoar and D. J. Randall, ed.), Vol. 2, pp. 241–274. Academic Press, New York.
Gorden, J. H., and Shellenburger, M. K. (1974). *Neuropharmacology* **13**, 129–133.
Grande, F. (1968). *Proc. Soc. Exp. Biol. Med.* **128**, 532–536.
Grant, W. C., and Grant, J. A. (1958). *Biol. Bull. (Woods Hole, Mass.)* **114**, 1–9.
Gras, J., Perrier, H., Perrier, C., and Gudefin, Y. (1971). *Comp. Biochem. Physiol. A* **39**, 45–51.
Greer, G. L., and Gardner, D. R. (1970). *Science* **169**, 1220–1222.
Gregory, R. W., and Fields, P. E. (1962). *Tech. Rep. Sch. Fish., Univ. Wash.* **52**, 1–58.
Groot, C. (1965). *Behaviour* **14**, 1–198.
Groot, C. (1972). *J. Fish. Res. Board Can.* **29**, 1431–1444.
Gualtierotti, S. (1964). *Ed. Med.* **3**, 112–134.
Gunter, G. (1945). *Publ. Inst. Mar. Sci., Univ. Tex.* **1**, 1–190.
Gunter, G. (1967). *In* "Estuaries" (G. H. Lauff ed.), Publ. No. 83, pp. 621–638. Am Assoc. Adv. Sci., Washington, D.C.
Gwinner, E. (1967). *Naturwissenschaften* **54**, 447.
Gwinner, E. (1968). *J. Ornithol.* **109**, 70–95.
Gwinner, E. (1972). *Proc. Int. Ornithol. Congr., 15th, 1970* pp. 218–236.
Gwinner, E. (1974). *Science* **185**, 72–74.
Gwinner, E. (1975). *In* "Avian Biology" (D. S. Farner, J. R. King, and K. C. Parkes, eds.), Vol. 5, pp. 221–274. Academic Press, New York.
Gwinner, E. (1977). *Annu. Rev. Ecol. Syst.* **8**, 381–405.
Gwinner, E., Turek, F., and Smith, S. D. (1971). *Z. Vergl. Physiol.* **75**, 323–331.
Hardisty, M. W. (1972). *Gen. Comp. Endocrinol.* **18**, 501–514.
Haskell, P. T., and Moorhouse, J. E. (1963). *Nature (London)* **197**, 56–58.
Hasler, A. D. (1966). "Underwater Guideposts." Univ. of Wisconsin Press, Madison.
Hasler, A. D. (1971). *In* "Fish Physiology" (W. S. Hoar and D. J. Randall, eds.), Vol. 6, pp. 429–506. Academic Press, New York.
Hasler, A. D., Horrall, R. M., Wisby, W. J., and Braemer, W. (1958). *Limnol. Oceanogr.* **3**, 353–361.
Haufe, W. O. (1946). *Bull. Entomol. Res.* **45**, 507–526.
Healy, M. C. (1967). *J. Fish. Res. Board Can.* **24**, 2321–2338.
Helms, C. W. (1968). *Am. Zool.* **8**, 151–167.
Helms, C. W., and Drury, W. H., Jr. (1960). *Bird-Banding* **31**, 1–40.

Helms, C. W., Aussiker, W. H., Bower, E. B., and Fretwell, S. D. (1967). *Condor* **69**, 560–578.

Herrington, L. P., and Smith, K. L. (1935). *J. Ind. Hyg.* **17**, 283–288.

Hery, F., Rover, E., and Glowinski, J. (1972). *Brain Res.* **43**, 445–465.

Hicks, D. L. (1967). *Condor* **69**, 387–399.

Higginbotham, A. C. (1939). *Ecology* **20**, 58–70.

Hinks, C. F. (1967). *Nature (London)* **214**, 386–387.

Hirano, T., Johnson, D. W., and Bern, H. A. (1971). *Nature (London)* **230**, 469–471.

Hirano, T., Johnson, D. W., Bern, H. A., and Utida, S. (1973). *Comp. Biochem. Physiol. A* **45**, 529–540.

Hoar, W. S. (1939). *J. Morphol.* **65**, 257–295.

Hoar, W. S. (1951). *Publ. Ont. Fish. Res. Lab.* **71**, 1–51.

Hoar, W. S., Keenleyside, M. H. A., and Goodall, R. G. (1955). *Can. J. Zool.* **33**, 428–439.

Hoar, W. S., Keenleyside, M. H. A., and Goodall, R. G. (1957). *J. Fish. Res. Board Can.* **14**, 815–830.

Hochachka, P. W., and Somero, G. N. (1973). "Strategies of Biochemical Adaptation." Saunders, Philadelphia, Pennsylvania.

Hoffman, K. (1971). *In* "Biochronometry" (M. Menaker, ed.), pp. 134–146. Natl. Acad. Sci. Washington, D.C.

Honma, Y., and Yoshie, S. (1974). *Arch. Histol. Jpn.* **36**, 237–250.

Horseman, N., and Meier, A. H. (1978). *Life Sci.* **22**, 1485–1490.

Houben, N. M. D., and Beenakkers, A. M. T. (1973). *J. Endocrinol.* **57**, liv.-lv.

Hughes, D. A. (1972). *Biol. Bull. (Woods Hole, Mass.)* **142**, 271–280.

Ibuka, N., and Kawamura, H. (1975). *Brain Res.* **96**, 76–81.

Idler, D. R., and Tsuyuki, H. (1958). *Can. J. Biochem. Physiol.* **36**, 783–791.

Jacobs, B. L., Wise, W. D., and Taylor, K. M. (1975). *Neuropharmacology* **14**, 501–506.

Jahn, L. A. (1969). *J. Fish. Res. Board Can.* **26**, 1243–1261.

Jenner, C. E., and Engels, W. L. (1952). *Biol. Bull. (Woods Hole, Mass.)* **103**, 345–355.

John, T. M., Meier, A. H., and Bryant, E. E. (1972). *Physiol. Zool.* **45**, 34–42.

John, T. M., McKeown, B. A., and George, J. C. (1973). *Comp. Biochem. Physiol. A* **46**, 497–504.

Johnson, A. M., and Hendrickson, G. O. (1958). *Proc. Iowa Acad. Sci.* **65**, 554–558.

Johnson, B., and Birks, P. R. (1960). *Entomol. Exp. Appl.* **3**, 327–339.

Johnson, C. G. (1969). "Migration and Dispersal of Insects by Flight." Methuen, London.

Johnson, C. G. (1974). *In* "The Physiology of Insecta" (M. Rockstein, ed.), Vol. 3, pp. 279–334. Academic Press, New York.

Johnson, D. W. (1973). *Am. Zool.* **13**, 799–818.

Johnson, W. E., and Groot, C. (1963). *J. Fish. Res. Board Can.* **20**, 919–938.

Johnston, D. W. (1968). *Auk* **85**, 13–18.

Johnston, D. W. (1973). *Condor* **75**, 108–113.

Joseph, M. M., and Meier, A. H. (1971). *J. Exp. Zool.* **178**, 59–62.

Kamberi, T. A., Mical, R. S., and Porter, J. C. (1970). *Fed. Proc., Fed. Am. Soc. Exp. Biol.* **29**, 378.

Kamiya, M. (1972). *Comp. Biochem. Physiol. B* **43**, 611–617.

Karnaky, K. J. Jr., Degnan, K. J., and Zadunaisky, J. A. (1977). *Science* **195**, 203–205.

Kashiwagi, M. (1971). *In* "Symposium sur l'étude physiologique de la migration. III Salmonides " (S. Utida, ed.), p. 11. Inst. Oceanogr. Res., University of Tokyo, Tokyo.

Kastin, A. J., Miller, L. H., Nockton, R., Sandman, C. A., Schally, A. V., and Stratton, L. O. (1973). *Prog. Brain Res.* **39**, 461–470.

Katada, M., Zama, K., and Igarashi, H., (1960). *Bull. Jpn. Soc. Sci. Fish.* **26**, 425–429.

Kerstetter, T. H., and Kirschner, L. B. (1974). *Comp. Biochem. Physiol. B* **48**, 581–589.

Keys, A., and Bateman, J. B. (1932). *Biol. Bull. (Woods Hole, Mass.)* **63**, 327–336.
King, J. R. (1961). *Condor* **63**, 128–142.
King, J. R. (1963). *Proc. Int. Ornithol. Congr., 13th, 1962* pp. 940–949.
King, J. R. (1968). *Comp. Biochem. Physiol.* **24**, 827–837.
King, J. R. (1972). *Proc. Int. Ornithol. Congr., 15th, 1970* pp. 200–217.
King, J. R., and Farner, D. S. (1956). *Proc. Soc. Exp. Biol. Med.* **93**, 354–359.
King, J. R., and Farner, D. S. (1963). *Condor* **65**, 200–223.
King, J. R., and Farner, D. S. (1965). *Ann. N.Y. Acad. Sci.* **131**, 422–440.
King, J. R., Mewaldt, L. R., and Farner, D. S. (1960). *Auk* **77**, 89–92.
King, J. R., Barker, S., and Farner, D. S. (1963). *Ecology* **44**, 513–521.
Kinne, O. (1975). In "Marine Ecology" (O. Kinne, ed.), Vol. 2, pp. 709–852. Wiley, New York.
Kiselevich, K. A. (1924). *Tr. Astrakh. Ikhtiol. Lab.* **6**, No. 1.
Kleerekoper, H., Taylor, G., and Wilton, R. (1961). *Trans. Am. Fish. Soc.* **90**, 73–78.
Kleimenov, I. V. (1962). *Rybn. Khoz. (Moscow)* p. 35–44.
Knepton, J. C. (1952). *Q. J. Fla. Acad. Sci.* **14**, 255–265.
Knowles, F., and Vollrath, L. (1966). *Z. Zellforsch. Mikrosk. Anat.* **69**, 474–479.
Kook, Y., Cho, K. B., and Yun, L. O. (1964). *Nature (London)* **204**, 385–386.
Kramer, G. (1951). *Proc. Int. Ornithol. Congr., 10th, 1950* pp. 271–280.
Kramer, G. (1953). *Verh. Dtsch. Zool. Ges., Suppl.* **17**, 72–84.
Krieger, D. T., and Krieger, H. P. (1967). *Science* **155**, 1421–1422.
Krieger, D. T., and Rizzo, F. (1969). *Am. J. Physiol.* **217**, 1703–1707.
Krieger, D. T., Silverberg, A. I., Rizzo, F., and Krieger, H. P. (1968). *Am. J. Physiol.* **215**, 959–967.
Krueger, A. P., Andriese, P. C., and Kotaka, S. (1968). *Int. J. Biometeorol.* **12**, 225–239.
Krulich, L., Hefko, E., and Aschenbrenner, J. E. (1975). *Endocrinology* **96**, 107–118.
Kubo, T. (1960). *Bull. Fac. Fish. Hokkaido Univ.* **11**, 15–19.
Lahlou, B., and Giordan, A. (1970). *Gen. Comp. Endrocrinol.* **14**, 491–509.
Lam, T. J. (1969). *Comp. Biochem. Physiol.* **31**, 909–913.
Lam, T. J. (1972). *Gen. Comp. Endocrinol., Suppl.* **3**, 328–338.
Landgrebe, F. W. (1941). *J. Exp. Biol.* **18**, 162–169.
Langslow, D. R., and Hales, C. N. (1969). *J. Endocrinol.* **43**, 285–294.
Larsson, A. L. (1973). *Gen. Comp. Endocrinol.* **20**, 155–167.
Larsson, A. L., and Lewander, K. (1972). *Comp. Biochem. Physiol. A* **43**, 831–836.
Leatherland, J. F., and McKeown, B. A. (1973). *J. Interdiscip. Cycle Res.* **4**, 137–143.
Leatherland, J. F., McKeown, B. A., and John, T. M. (1974). *Comp. Biochem. Physiol. A* **47**, 821–828.
Lee, R. W., and Meier, A. H. (1967). *J. Exp. Zool.* **166**, 307–316.
Lees, A. D. (1961). *Symp. R Entomol. Soc. London* **1**, 68–79.
Lees, A. D. (1967). *J. Insect Physiol.* **13**, 289–318.
Lefebvre, P. (1975). *Biochem. Pharmacol.* **24**, 1261–1266.
Leibson, L., Plisetskaya, E. M., and Mazina, T. I. (1968). *Zh. Evol. Biokhim. Fiziol.* **4**, 121–127.
Leloup, J., and Fontaine, M. (1960). *Ann. N.Y. Acad. Sci.* **86**, 316–353.
Leloup-Hatey, J. (1964). *Ann. Inst. Oceanogr. (Paris)* **42**, 224–337.
Levanidov, I. P. (1932). *Byull. Vsekaspiisk. Nauchn. Rybokh. Eksped. Nos. 5–6.*
Lewis, T. L., and Epple, A. (1972). *Science* **178**, 1286–1288.
Linko, R. (1964). *Suom. Kemistil.* **37**, 90–92.
Lobanov-Rostovsky, I. (1969). *Oceans Mag.* **1**, 4–16.
Lofts, B., Marshall, A. J., and Wolfson, A. (1963). *Ibis* **105**, 99–105.
Lotrich, V. A. (1975). *Ecology* **56**, 191–198.

Lovern, J. A. (1964). *Oceanogr. Mar. Biol.* **2**, 169–191.

McCleave, J. (1967). *J. Fish. Res. Board Can.* **24**, 2011–2044.

McCleave, J., and LaBar, G. W. (1972). *Trans Am. Fish. Soc.* **101**, 44–54.

McGreal, R. D., and Farner, D. S. (1956). *Northwest Sci.* **30**, 12–23.

McInerney, J. E. (1964). *J. Fish. Res. Board Can.* **21**, 995–1018.

McKeown, B. A., and van Overbeeke, A. P. (1972). *J. Fish. Res. Board Can.* **29**, 303–309.

McKeown, B. A., John, T. M., and George, J. C. (1974). *Arch. Int. Physiol. Biochim.* **82**, 55–62.

McMillan, J. (1972). *J. Comp. Physiol.* **79**, 105–112.

McMillan, J. P., Gauthreaux, S. A., and Helms, C. W. (1970). *BioScience* **20**, 1259–1260.

Maetz, J. (1969). *Gen. Comp. Endocrinol., Suppl.* **2**, 299–316.

Mangum, C., and Towle, D. (1977). *Am. Sci.* **65**, 67–75.

Marshall, A. J. (1961). *In* "Biology and Comparative Physiology in Birds" (A. J. Marshall, ed.), Vol. 2, pp. 307–339. Academic Press, New York.

Martin, D. D., and Meier, A. H. (1973). *Condor* **75**, 369–374.

Martin, H., and Lindauer, M. (1973). *Fortschr. Zool.* **21**, 211–228.

Matthews, G. V. T. (1955). *J. Exp. Biol.* **32**, 39–58.

Mayer, N., Maetz, J., Chan, D. K. O., Forster, M., and Chester Jones, I. (1967). *Nature (London)* **214**, 1118–1120.

Mayer, R. J., and Candy, D. J. (1969). *J. Insect Physiol.* **15**, 611–620.

Mazeaud, F. (1964). *C. R. Hebd. Seances Acad. Sci.* **258**, 36–40.

Mazeaud, F. (1973). Thesis, Science Faculty, Paris (cited in Y. A. Fontaine, 1975).

Mazeaud, M. (1972). *Comp. Gen. Pharmacol.* **3**, 457–468.

Mehrle, P. M., and Fleming, W. R. (1970). *Comp. Biochem. Physiol.* **36**, 597–603.

Meier, A. H. (1969). *Gen. Comp. Endocrinol., Suppl.* **2**, 55–62.

Meier, A. H. (1973). *Am. Sci.* **61**, 184–187.

Meier, A. H. (1975). *In* "Hormonal Correlates of Behavior" (B. E. Elefthériou and R. L. Sprott, eds.), Vol. 2, pp. 469–550. Plenum, New York.

Meier, A. H. (1976). *Proc. Int. Ornithol. Congr., 16th, 1974* pp. 355–368.

Meier, A. H. (1977). *Am. J. Physiol.* **232**, E193–E196.

Meier, A. H., and Burns, J. T. (1976). *Am. Zool.* **16**, 649–659.

Meier, A. H., and Davis, K. B. (1967). *Gen. Comp. Endocrinol.* **8**, 110–114.

Meier, A. H., and Farner, D. S. (1964). *Gen. Comp. Endocrinol.* **4**, 584–595.

Meier, A. H., and Fivizzani, A. J. (1975). *Proc. Soc. Exp. Biol. Med.* **150**, 356–362.

Meier, A. H., and MacGregor, R., III (1972). *Am. Zool.* **12**, 257–271.

Meier, A. H., and Martin, D. D. (1971). *Gen. Comp. Endocrinol.* **17**, 311–318.

Meier, A. H., Burns, J. T., and Dusseau, J. W. (1969). *Gen. Comp. Endocrinol.* **12**, 282–289.

Meier, A. H., Garcia, L. E., and Joseph, M. M. (1971a). *Biol. Bull. (Woods Hole, Mass.)* **141**, 331–336.

Meier, A. H., Martin, D. D., and MacGregor, R., III (1971b). *Science* **173**, 1240–1242

Meier, A. H., Trobec, T. N., Joseph, M. M., and John, T. M. (1971c). *Proc. Soc. Exp. Biol. Med.* **137**, 408–415.

Meier, A. H., Trobec, T. N., Haymaker, H. G., MacGregor, R., and Russo, A. C. (1973). *J. Exp. Zool.* **184**, 281–288.

Menaker, M. (1961). *J. Cell. Comp. Physiol.* **57**, 81–86.

Menaker, M. (1968). *Proc. Nat. Acad. Sci. U.S.A.* **59**, 414–421.

Merkel, F. W. (1938). *Ber. Ver. Schles. Ornithol.* **23**, 1–72.

Merkel, F. W. (1960). *Umschau* **60**, 243–246.

Merkel, F. W. (1963). *Proc. Int. Ornithol. Congr., 13th, 1962* pp. 950–959.

Michel, R. (1972). *Gen. Comp. Endocrinol.* **19**, 96–101.

Miles, S. G. (1968). *J. Fish. Res. Board Can.* **25**, 1591–1602.

Millar, J. B. (1960). Ph.D. Dissertation, University of Wisconsin, Madison.
Miller, L. J., and Weise, C. M. (1978). *Condor* **80**, 94–96.
Miller, R. E., and Ogawa, N. (1962). *J. Comp. Physiol. Psychol.* **55**, 211–213.
Milne, K. P., Ball, J. N., and Chester Jones, I. (1971). *J. Endocrinol.* **49**, 177–178.
Minick, M. C., and Chavin, W. (1972). *Comp. Biochem. Physiol. A* **41**, 791–804.
Mittler, T. E., and Dadd, R. H. (1966). *Ann. Entomol. Soc. Am.* **59**, 1162–1166.
Moore, R. Y., and Eichler, V. B. (1972). *Brain Res.* **42**, 201–206.
Morawa, F. W. (1955). *Z. Fisch. Deren Hilfswiss.* [N.S.] **4**, 101–136.
Morton, M. L. (1967). *Condor* **69**, 491–512.
Morton, M. L., and Liebman, H. A. (1974). *Comp. Biochem. Physiol. A* **48**, 329–335.
Morton, M. L., and Mewaldt, L. R. (1962). *Physiol. Zool.* **35**, 237–247.
Müller, K. (1973). *J. Comp. Phsyiol.* **84**, 357–359.
Nagahama, Y., and Yamamoto, K. (1970). *Bull. Fac. Fish. Hokkaido Univ.* **21**, 169–177.
Nakamura, T. (1962). *Misc. Rep. Yamashina Inst. Ornithol. Zool.* **3**, 185–188.
Nakamura, T. (1969). *Tori Bull. Ornithol. Soc. Jpn.* **19**, 87–108.
Nakono, T., and Tomlinson, N. (1967). *J. Fish. Res. Board Can.* **24**, 1701–1715.
Naylor, E., and Atkinson, R. J. A. (1972). *Symp. Soc. Exp. Biol.* **26**, 395–415.
Naylor, E., Smith, G., and Williams, B. G. (1973). In "Neurobiology of Invertebrates" (J. Salanki, ed.), pp. 423–429. Akadémiai Kiadó, Budapest.
Nice, M. M. (1937). *Trans. Linn. Soc. N.Y.* **4**, 61–247.
Nikolskii, G. V. (1963). *Isr. Prog. Sci. Transl. Cat.* No. 233
Noeske, T. A., and Meier, A. H. (1977). *J. Exp. Zool.* **202**, 97–102.
Oakeson, B. (1953). *Condor* **55**, 3–16.
Odhiambo, T. R. (1965). *Nature (London)* **207**, 1314–1315.
Odhiambo, T. R. (1966). *J. Exp. Biol.* **45**, 51–63.
Odum, E. P. (1960a). *Proc. Int. Ornithol. Congr. 12th, 1958* pp. 563–576.
Odum, E. P. (1960b). *Am. J. Clin. Nutr.* **8**, 621–629.
Odum, E. P., and Major, J. C. (1956). *Condor* **58**, 222–228.
Odum, E. P., and Perkinson, J. D. (1951). *Physiol. Zool.* **24**, 216–230.
Odum, E. P., Rogers, D. T., and Hicks, D. L. (1964). *Science* **143**, 1037–1039.
Ogawa, M. (1974). *Comp. Biochem. Physiol. A* **49**, 545–553.
Ogawa, M. (1975). *Comp. Biochem. Physiol. A* **52**, 539–543.
Ogawa, M. (1977). *Can. J. Zool.* **55**, 872–876.
O'Hea, E. K., and Leveille, G. A. (1968). *Comp. Biochem. Physiol.* **26**, 111–120.
O'Hea, E. K., and Leveille, G. A. (1969). *Comp. Biochem. Physiol.* **30**, 149–159.
Oksche, A., and Hartwig, H. G. (1975). In "Brain-Endocrine Interaction" (K. M. Knigge, ed.), Vol. II, pp. 40–53. Karger, Basel.
Oliver, J. A. (1955). "The Natural History of North American Amphibians and Reptiles." Van Nostrand-Reinhold, Princeton, New Jersey.
Olivereau, J. M. (1971). *Z. Vergl. Physiol.* **72**, 435–441.
Olivereau, M. (1954). *Ann. Inst. Oceanogr. (Paris)* **29**, 95–296.
Olivereau, M. (1975). *Gen. Comp. Endocrinol.* **27**, 9–27.
Olla, B. L., and Studholme, A. L. (1972). In "Behavior of Marine Animals" (H. E. Winn and B. L. Olla, eds.), Vol. II, pp. 303–325. Plenum, New York.
Orr, R. T. (1954). *Proc. Calif. Acad. Sci.* **28**, 165–246.
Otto, R. G., and McInerney, J. E. (1970). *J. Fish. Res. Board Can.* **27**, 793–800.
Page, T. C., and Larimer, J. L. (1972). *J. Comp. Physiol.* **78**, 107–120.
Palmer, J. D. (1973). *Biol. Rev. Cambridge Philos. Soc.* **48**, 305–372.
Palmgren, P. (1949). *Ibis* **91**, 561–576.
Papi, F. (1955). *Experientia* **11**, 201.
Pardi, L., and Papi, F. (1952). *Naturwissenschaften* **39**, 262–263.

Pardo, R. J., and de Vlaming, V. L. (1976). *Copeia* pp. 563–573.
Patent, G. (1970). *Gen. Comp. Endocrinol.* **14**, 215–242.
Pentegov, B. P., Mentov, Y. N., and Kurnaev, E. F. (1928). *Izv. Tikhookean. Nauchno-Promysl. Stn.* **2**, No. 1.
Perrett, W. S., Barrett, B. B., Latapie, W. R, Pallard, J. F., Mock, W. R., Adkins, G. B., Guidry, W. J. and White, C. J. (1971). "Cooperative Gulf of Mexico Estuarine Inventory and Study, Louisiana" Vol. I. Louisiana Wildlife and Fisheries Commission, Baton Rouge.
Perrier, C. (1971). Ph.D. Dissertation, University of Claude Bernard, Lyon, France.
Petterson, D. A. (1972). *Prog. Fish Cult.* **34**, 110–112.
Pettingill, O. S. (1972). "Ornithology in Laboratory and Field." Burgess, Minneapolis, Minnesota.
Peyraud-Waitzenegger, M. 1972. *J. Physiol. (Paris)* **64**, 685–694.
Pfeiler, E., and Kirschner, L. B. (1972). *Biochim. Biophys. Acta* **282**, 301–310.
Pickford, G. E., Griffith, R. W., Torretti, J., Hendlez, E., and Epstein. F. H. (1970a). *Nature (London)* **228**, 378–379.
Pickford, G. E., Pang. P. K. T., Weinstein, E., Torretti, J., Hendler, E., and Epstein, F. H. (1970b). *Gen. Comp. Endrocrinol.* **14**, 524–534.
Plisetskaya, E. M., and Mazina, T. I. (1969). *Zh. Evol. Biokhim. Fiziol.* **5**, 457–463.
Potts, W. T. W., and Fleming, W. R. (1970). *J. Exp. Biol.* **53**, 317–327.
Potts, W. T. W., and Fleming, W. R. (1971). *J. Exp. Biol.* **54**, 63–75.
Preiser, R. (1957). Ph.D. Dissertation, Hoheheim University.
Preston, F. W. (1966). *Ecology* **47**, 375–392.
Putzig, P. (1937). *Vogelzug* **8**, 116–130.
Putzig, P. (1938). *J. Ornithol.* **86**, 123–165.
Quay, W. B. (1965). *Life Sci.* **4**, 379–384.
Ramwell, P. W., and Shaw, J. (1970). *Recent Prog. Horm. Res.* **26**, 139–187.
Randall, D. J., and Stevens, D. E. (1967). *Comp. Biochem. Physiol.* **21**, 415–424.
Ranje, C., and Ungerstedt, U. (1977). *Brain Res.* **134**, 83–93.
Rankin, M. A. (1974). *In* "Experimental Analysis of Insect Behaviour" (L. B. Browne, ed.), pp. 319–329. Springer-Verlag, Berlin and New York.
Rankin, M. A., and Riddiford, L. M. (1977). *Gen. Comp. Endocrinol.* **33**, 309–321.
Reite, O. B. (1969). *Acta Physiol. Scand.* **75**, 221–239.
Rice, D. W., and Wolman, A. A. (1971). Special Publication No. 3. Am. Soc. Mammal. Stillwater, Oklahoma.
Richardson, N. E., and McCleave, J. D. (1974). *Biol. Bull. (Woods Hole, Mass.)* **147**, 422–432.
Richkus, W. A. (1974). *J. Fish. Res. Board Can.* **31**, 1485–1497.
Ridgway, S. H. (1972). *In* "Mammals of the Sea: Biology and Medicine" (S. H. Ridgway, ed.), pp. 590–747. Thomas, Springfield, Illinois.
Roberts, T. W. (1941). *Hagen Anat. Rec.* **81**, 46–47.
Robertson, O. H. (1948). *Physiol. Zool.* **21**, 282–295.
Robertson, O. H. (1955). *Trans. Assoc. Am. Physicians* **68**, 33–41.
Robertson, O. H., and Wexler, B. C. (1960). *Endocrinology* **66**, 222–239.
Robertson, O. H., Krupp, M. A., Thomas, S. F., Favour, C. B., Hane, S., and Wexler, B. C. (1961). *Gen. Comp. Endocrinol.* **1**, 473–484.
Robiller, F. (1975). *Beitr. Vogelkd.* **21**, 1–17.
Robinson, N. L., and Goldsworthy, G. J. (1977). *J. Insect Physiol.* **23**, 153–158.
Rodriques, G., and Naylor, E. (1972). *J. Mar. Biol. Assoc. U.K.* **52**, 81–95.
Rommel, S. A., and McCleave, J. D. (1971). *IEEE Trans Biomed. Eng.* **18**, 421–424.
Rommel, S. A., and McCleave, J. D. (1973). *J. Fish. Res. Board Can.* **30**, 657–662.
Rowan, W. (1926). *Proc. Boston Soc. Nat. Hist.* **38**, 147–189.

Rowan, W. (1931). "The Riddle of Migration." Williams & Wilkins, Baltimore, Maryland.
Rusak, B. (1977). *J. Comp. Physiol. A* **118**, 145–164.
Ryapolova, N. I. (1964). *Tr. Vses. Nauchno-Issled. Inst. Morsk. Rybr. Khoz. Okeanogr.*, 66–69.
Sandman, C. A., Kastin, A. J., and Schally, A. V. (1969). *Experientia* **25**, 1001–1002.
Sansum, E. L., and King, J. R. (1976). *Physiol. Zool.* **49**, 407–416.
Saunders, D. S. (1976). "Insect Clocks." Pergamon, Oxford.
Schaefers, G. A., and Judge, F. D. (1971). *J. Insect Physiol.* **17**, 365–379.
Schildmacher, H., and Rautenberg, W. (1952). *Biol. Zentralbl.* **71**, 397–405.
Schooley, J. P., Riddle, O., and Bates, R. W. (1941). *Am. J. Anat.* **69**, 123–154.
Segawa, T., Kawasaki, K., and Yajima, H. (1973). *Jpn. J. Pharmacol.* **23**, 121–123.
Seigmund, R., and Wolff, D. L. (1973). *Experientia* **29**, 54–58.
Serventy, D. L., and Marshall, A. J. (1959). *Emu* **57**, 99–126.
Seshadri, B. (1959). *Curr. Sci.* **28**, 121–122.
Shank, M. C. (1959). *Auk* **76**, 44–54.
Shulman, G. E. (1960). *Tr. Azovo-Chernomorsk. Nauchno-Issled. Inst. Morsk. Rybn. Khoz. Okeanogr.* **18.**
Shulman, G. E. (1974). "Life Cycles of Fish, Physiology and Biochemistry." Keter Publishing House, Jerusalem.
Singley, J. A., and Chavin, W. (1971). *Am. Zool.* **11**, 653.
Sklower, A. (1930). *Forsch. Fortschr.* **6**, 435–436.
Slijper, E. J. (1962). "Whales." Hutchinson, London.
Southwood, T. R. E. (1961). *Proc. R. Entomol. Soc. London, Ser. A* **36**, 63–66.
Spence, B., Meier, A. H., and Dietz, T. H. (1977). *Copeia* pp. 557–560.
Spieler, R. E. (1975). Ph.D. Dissertation, Louisiana State University, Baton Rouge.
Spieler, R. E., and Meier, A. H. (1975). *J. Fish. Res. Board Can.* **33**, 183–186.
Spieler, R. E., Meier A. H., and Loesch, H. C. (1976a). *Copeia* pp. 605–608.
Spieler, R. E., Meier, A. H., and Loesch, H. C. (1976b) *Gen. Comp. Endocrinol.* **29**, 156–160.
Spieler, R. E., Meier, A. H., and Noeske, T. A. (1978). *Nature (London)* **271**, 469–471.
Sprott, R. L. (1967). *Science* **157**, 1206–1207.
Srivastava, A. K., and Meier, A. H. (1972). *Science* **177**, 185–187.
Steen, J. B., and Kruysse, A. (1964). *Comp. Biochem. Physiol.* **12**, 127–142.
Stengel, M. M. (1974). *In* "Experimental Analysis of Insect Behaviour" (L. B. Browne, ed.), pp. 297–303. Springer-Verlag, Berlin and New York.
Stephan, F. K., and Nunez, A. A. (1977). *Behav. Biol.* **20**, 1–16.
Stephan, F. K., and Zucker, I. (1972). *Porc. Natl. Acad. Sci. U.S.A.* **69**, 1583–1586.
Stetson, M. H., and Erickson, J. E. (1972). *Gen. Comp. Endocrinol.* **19**, 355–362.
Stickney, A. P. (1972). *Ecology* **53**, 438–445.
Strong, L. (1968). *J. Insect Physiol.* **14**, 1685–1692.
Sundararaj, B. I., Vasal, S., and Halberg, F. (1973). *Int. J. Chronobiol.* **1**, 362–363.
Sutherland, O. R.W., and Mittler, T. E. (1969). *Entomol. Exp. Appl.* **12**, 240–241.
Swallow, R.L., and Fleming, W. R. (1966). *Am. Zool.* **6**, 562.
Swift, D. R. (1955). *J. Exp. Biol.* **32**, 751–764.
Swift, D. R. (1964). *J. Fish. Res. Board Can.* **21**, 133–138.
Szulc-Olech, B. (1965). *Bird Study* **12**, 1–7.
Tabb, D. C., Dubrow, D. L., and Jones, A. E. (1962). *Tech. Serv. Fla. State Board Conserv.* **37**, 1–30.
Taranenko, N. F. (1966). *Tr. Azovo-Chernomorsk-Nauchno-Issled. Inst. Morsk. Rybn. Khoz. Okeanogr.* **24.**
Tashima, L., and Cahill, G. F. (1968). *Gen. Comp. Endocrinol.* **11**, 262–271.

Tashiro, M., Yamagishi, Y., and Suzuuchi, T. (1972). *Hokkaidoritsu Suisan Shikenjo Hokoku* **14**, 1-16.
Tashjian, A., Voelkel, E., McDonough, L., and Levine, L., (1975). *Nature (London)* **258**, 739-741.
Taylor, D. H., and Ferguson, D. E. (1970). *Science* **168**, 390-392.
Taylor, J. L. (1962). *Rep. Conf. Soc. Biol. Rhythms, 7th, 19* p. 153.
Tesch, F. W. (1967). *Helgol. Wiss. Meeresunters.* **16**, 92-111.
Tesch, F. W. (1974). *ICES Anadromous Catadromous Fish Comm.* M5, 1-27.
Tesch, F. W. (1975). *In* "Marine Ecology" (O. Kinne, ed.), Vol. 2, pp. 657-707. Wiley, New York.
Thapliyal, J. P., Tewari, P. D., and Bhatnagar, G. (1963). *Proc. Summer Sch. Zool., 1st, 1961* pp. 391-401.
Thomas, V. G., and George, J. C. (1975). *Physiol. Zool.* **48**, 157-167.
Thomson, A. J., and Sargent, J. R. (1977). *J. Exp. Zool.* **200**, 33-40.
Thorpe, A., and Ince, B. W. (1974). *Gen. Comp. Endocrinol.* **23**, 29-44.
Timiras, P. S., and Woodbury, D. M. (1956). *Endocrinology* **58**, 181-192.
Tinbergen, L., and Verwey, J. (1945). *Arch. Neerl. Zool.* **7**, 213-286.
Toba, H. H., Paschke, J. E., and Friedman, S. (1967). *J. Insect Physiol.* **13**, 381-396.
Towle, D. W., and Taylor, D. D. (1976). *Am. Zool.* **16**, 224.
Trobec, N. (1974). *In* "Chronobiology" (L. E. Scheving, F. Halberg, and J. E. Pauly, eds.), pp. 147-151. Igaku Shoin Ltd., Tokyo.
Truchan, L. C., and Boyer, S. D. (1972). *Physiol. Zool.* **45**, 204-214.
Twente, J. W., Jr. (1955). *Ecology* **36**, 706-732.
Underhill, G. W. (1940). *J. Econ. Entomol.* **33**, 915-917.
Utida, S., Hirano, T., Oide, H., Ando, M., Johnson, D. W., and Bern, H. A. (1972). *Gen. Comp. Endocrinol., Suppl.* **3**, 317-327.
Van Oordt, G. J. (1959). *Ostrich, Suppl.* **2**, 342-345.
Van Veen, T., Hartwig, H. G., and Müller, K. (1976). *J. Comp. Physiol.* **111**, 209-219.
Varanelli, C. C., and McCleave, J. D. (1974). *Anim. Behav.* **22**, 178-186.
Vasiliev, A. S., and Gleiser, S. Y. (1973). *Vopr. Ikhtiol.* **13**, 381-383.
Vepsalainen, K. (1971). *Acta Entomol. Fenn.* **28**, 101-102.
Vollrath, L. (1966). *Z. Zellforsch. Mikrosk. Anat.* **73**, 107-131.
von Lucanus, F. (1923). "Die Rätsel des Vogelzuges." Beyer & Mann, Langensalza.
Wagner, H. O. (1930). *Z. Vergl. Physiol.* **12**, 703-723.
Wagner, H. O. (1956). *Z. Vergl. Physiol.* **38**, 355-373.
Wagner, H. O. (1961). *Z. Tierpsychol.* **18**, 302-319.
Wagner, H. O., and Muller, C. (1963). *Z. Morphol. Oekol. Tiere* **53**, 107-151.
Wagner, H. O., and Thomas, I. (1957). *Z. Vergl. Physiol.* **40**, 73-84.
Wajc, E., and Pener, M. P. (1971). *Gen. Comp. Endocrinol.* **17**, 327-333.
Walker, A. T. (1964). *Physiol. Zool.* **37**, 57-64.
Wallgren, H. (1954). *Acta Zool. Fenn.* **84**, 1-110.
Weisbart, M., and Idler, D. R. (1970). *J. Endocrinol.* **46**, 29-43.
Weise, C. M. (1956). *Ecology* **37**, 275-287.
Weise, C. M. (1967). *Condor* **69**, 49-68.
Weis-Fogh, T. (1952). *Philos. Trans. R. Soc. London, Ser. B* **237**, 1-36.
Wellington, W. G. (1954). *Can. Entomol.* **86**, 312-333.
West, G. C., and Peyton, L. J. (1972). *Bird-Banding* **43**, 241-256.
Wheeland, R. A., Martin, R. J., and Meier, A. H. (1976). *Comp. Biochem. Physiol. B* **53**, 379-385.
Williams, B. G., and Naylor, E. (1969). *J. Exp. Biol.* **51**, 715-725.

Wilson, A. C., and Farner, D. S. (1960). *Condor* **62**, 414–425.

Wolfson, A. (1945). *Condor* **47**, 95–127.

Wolfson, A. (1954). *J. Exp. Zool.* **125**, 353–376.

Wolfson, A. (1958). *J. Exp. Zool.* **139**, 349–379.

Wolfson, A. (1959). *In* "Photoperiodism and Related Phenomena in Plants and Animals" (R. B. Withrow, ed.), Publ. No. 55, pp. 679–716. Am. Assoc. Adv. Sci., Washington D.C.

Woodhead, A. D. (1975). *Oceanogr. Mar. Biol.* **13**, 287–368.

Woodhead, P. M. J. (1970). *J. Fish. Res. Board Can.* **27**, 2337–2338.

Wurtman, R. J., and Fernstrom, J. D. (1975). *In* "Biochemistry and Behavior" (S. H. Snyder, ed.), pp. 685–693. MIT Press, Cambridge, Massachusetts.

Yablonkevich, M. L. (1975). *J. Evol. Biochem. Physiol. (Engl. Transl.)* **10**, 529–534.

Yokoyama (1977). *Cell Tissue Res.* **176**, 91–108.

Zaks, M. G., and Sokolova, M. M. (1961). *Vopr. Ikhtiol.* **1**, 333–346.

Zaugg, W. S., and McLain, L. R. (1972). *J. Fish. Res. Board Can.* **29**, 167–171.

Zimmerman, J. L. (1966). *Condor* **68**, 377–387.

5 Mechanisms of Orientation, Navigation, and Homing*

KENNETH P. ABLE

Real advances in understanding a subject like bird migration almost always come as partial or complete surprises. A generation ago neither I nor any other scientist would have predicted with the slightest confidence several of the (recent) discoveries. . . .

Donald R. Griffin, 1964

*This chapter is dedicated to the memory of William T. Keeton, in recognition of his outstanding contributions to the study of avian homing.

Animal Migration, Orientation, and Navigation
Copyright © 1980 by Academic Press, Inc.
All rights of reproduction in any form reserved.
ISBN 0-12-277750-6

I. Introduction

A. General Background

The spectacular long-distance migrations of birds, fishes, insects, and mammals, and the mysterious homing abilities of many kinds of animals have intrigued layman and scientist alike. Research in the past decade has produced major advances in our understanding of the mechanisms by which animals perform these feats. Griffin's closing remarks in *Bird Migration* (1964) were prophetic and yet remain as true today as they were 15 years ago. New data have more clearly defined the behavioral capabilities of various species, new sensory modalities have been discovered, and new hypotheses advanced. Yet as of this writing we have no viable hypothesis to explain the Manx shearwater *(Puffinus puffinus)* that homed nearly 5000 km across the Atlantic Ocean in only 12.5 days (Matthews, 1953b), the young pigeons that vanish toward the home loft a few seconds after being released for the first time tens of kilometers away (see Keeton, 1974a), or the Pacific salmon *(Oncorhynchus* spp.) that returns to its natal stream following 2 or more years and thousands of kilometers at sea. For many species, especially the insects that migrate long distances, reptiles, and terrestrial mammals, the descriptive data are not yet sufficient to define precisely the abilities that we wish to explain.

In this chapter I will attempt to review selected recent developments in the study of orientation and navigation mechanisms in a wide variety of vertebrates and invertebrates. While emphasizing what I believe to be the most important recent advances in knowledge, I hope to draw attention to the salient similarities and differences among groups as well as to point out the most critical gaps in our knowledge.

Two recent reviews of homing and migratory orientation in birds alone by Keeton (1974a) and Emlen (1975), respectively, were both nearly the length of this chapter. Therefore, I have made no attempt to be exhaustive. More detailed recent reviews have been published on fish (Waterman, 1972; Legget, 1977), terrestrial mammals (Joslin, 1977), and shore-living arthropods (Herrnkind, 1972); Schmidt-Koenig (1975) briefly discussed orientation and homing in a variety of species and Griffin (1978) reviewed the sensory implications of several particularly intriguing orientation phenomena. Johnson's (1969) classic volume gives a comprehensive treatment of insect migration and orientation. Baker (1978) has discussed orientation and navigation in his treatise on the evolution and ecology of animal migration. Wherever possible, I shall further restrict this review to orientation in the context of seasonal or other long-distance movements. Orientation with respect to very local environmental gradients (taxes and kineses) (see recent review by Jander, 1975) will not be discussed except in cases in which it appears to be pertinent to long-distance movements.

Undoubtedly in large part because their migrations are the most extensive and spectacular, our understanding of orientation mechanisms in birds greatly exceeds that in other groups (the honeybee, *Apis mellifera*, is a notable exception). Recent work with both migrant bird species and homing pigeons has revealed greater complexity than was previously expected: no single cue system or mechanism is likely to explain bird navigation. Birds have access to and apparently make use of a variety of sources of orientational information. Because of the rigors of their great migrations, natural selection has probably favored redundancy in avian orientation systems and additional complexity can be expected to be found as we begin to explore how orientation behavior varies among species, age classes, sexes, and in time and space.

This change in outlook on the part of students of bird orientation should be applied to other animal groups, in many of which research still seems preoccupied with a search for unitary mechanisms (some important exceptions will be pointed out in the following pages). If the past history of orientation research is a valid predictor, we should expect some surprises, and as Griffin (1978) warns, we should not constrain our thinking to simple explanations. On the other hand, it must be borne in mind that the orientation and navigation mechanisms of a species have evolved in concert with its life history mileu. Whereas many discoveries made with birds have proved prophetic for other groups, one must beware not to overgeneralize among species. My approach in this chapter will be to attempt to explain the existing data by means of the simplest possible hypothesis, while bearing in mind that additional known or unknown factors may be involved.

In the following pages I will first review visual and nonvisual compass orientation mechanisms. The approach will be to examine each capability in turn, enumerating for all animal groups the relevant data documenting its occurrence. Following these two sections, I will discuss the interaction of compass cues within species and groups. Of necessity, this discussion will involve a detailed analysis of the few cases in which the integration of several cues has been explored. Throughout, the reader should consider the potential adaptive value of various capabilities *vis-à-vis* the orientation tasks performed by the animals: most models to explain bird navigation will not be applicable to the shorter range homing performed by salamanders.

The chapter will conclude with a consideration of homing and navigation. I have adopted a conservative approach to the problem of navigation, preferring to assume that homing in most species can be explained without invoking true navigation (see Section I,B). As in the case with birds, the simpler hypotheses may of necessity be rejected, but I believe it is prudent to heed Occams' Razor until the data force us to take that course. As a result, I find that only birds have been shown unequivocally capable of homing from unfamiliar localities over distances and at speeds that preclude piloting or various search strategies.

B. Some Problems of Definition

The terms migration and navigation are used with a variety of different meanings even in recent literature. Among ornithologists, migration has been restrictively defined to refer only to those long-distance movements that involve a subsequent return trip by the surviving members of the population. Defined in this way the term is not appropriate for most long-range flights of insects which are one-way movements, and use of the term has been debated, especially among entomologists (Williams, 1958; Johnson, 1969; Urquhart, 1958; Kennedy, 1961; Dingle, 1972). Such difficulties in formulating a rigid definition are not surprising inasmuch as various forms of short-distance dispersal, one-way migration, and round trip migration may form an evolutionary continuum (see Chapter 2 by Gauthreaux). For the purposes of this chapter, I shall use migration in a rather broad sense, referring to any oriented, long-distance, seasonal movement of individuals.

Discussion of navigation and related phenomena require a somewhat more rigorous definition. Griffin's (1955) scheme still has heuristic value and I will use it here. He defined three types of homing ability:

1. *Type I or piloting.* To find a goal by referring to familiar landmarks. Type I homing may also entail random or systematic search until familiar landmarks are encountered. Griffin originally defined these homing types with reference to birds and in subsequent use (e.g., Schmidt-Koenig, 1965) the landmarks have been assumed to be visual. However, the term should also encompass cases of homing in which animals use other types of cues as guideposts or maintain sensory contact with the goal (e.g., by olfaction in salamanders).

2. *Type II or compass orientation.* The ability to orient in a given compass direction without reference to landmarks when released in unfamiliar territory. This behavior will result in homing only when the given compass direction leads to home or a familiar region.

3. *Type III or true navigation.* The organism is able to select the direction toward a goal (e.g., home) when released in unfamiliar territory in a new and unfamiliar direction. This is the most complex of homing capabilities; it occurs in the absence of familiar landmarks and any form of sensory contact with the goal. The use of the term implies no particular mechanism (several quite different models are possible, see Section V,B) and it has been convincingly demonstrated in only a small number of species.

II. Visual Orientation Mechanisms

This section discusses compass orientation mechanisms based on the preception of light. This will include light stimuli perceived by eyes and extraoptic structures and the specific cues considered will be the sun, stars, the moon, landmarks, and the plane of polarization of polarized light.

A. THE SUN AS AN ORIENTATION CUE

Because the sun is such an overwhelming stimulus in the environment of organisms with form vision, it is not surprising that it is used for orientation by many animals. Santschi (1911) was the first to demonstrate this when he predictably shifted the homing movements of the myrmicine ant, *Messor barbarus*, by reversing the apparent position of the sun with a mirror. Since that time, a large number of elegant experiments have explored sun orientation. It appears that some animals use the sun essentially as a fixed reference point whereas others possess a compass system that compensates for the changing position of the sun during the day.

1. *Sun Orientation without Time Compensation*

Animals may orient their bodies or movements with respect to the sun for a variety of reasons. If it is not necessary for a constant compass direction to be maintained we might expect individuals to orient at some fixed angle to the sun's azimuth regardless of the time of day. Although the behavior is probably unrelated to migration, dragonflies(*Sympetrum* spp.) alighting on stationary objects were found to orient consistently at right angles to the sun (Hisada, 1972). Once landed, they did not change orientation as sun position changed. This behavior resulted in a bimodal distribution of facing directions which could be predictably changed with a mirror. Unlike superficially similar bimodal orientation in other species, the dragonflies were apparently responding to the sun itself rather than to patterns of polarized light (see Section II,B).

Baker (1968a,b, 1969) has discussed the flight orientation of several species of short-distance migrant butterflies. At any given time individuals may be observed flying on constant tracks in many directions. However, each individual, if captured and released repeatedly, maintains the same compass direction over a short period of time. These data were interpreted to mean that within a population many different flight directions were represented, but that each individual oriented in only one direction. This system was hypothesized to be adaptive for individuals of species keyed to ephermeral and patchily distributed food plants, resting, or hibernation sites. Observation of flying butterflies throughout the day revealed that flight direction shifted in phase with the azimuth of the sun, implying that they were orienting at a constant angle with respect to the sun.

Whereas the data for some of the species (especially *Pieris rapae*) seem to support Baker's hypothesis, his results and interpretations have been questioned (e.g., Johnson, 1969, pp. 703–704). Some of the species he studied are not known to migrate and in the others it is not clear that he was always observing insects in migration. Flight orientation such as Baker describes, if maintained throughout the day, would cause a butterfly to move over a curved track. This might well be a satisfactory means of dispersal over short distances if suitable habitat was widely spaced. For a long-distance migrant in which highly directional movement is necessary, it would be inefficient. Cases in which migrating butterflies maintained a given compass direction over many hours are numerous (including *Pieris brassicae*, one of the species studied by Baker; Blunck, 1954), especially in long-distance migrants (e.g., *Danaus, Vanessa, Ascia*) (Johnson, 1969; Schmidt-Koenig, 1979a). Whereas such data do not demonstrate a time-compensated sun compass, they suggest its existence and do not support Baker's hypothesis. Furthermore, his arguments (Baker, 1978) regarding the evolution and maintenance in a population of fixed-angle sun orientation resort to group selection and are unconvincing.

2. Sun Orientation with Time Compensation

For an organism to maintain a constant compass direction using only the sun as a cue, it must correct for the apparent movement of the sun across the sky. Animals that do this are said to possess a time-compensated sun compass or simply a sun compass (time compensation will be implicit in my use of this term). Demonstration of sun compass orientation is usually based on one of two types of evidence: (1) predicted changes in orientation occur when the animal's internal clock is phase-shifted relative to sun time; or (2) a constant compass orientation is maintained at different times of day and therefore in relation to different sun positions. Clock shift experiments provide more unequivocal evidence of sun compass orientation , although many studies, especially with invertebrates, fail to distinguish between orientation based on the sun itself versus that based on the plane of light polarization (Section II,B).

The sun compass was discovered almost simultaneously in the honeybee by von Frisch (1950) and in a bird by Kramer (1950). Subsequent work by von Frisch and colleagues showed clearly that sun compass orientation is an integral part of the complex foraging behavior of the bees, that they compensate for the changing position of the sun, and that polarization patterns are used when the sun itself is not directly visible (reviewed in von

Frisch, 1967). Other recent developments in our understanding of honeybee orientation and communication are contained in Gould's papers (1975, 1976).

The classic experiments of Kramer (1950, 1951) were made possible by the discovery that caged starlings *(Sturnus vulgaris)* exhibited spontaneous migratory restlessness that was oriented in the proper direction when the birds were able to see the sun. Kramer then shifted the sun's apparent position with mirrors and the starlings' hopping was shifted by the predicted amount. Further experiments with starlings, pigeons *(Columba livia)*, and western meadowlarks *(Sturnella neglecta)* trained at one time of day and tested at another showed constant compass directions, implying time compensation (Kramer and von St. Paul, 1950; Kramer, 1952; von St. Paul, 1956). If tested in the presence of a stationary light source, the birds' orientation shifted gradually during the course of the day (Kramer, 1953).

Hoffmann (1954) was the first to apply the rigorous technique of phase-shifting an animal's chronometer to the analysis of sun orientation. His classic results proved conclusively that starlings compensated for the apparent movement of the sun via their internal clocks at the expected rate of about 15° per hour. Schmidt-Koenig (reviewed 1965) was able to similarly shift the vanishing directions of homing pigeons and Matthews (1961, 1963) altered the fixed northwesterly departure directions of released mallards *(Anas platyrhynchos)* by using clock shifts.

Similar types of experiements have confirmed the existence of sun compass orientation in a variety of other invertebrate and vertebrate species including the following: Arthropoda—the lycosid spiders *Arctosa variana* and *Lycosa fluviatilis* (Papi and Tongiorgi, 1963); the amphipods *Talitrus saltator*, *Talorchestia* spp., *Orchestia mediterranea*, and *Orchestoidea* spp. (e.g., Pardi and Papi, 1952; Pardi, 1960; reviewed in Herrnkind, 1972); the isopod *Tylos latreillii* (Pardi, 1954); the decapods *Goniopsis cruentata* (Schone, 1963) and *Uca* spp. (Altevogt and von Hage, 1964; Herrnkind, 1972); the beetle *Paederus rubrothoracicus* (Ercolini and Badino, 1961); the ants *Formica rufa*, *Lasius niger* (Jander, 1957), and *Cataglyphis bicolor* (Wehner, 1972); vertebrates—numerous migratory and nonmigratory fish (see Waterman, 1972; Leggett, 1977); seven species of anuran amphibians and the newt *Taricha granulosa* (see Ferguson, 1971); the turtle *Chrysemys picta* (De Rosa and Taylor, 1978); the snakes, *Natrix sipedon* and *Regina septemvittata* (Newcomer *et al.*, 1974); the nocturnal bird migrants *Lanius collurio*, *Sylvia nisoria* (von St. Paul, 1953), and *Zonotrichia albicollis* (Able and Dillon, 1977); and the rodent *Apodemus agrarius* (Luters and Birukow, 1963). Other reported cases appear less certain, including the widely cited work on the insect *Velia* by Birukow (1960; see Heran, 1962; Schmidt-Koening, 1975).

3. Role of the Sun in Oriented Movements

Because of its widespread occurrence among mobile animals, the sun compass is often assumed to be ubiquitous and important in oriented dispersal. However, this assumption is neither obviously correct nor easy to demonstrate. The extensive redundancy in orientation systems in several well-studied species (see Section IV) makes the design of experiments to evaluate the role of a single cue very difficult (see Emlen, 1975). Showing that an animal can orient in the absence of the sun in no way precludes its use when available.

Furthermore, we should not expect even closely related species to employ identical means of orientation. For example, the beach-dwelling isopods *Tylos latreilli* and *T. punctatus* perform daily movements correlated with tides. *Tylos latreilli* is nocturnal and moves from burrows in moist areas to higher zones up to 50 m away. Before morning the isopods return to their burrows near the water (Tongiorgi, 1962, 1969). *Tylos latreilli* orient by means of a sun compass, also using polarized light (Pardi, 1954; Papi, 1960). Its congener, *punctatus,* migrates at night from the high tide line seaward, returning to the upper beach before dawn. It appears not to possess a sun compass, but apparently orients with respect to gravity. Clear orientation was found on slopes as small as 3° (Hamner *et al.,* 1968). Inhabiting the lower beach where a consistent slope exists, gravity orientation may be sufficient in this species, whereas it would be useless for *T. latreilli* on the uneven terrain above the high tide line (see Herrnkind, 1972). Even though comparative studies of orientation are in an embryonic stage, similar examples have been found in other groups.

In a number of cases, it has been possible to show that the sun is the primary, if not the only, orientation cue. Generally, the criterion is that orientation degenerates to random movement in the absence of the sun and patterns of sky polarization. However, recent discoveries concerning bird orientation in magnetic fields (Section III,A) warn us not to regard this criterion too seriously; subtle directionality might be revealed by second-order analysis.

Among invertebrates, especially shore-dwelling arthropods, the sun seems to provide the primary orientation information. Talitrids tested in the experimental chambers of Pardi and Papi (1952) moved randomly under solid overcast skies. In bees the sun is certainly the primary cue (von Frisch, 1967), but both bees and ants have back-up cues that probably include landmarks, wind direction, and odors (Wellington, 1974; Wehner, 1972). Butterflies (e.g., the monarch, *Danaus plexippus,* and *Ascia monuste)* often appear to avoid migrating under solid overcast, but are well oriented when they do fly under these conditions (Johnson, 1969; Schmidt-Koenig, 1975). Landmarks appear to be important in the maintenance of their flight direction (Nielsen, 1961), but how the correct direction is determined initially is unknown.

The sun also appears to be the primary cue involved in several types of orientation in fish. Hasler *et al.* (1958) displaced white bass *(Roccus chrysops)* from spawning grounds to several release points in a large fresh-water lake. The fish, followed by means of small attached floats, moved consistently northward. Harden Jones (1968) questioned whether the directions taken by these fish were significantly oriented. I have reanalyzed the data as presented in his polar diagrams (Fig. 77, p. 240) and found the uncorrected data significantly oriented by the Rayleigh test $(p < .001)$. Twelve fish fitted with opaque eye-caps and three released under overcast skies were not oriented, implying an important role for the sun. Hasler *et al.* (1958) applied an unspecified correction for the direction of drift and replotted the data. As Harden Jones noted, seven fish were omitted from the corrected data, but that distribution is also significantly oriented $(p < .001)$ and is homeward directed by the V test. However, it is not clear that anything more than type II orientation was involved.

Two species of parrot fish *(Scarus)* were studied in shallow waters off Bermuda (Winn *et al.*, 1964). During the day these fish feed along the shore and in bays; at night they move to offshore caves. Displaced individuals showed a consistent southeast orientation (apparently type II orientation) which appeared to be disrupted under overcast (only seven releases) and at night (12 releases). A 6-hr phase delay applied to seven fish produced a change in departure directions, but the magnitude of the shift was apparently greater than predicted. Nonetheless, these data appear to be the most convincing evidence that fish employ sun-compass orientation in natural situations.

Many animals that live near shorelines perform movements oriented by the sun. In general, the organisms tend to move on an axis perpendicular to the nearest shore, a direction apparently innate in talitrid amphipods (Pardi, 1960), but learned in wolf spiders *(Arctosa)* (Papi and Tongiorgi, 1963), fiddler crabs *(Uca)* (Herrnkind, 1972), fish (Goodyear, 1970), and amphibians. In the latter group, new shorelines can be learned by individuals, the behavior has been termed *y*-axis orientation and studied extensively by Ferguson and students (see Ferguson, 1967).

Figure I shows schematically the operation of *y*-axis orientation in vertebrates. Individuals displaced to a central location or a distant unfamiliar place and tested in a circular arena from which they can see only the sky take up a direction toward or away from the shore near which they had been living. Whether the animals move in a direction corresponding to shore or deeper water can be predicted on the basis of the species involved, season, or stage in the life cycle. When individuals are transplanted to a new shoreline with a different orientation, the new direction may be learned within a few days (Ferguson, 1971).

The mechanisms of this behavior have been studied extensively and it appears to be due entirely or primarily to celestial orientation. Typical data

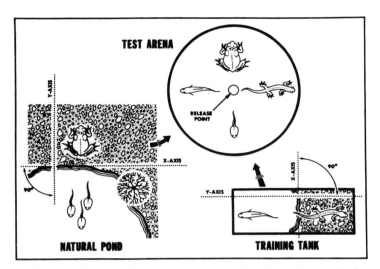

Fig. 1. Schematic diagram of *y*-axis orientation in selected vertebrates. Toads and tadpoles collected near a natural shore (lower left) move in opposite directions parallel to the natural *y*-axis when released in the center of an aquatic test arena from which they can see only the sky (upper right). Animals may also be trained to new shorelines as illustrated by the fish and salamander at the lower right. Again, aquatic and terrestrial forms move in opposite directions. Animals may be displaced over any distance to the test arena and their orientation is not related to the homeward direction. (From Adler, 1970.)

from the newt, *Taricha granulosa*, are shown in Fig. 2 (Landreth and Ferguson, 1967). Animals from shorelines in the four cardinal directions oriented as predicted by the *y*-axis hypothesis (Fig. 2a–d) under sunny skies. They also oriented in the predicted direction at night (Fig. 2e), but were tested in the presence of the moon. Under cloudy skies no orientation was observed (Fig. 2f). Tests of newts and other species transported long distances from the home ponds yielded similar results, precluding orientation to local cues (e.g., odors).

Further studies of *y*-axis orientation in cricket frogs *(Acris gryllus)* led to the surprising discovery that the behavior may be mediated through the eyes or an extraoptic photoreceptor (Taylor and Ferguson, 1970). Figure 3 shows that the frogs swam in the expected directions when the tops of their heads were covered with opaque Teflon inserted under the skin, or when blinded and the top of the head covered with clear plastic. Orientation disappeared when both eyes and the presumed extraoptic photoreceptive site were blocked. Blind frogs could also be clock-shifted and could learn a new *y*-axis direction. The fact that the data from blind frogs were unimodal (Fig. 3D) suggested that their extraoptic orientation was based on the sun per se, rather than polarized light (see Section II,B).

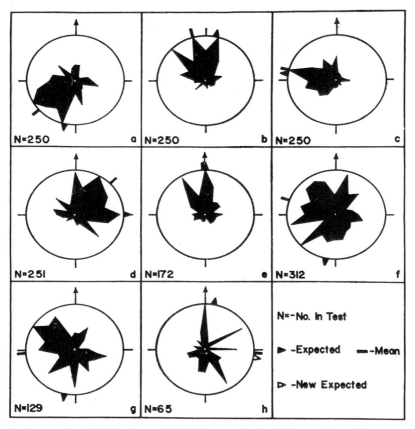

Fig. 2. y-axis orientation in newts *(Taricha granulosa)* which move toward deeper water. (a), (b), and (c) show the orientation under sunny skies of newts taken from near shorelines with water located southward, northward, and westward, respectively. The animals oriented as predicted at night under moonlight (e), but were random under overcast skies during the day (f). After 7 days in constant darkness the animals failed to orient in the predicted direction (g). A 6-hr clock shift produced the predicted shift in orientation. All mean directions significant by the Rayleigh test ($p \leq .01$) except (f) and (g). (From Landreth and Ferguson, 1967; copyright 1967 by the American Association for the Advancement of Science.)

These studies assumed that the relevant extraocular photoreceptor was located in the head and the pineal organ was suspected. This has been tested directly in the tiger salamander *(Ambystoma tigrinum)* by Taylor and Adler (1978). Adult salamanders were trained to a particular compass direction in outdoor tanks with an artificial shore at one end. Tests were performed in circular aquatic arenas outdoors under clear skies. Figure 3F–J shows that pinealectomy resulted in the disappearance of the shoreward orientation in blind animals. The unimodal orientation of the blind, clear

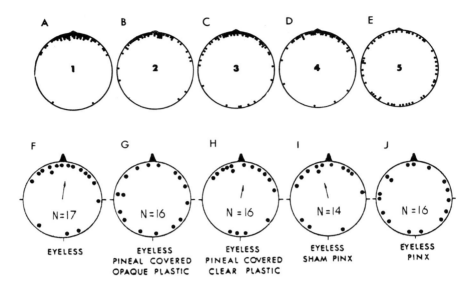

Fig. 3. Extraoptic y-axis orientation in amphibians. A–E, cricket frogs *(Acris gryllus)*; F–J, tiger salamanders *(Ambystoma tigrinum)*. Cricket frogs orient in the predicted y-axis direction (indicated by pointer outside of circle) when they have normal vision and the tops of their heads are not covered (A), when they have normal vision, but the tops of their heads are covered with opaque Teflon (B), when blinded and with clear plastic covering the presumed extraoptic photoreceptor site (C), and when blinded and with the top of the head uncovered (D). When the tops of the heads of blind frogs were covered with opaque Teflon, orientation disappeared (E). Blind tiger salamanders also orient as predicted (F, H), but orientation disappeared when the pineal region was covered (G). Pinealectomy resulted in disorientation in blind animals (J), but not in those with eyes (not shown). (From Taylor and Ferguson, 1970, and Taylor and Adler, 1978.)

plastic and sham groups again suggests that extraoptic orientation may be performed by reference to the sun's position per se.

In homing experiments with box turtles *(Terrapene carolina)*, displaced individuals initially oriented in a homeward direction under sun (see Section V,A), but moved in random directions under solid overcast (Gould, 1957). More clear-cut data show the importance of the sun in bird orientation and homing. Although not necessary, the sun compass plays a primary compass role in homing pigeons; departure flight directions are readily manipulated with clock shifts and initial orientation shows marked deterioration under solid overcast (see Keeton, 1974a). The fixed northwestward flight directions of several species of ducks (nonsense orientation of Matthews, 1961) can be clock-shifted in the daytime (but not at night, Matthews, 1963) and deteriorates under overcast skies (Bellrose, 1958, 1963; Matthews, 1963).

In his pioneering observations on the orientation of migratory restlessness *(Zugunruhe)* in birds placed in circular cages, Gustav Kramer (1949)
noticed that nocturnal migrants showed a strong tendency to hop toward
extraneous light sources (e.g., horizon glow, the moon). If he placed the
birds in their cages before sunset they appeared to be less distracted and
oriented in appropriate migratory directions after dark. Whereas these
casual observations suggested that the sun might be used in migratory
orientation even by species that migrate exclusively at night, no relevant
data were available until very recently. Solid overcast skies of several days
duration sometimes result in a drastic increase in the spread of flight directions of migrants observed in the field (Hebrard, 1971; Emlen and Demong,
1978; K. P. Able, unpublished radar tracks).

More direct evidence implicating the role of the sun comes from studies
of orientation in the night migrant savannah sparrow *(Passerculus sandwichensis)* (Moore, 1978). Figure 4 shows the oriented hopping of sparrows
in spring in funnel orientation cages (Emlen and Emlen, 1966). When allowed to see the sun near the time of sunset only, or sunset and stars, the
birds oriented in the appropriate migratory direction. When allowed to
view only the stars after dark, orientation was much poorer and, in fact,
many individuals failed to orient under this condition. Similar results have
been obtained with another relatively short-distance nocturnal migrant, the
white-throated sparrow *(Zonotrichia albicollis)* (Bingman and Able, 1979).
Although this species possesses a time-compensated sun compass (Able and
Dillon, 1977), present evidence does not preclude the likelihood that night
migrants use the sun as a fixed reference point. The role of the sun relative
to other cues involved in migratory orientation is discussed in Section IV.

B. Polarized Light Orientation

Many animals are known to be able to perceive the plane of polarization
(*e*-vector) of linearly polarized light. Since von Frisch's (1949) discovery of
the use of polarized light in honeybee foraging flight orientation, the ability
has been demonstrated in many arthropods and one cephalopod *(Octopus)*
(see Waterman, 1966). Only recently has the perception of polarization
patterns in vertebrates other than man been demonstrated. Waterman and
Forward (1970, 1972) and Forward *et al.* (1972) presented the first fully
convincing data on the marine teleost *Zenarchopterus*. Shortly thereafter,
polarized light perception was found in the goldfish *(Carassius auratus)*
(Kleerekoper *et al.*, 1973; Waterman and Hashimoto, 1974), the tiger
salamander *(Ambystoma tigrinum)* (Adler and Taylor, 1973; Taylor and
Adler, 1973), and the pigeon *(Columba livia)* (Kreithen and Keeton, 1974).

The *e*-vector of polarized light can provide an axis of orientation for
those animals that can perceive it. Theoretically, it can also be used to
determine the position of the sun and might therefore by used as a means of

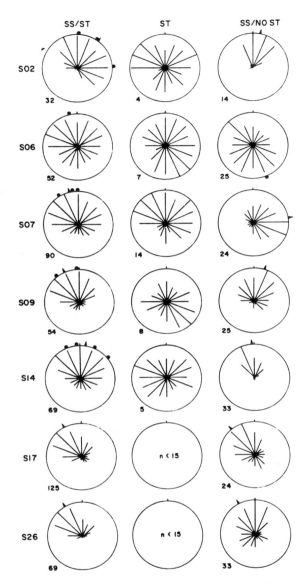

Fig. 4. Cage orientation of savannah sparrows (*Passerculus sandwichensis*) in spring at Clemson, South Carolina. Each row shows the behavior of one individual under the three test conditions: SS/ST birds saw sunset in the cages and stars after dark; ST birds were placed in the cages after all traces of sunset had disappeared; SS/NO ST birds saw sunset but their cages were covered with an opaque board before stars were visible. The numbers at the lower left of each circle denote the greatest number of activity units in any 15° sector and define the length of the longest vector. Arrows indicate significant mean directions ($p < .05$, Rayleigh tests) of the pooled activity and dots show significant means for single nights. (From Moore, 1978.)

performing sun compass orientation when the sun is not visible (e.g., when the sky is partly cloudy). For an unambiguous determination of the sun's position or azimuth in air, e-vector information from at least two widely separated points in the sky or some other second piece of information (e.g., degree of polarization, mean light intensity, altitude of the sun) must be available (Kirschfeld et al., 1975). Underwater polarization patterns are produced by primary scattering of directional light in the water and are a function of sun azimuth, sun altitude, and depth (Waterman and Westell, 1956). They provide a means of directly determining the sun's position to depths well below those at which it could be localized by direct form vision (only near the surface) or by radiance distribution (see Harden Jones, 1968; Waterman, 1972).

Polarized light orientation in invertebrates has been studied most intensively in bees and ants. Both species use exclusively ultraviolet wavelengths (410–430 nm) for e-vector detection (von Frisch, 1967; Duelli and Wehner, 1973; von Helverson and Edrich, 1974) and both can detect the polarization axis when only a very small area of blue sky is visible [10°–15° or a field of < 25 ommatidia (von Frisch, 1967; Duelli, 1975); a much smaller effective field, 1° diameter or approximately 3–7 ommatidia has been reported for the honeybee by Edrich and von Helverson (1976)]. Faultless orientation appears to require a somewhat larger area of exposure [150–200 ommatidia in bees (Zolotov and Frantsevich, 1973)] to the belt of sky with maximum polarization (30–40%). Increasing the degree of polarization above natural levels yielded only a slight improvement in orientation (Zolotov and Frantsevich, 1973).

The desert ant *Cataglyphis bicolor* is a solitary, predatory species that returns to its burrow from foraging excursions without aid of scent trails. Wehner and his colleagues have studied the orientation of the species extensively in its uniform desert environment. The standard procedure (see Wehner, 1972) involves training the ants to a feeding station and then testing their homeward orientation in an arena at a remote site. Training directions are maintained by individuals for many days.

To examine the role of polarized light in orientation, Duelli and Wehner (1973) used a test area on a sandy plain over 600 m from the nest site. Using a frame mounted on wheels they were able to place filters over the homing ants and manipulate the polarization of skylight. A black skirt around the vehicle prevented the ants from seeing landmarks. *Cataglyphis bicolor* is capable of using several cues for retracing the route home (see Section IV), but polarized light appeared to be the dominant one. Figure 5B shows the orientation of ants moving under Plexiglass (control) just prior to sunset. The sun was not visible to them, but polarization patterns were normal and the ants were strongly oriented. Under a configuration of filters that

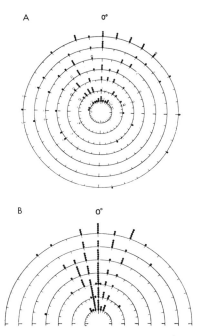

Fig. 5. The orientation of homing ants *(Cataglyphis bicolor)* by solar cues. (A) Distribution of return directions plotted at 1-m intervals with distances >7 m lumped on the periphery. These runs were performed under filters that eliminated the polarization pattern of the sky. Closed circles, sun visible to the ants; open circles, sun not visible. (B) Return runs under plexiglass between 11–40 min before sunset. The sun was not visible to the ants. (From Duelli and Wehner, 1973.)

eliminated the polarization pattern of the sky they were correctly oriented only when the sun was visible (Fig. 5A). Wavelengths in the range 380–410 nm are necessary for homeward orientation, and polarized light orientation also interacts with orientation to wind direction.

The adaptiveness of the ability to use polarized light for orientation is obvious. Polarization patterns can be seen so long as some blue sky is visible, and can give information about the position of the sun. Furthermore, the e-vector is available during dawn and dusk when the sun is below the horizon. Wellington (1974) has suggested that bumblebees *(Bombus terricola)* are able to forage longer because polarization perception by the ocelli allows them to be active when it is too dark for orientation to landmarks via the compound eyes.

Waterman and Forward (1971, 1972) demonstrated consistent orientation to the e-vector in the hemiramphid fish *Zenarchopterus.* The fish were collected in an interisland channel in the Palau Islands and tested in an underwater apparatus 3–5 m below the surface. The orientation of the fish was statistically significant only if selected data were used, and because little is known about the natural behavior of this species, the observed orientation patterns were not readily interpretable. It is known, however, that the species normally swims at the water surface, and Forward *et al.* (1972) performed experiments in a shallow experimental vessel on land. Swimming at the water surface, the fish showed a significant tendency to orient parallel to the *e*-vector even when the vector was manipulated so that it no longer had the normal perpendicular relationship to the sun's azimuth. Therefore, polarotaxis seemed to dominate over the sun although there were indications of interaction between the cues.

Many amphibians are nocturnal and/or crepuscular and live in habitats in which, because of their limited locomotor abilities, the sun may not always be visible. Polarized light orientation could be useful in these situations, and two quite distantly related forms have been shown to exhibit adaptive responses to polarization patterns.

The familiar *y*-axis orientation behavior was used as a paradigm to examine *e*-vector orientation in adult tiger salamanders *(Ambystoma tigrinum)* (Adler and Taylor, 1973; Taylor and Adler, 1973). The salamanders were trained indoors in a metal tank filled to 9 cm with water and containing an artificial shore of bricks. The tank was illuminated with linearly polarized light with the *e*-vector aligned parallel to the shore. After daily training for 21 days, the animals were tested in water in an arena under polarized light (*e*-vector rotated 90° from true compass bearing used during training). All sighted as well as blinded animals showed a significant bidirectional orientation as expected (the *e*-vector has no polarity and therefore provides only a bidirectional axis). Opaque plastic inserted under the skin over the skull caused a disappearance of orientation, whereas animals with clear plastic were able to orient, albeit poorly relative to both blind and sighted groups. The important fact that salamanders with eyes failed to orient when the tops of their heads were covered demonstrated that polarized light is apparently used only via an extraoptic receptor.

Similar procedures revealed *e*-vector orientation in larval bullfrogs *(Rana catesbeiana)* tested under both artificial light and outdoors at sunrise and sunset (Auburn and Taylor, 1979). The tadpoles were not oriented outdoors under solid overcast. Untrained tadpoles taken directly from a pond also swam toward deep water when tested at sunset, suggesting a naturally occurring analogue of the experimentally induced behavior. Unlike salamanders, frogs have two discrete intracranial photoreceptors, the pineal body and its derivative, the frontal organ. The relationships of these structures and the eyes in polarized light orientation were explored by electrocauterizing the frontal organ, surgically removing the eyes, and covering the pineal with opaque plastic inserted under the skin atop the skull (Auburn and Taylor, 1979). As with salamanders, the eyes appear not to be involved in polarotaxis, which can be performed via the pineal and perhaps the frontal organ.

These recent developments suggest that the polarized light compass may turn out to be quite widespread even in vertebrates. However, except in bees, the widely hypothesized use of polarization patterns as a means of locating the position of the obscured sun has not been convincingly demonstrated in any group. Whereas insects appear to use only ultraviolet light in polarotaxis, the spectral sensitivity of the behavior in vertebrates is entirely unknown. Recently, two species of amphibians (Dietz, 1972; Kimeldorf and Fontanini, 1974), a lizard (Moehn, 1974), and two birds (Huth and Burkhardt, 1972; Kreithen and Eisner, 1978) have been found able to perceive near ultraviolet wavelengths (320–400 nm). However, the mechanisms of detection and behavioral significance of the capability are unknown. Much additional work is needed to determine the ways in which polarized light orientation interacts with other compass systems.

C. STELLAR ORIENTATION

If diurnal animals use the sun as a compass for orientation, it seemed inescapable logic to suppose that species migrating at night used the stars in much the same way. As with the sun compass, Kramer (1949, 1951) performed the pioneering work by examining the orientation of the *Zugunruhe* characteristic of night-migrating birds during spring and fall. Kramer's finding that two species *(Sylvia atricapilla* and *Lanius collurio)* fluttered and hopped in the appropriate seasonal direction when placed in a circular cage under clear night skies has since been confirmed in many species (see Emlen, 1975, for a review). Typically, birds placed in outdoor cages after dark show reasonably consistent directionality when stars are visible. Overcast skies usually result in a drastic deterioration in orientation or diminution in activity. Results of this sort have led to the conclusion that birds can select their migratory direction from stellar information and that, in fact, visible stars are necessary for this process. Such rigid conclusions are not supported by recent data (see Section IV).

The classic experiments of the Sauers (Sauer, 1957, 1961; Sauer and Sauer, 1960) and Emlen (1967a,b) have shown that nocturnal bird migrants can select migratory directions based solely on stars. Under planetarium skies simulating natural star patterns several species of sylviid warblers and indigo bunting *(Passerina cyanea)* showed the predicted orientation when the star projector was on. They maintained this stellar direction even when tested under planetarium skies in which stellar north was reversed 180°. When stars were turned off and the planetarium dome was diffusely illuminated with dim light the nocturnal activity of the birds appeared to deteriorate to random or cease altogether. Whether the birds under these conditions really hopped randomly has become problematical (see Emlen, 1975, pp. 199–203), but it seems clear that the birds were relying heavily on stellar information. In many cases during these experiments, magnetic and stellar directions did not conform to their natural relationship.

The mechanism of star orientation in birds has been explored in detail only in the indigo bunting. By shifting planetarium skies out of phase with local time and selectively blocking certain stars and areas of the sky, Emlen (1967a) showed that the buntings apparently use star patterns, probably over large areas of the sky, as orientation cues. Because of the fixed geometric relationships among the stars, this method of star orientation can be used without precise knowledge of time, season, or geographic locality. Further evidence supporting this model of star orientation was obtained by simultaneously testing birds in spring and autumn physiological condition under the same star configuration (Emlen, 1969a). Under the spring planetarium sky the group of buntings in spring condition hopped northward while the birds accelerated to autumn condition hopped southward. This experiment has been repeated with white-throated sparrows (Miller and Weise, 1978) and a possible endocrine mechanism producing the reversal of orientation has been suggested by the work of Martin and Meier (1973; see Meier and Fivizzani, Chapter 4).

In an additional series of elegant experiments, Emlen (1969b, 1970) discovered how the star compass becomes set in young indigo buntings prior to their first autumn migration. First, a group of ten nestling buntings was taken into the laboratory and reared in visual isolation from stars or any other point sources of light. When they came into migratory condition for the first time, they were tested in orientation cages under a stationary planetarium sky with a normal configuration of stars. All showed nocturnal activity, but their orientation directions over many nights were random. These data suggested, contrary to the earlier assertions of Sauer (1957), that early visual experience with stars is necessary for the development of normal migratory orientation, at least in this species.

Exactly what comprises the relevant experience was explored by allowing two other groups of young buntings to see planetarium skies with normal seasonal star patterns during early development (22 nights prior to the postjuvenal molt). The sky that the two groups observed differed in only one respect: both rotated at the normal speed, but one group of birds observed a sky with a normal axis of rotation centered on Polaris, whereas the other group observed a sky manipulated to rotate around a different pole star, Betelgeuse in Orion. Emlen was thus able to test the hypothesis that it is the apparent axis of celestial rotation which provides the reference system by which the birds set their star compass. When the birds came into migratory condition, both groups were tested simultaneously under a *stationary* planetarium sky. As predicted, the two groups oriented in quite different stellar directions, but both oriented toward what would have been south as defined by the rotational axes to which they had been exposed (Fig. 6). As important as these results are, they leave open the question of why the young buntings selected a southerly heading at the time of their first migration.

This extensive analysis of the orientation system of the indigo bunting has unfortunately not been generalized to other species. Sauer's (1957) experiments with sylviid warblers differed in several important respects from the picture that has emerged from Emlen's studies. The circumpolar region of the night sky, of great importance to the indigo bunting, seemed to be of little significance to the warblers. In addition, there were indications that clock compensation was involved in star orientation in the warblers and that young birds were able to orient spontaneously without prior experience with stars or celestial rotation (see Sauer, 1971; Emlen, 1971). Only comparative studies in the future can help resolve these apparent inconsistencies.

Given the variety of species that perform long-distance movements at night, it is perhaps surprising that stellar orientation is not widespread among animals. However, aside from inconclusive results on nocturnal y-axis orientation in the frog *Acris gryllus* (Ferguson et al., 1965), star orientation has not been reported in any nonavian species.

D. Lunar Orientation

Like stars, the moon provides a potential orientation cue for species that move at night. However, its use as a compass presents many obstacles. First, it is visible at night for only a part of each lunar cycle. During the time that it is visible its image size and position in the sky change from night to night, albeit in a predictable manner. Most important, to use the moon as a compass an animal would have to possess an internal clock in phase with lunar time. For most organisms, this would mean two independent timing systems: the nearly ubiquitous sun clock entrained by the day–night cycle, and a moon clock that operates at a slightly slower rate and is synchronized by other unknown stimuli. All these problems apply in the case of the beach amphipods (*Talitrus*, *Orchestoideas*, and *Talorchestia*), which apparently perform both sun and moon compass orientation. They provide the only known cases of lunar orientation.

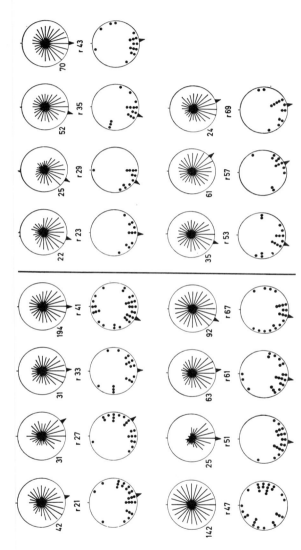

Fig. 6. The role of celestial rotation in the development of migratory orientation in the indigo bunting (*Passerina cyanea*). On the left is the autumn orientation of 8 birds permitted a regular view of a normal, rotating planetarium sky during early development. On the right are 7 birds exposed to a planetarium sky that rotated about an unnatural axis with Betelgeuse as the pole star. All tests were conducted under a stationary planetarium sky and the data are plotted with Polaris designating north in the left group, Betelgeuse at the top in the right group. Vector diagrams show the pooled activity of each bird over all test nights and the individual nightly means are shown below them. Arrow points indicate significant mean directions ($p < .05$, Rayleigh test). (From Emlen, 1970; copyright 1970 by the American Association for the Advancement of Science.)

Papi and Pardi (1953, 1959, 1963) first made the startling proposal that beachhoppers used the moon as a compass to orient their nocturnal movements. The experiments and data have been controversial (Enright, 1961) and attempts to repeat them have not always been successful. Nonetheless, in a review of all available data, including his own replications of the experiments, Enright (1972) concluded that the data generally support the moon compass hypothesis much as originally proposed by Papi and Pardi.

Several types of results bear on the question, but in general the experimental procedures involved are similar to those used in the analysis of sun compass orientation in these species. In circular chambers, the amphipods often show oriented movements when they can see only the moon and stars. In the absence of the moon they are usually not oriented. When the animals orient under the moon, the direction of their movement is that predicted on the basis of their usual seaward movement in observation chambers.

The conclusion that this orientation is based on the moon comes primarily from mirror experiments. While relatively few of these experiments have been performed and the results are not always consistent with predictions, shifting the position of the moon's image with a mirror usually yields major changes in the orientation directions of the amphipods. Furthermore, the weight of evidence supports the conclusion that the animals compensate for the changing position of the moon both during a single night and with less precision from day to day during the lunar cycle. Despite the inherent difficulties with the hypothesis, Enright (1972) concluded that the data could best be explained by a time-compensated moon compass.

E. LANDMARKS

Visual landmarks cannot, of course, provide animals with information on absolute compass direction. Landmarks could, however, be used in conjunction with compass cues to maintain directionality during oriented movement or be substituted for compass systems in familiar areas. In addition, they may well play an important role in at least the final stages of homing navigation in some species (Sections V,A and B).

Both walking and flying insects appear to use visual landmarks as a back-up cue system set by other references (e.g., the sun). In Wehner's (1972) experiments with the desert ant, *Cataglyphis bicolor*, the animals apparently used landmarks on the horizon in combination with their sun compass. If trained to a given direction with both sun and landmarks the ants oriented primarily by the sun if the two cues were altered so as to indicate conflicting directions. However, if rewarded only in the direction of the landmark (a small black screen), the ants began using it as the primary cue by the fifth trial. Bumblebees *(Bombus terricola)* also pilot by landmarks during foraging flights when light intensity is high.

Landmarks could be useful in a number of ways to long-distance bird migrants. Although there is little direct evidence that they do so, a migrating bird might be able to compensate for or at least gauge wind drift during flight by monitoring features on the ground below. Radar observations of wood pigeons *(Columba palumbus)* and cranes *(Grus grus)* migrating over Sweden and the Baltic Sea during the day indicated that the birds compensated for wind drift while flying over land. Over the sea they drifted slightly as if they were using the moving waves below them as stationary landmarks (Alerstam and Pettersson, 1976). Even at night, most areas of the earth are conspicuously dotted with lights and nocturnal migrants could also use these features to monitor wind drift.

Most diurnal bird migrants respond to major topographic features, often following coastlines, mountain ridges, etc., especially when these are oriented more or less in the direction of their movement. Nocturnal migrants generally seem less influenced by features on the ground. They often pass over coastlines, islands, or other major landmarks without changing course. There are, however, some exceptions to this general pattern. Lowery and Newman (1966), Richardson (1978b), and S. A. Gauthreaux (personal communication) have all observed night migrants apparently turning or landing in response to shorelines of large bodies of water. That migrants often do not respond to such features may simply reflect species differences or differences in flight strategy at different points along the migratory route.

Pilotage by familiar landmarks of one kind or another may explain the homing success of many animals, as discussed in Section V,A. However, it cannot account for homing navigation in pigeons and some other birds although it may play some role in their normal behavior. Wagner (1972) has found that the initial orientation of homing pigeons in Switzerland was often influenced by lakes, valleys, snow fields, and villages. This effect was short-lived, however, disappearing once the home directions had seemingly been determined. Michener and Walcott (1967) and Walcott and Michener (1967) found no evidence of strict pilotage during homing flights by pigeons, but did note changes in directions when some familiar tall buildings came into view. Even very close to the home loft familiar landmarks are often ignored and clock shifts are sometimes effective even within sight of the loft (reviewed by Keeton, 1974a).

Recent important experiments by Schlichte and Schmidt-Koenig (1971), Schmidt-Koenig and Schlichte (1972), and Schlichte (1973) have shown that pigeons are initially homeward oriented and can, in fact, return to the home loft wearing frosted contact lenses that preclude form vision of landmarks more than a few meters away. Landmarks, even those immediately associated with the loft, are obviously not essential for successful homing or initial orientation although they may well be used when available.

In other birds topographic features may be considerably more important. Two studies on bank swallows *(Riparia riparia)* suggest that landmarks may play a fairly major role in homing (Sargent, 1962; Downhower and Windsor, 1971) and the same may be true of the related purple martin *(Progne subis)* (Southern, 1968). Furthermore, early airplane tracking studies of homing (Griffin and Hock, 1949; Griffin, 1955) suggested that birds might search, perhaps systematically, until familiar terrain was reached and then employ type I orientation the rest of the way home. Whereas piloting and search plus piloting hypotheses cannot explain the homing behavior of pigeons, Manx shearwaters *(Puffinus puffinus)* (Matthews, 1953b), or Laysan albatrosses *(Diomedia immutabilis)* (Kenyon and Rice, 1958) we should perhaps be cautious in generalizing these capabilities to other species.

III. Nonvisual Orientation Mechanisms

Most animals that perform long-distance migrations or dispersal have well-developed abilities for form vision. Even for these species, however, movements sometimes occur under conditions in which visual cues are absent. Natural selection should favor individuals possessing back-up or additional systems and in recent years it has become increasingly apparent that the orientation capabilities of many forms include nonvisual cues, especially sensitivity to the earth's magnetic field.

A. ORIENTATION BASED ON THE EARTH'S MAGNETIC FIELD

The idea that animals might use terrestrial magnetism as a compass in much the same way that we do dates back nearly a century (reviewed by Keeton, 1974a). The hypothesis was first applied to birds and, combined with a postulated detection of the Coriolis force, was elaborated into a system for bicoordinate navigation by Yeagley (1947, 1951). Yeagley's hypothesis was vigorously attacked both on theoretical and empirical grounds, and aside from the work of Brown on a diverse array of invertebrates (reviewed in Brown, 1971), the possibility of magnetic orientation was generally regarded as quite unlikely for nearly a decade.

Nonetheless, working with the European robin *(Erithacus rubecula)*, Merkel and his colleagues (Merkel and Fromme, 1958; Merkel et al., 1964; Merkel and Wiltschko, 1965) reported oriented *Zugunruhe* in visually cueless situations. These reports were met with considerable skepticism because the behavior of the birds was extremely variable. Significant directionality was obtained only on second-order analysis in which the mean directions of individual bird-nights, most of which did not themselves differ significantly from random, were used as data points.

There were, however, several suggestive features of these earlier experiments at Frankfurt. Most important, the pooled data for spring and fall yielded nearly opposite directions that corresponded with the expected migratory directions of robins. Second, the system appeared to function only at magnetic field intensities very close to that of the earth (approximately 0.5 G). Third, the orientation was clearest in the type of cage used by Merkel and the Wiltschkos (see Wallraff, 1972, Emlen et al., 1976). Finally, by enclosing the orientation cages in Helmholtz coils and changing the direction of the horizontal component of the resultant magnetic field they were able to predictably shift the orientation of the birds (Merkel and Wiltschko, 1965; Wiltschko and Merkel, 1965; Wiltschko, 1968; Merkel, 1971). Since these earlier reports, a large volume of data has accumulated on a number of species of nocturnal migrants (Sylvia communis, Wiltschko and Merkel, 1971; S. borin, Wiltschko, 1974; S. cantillans, Wiltschko and Wiltschko, 1975a; S. atricapilla, Viehmann and Wiltschko, 1977; Viehmann, 1979; Passerina cyanea, Emlen et al., 1976). When the data are subjected to the type of analysis customarily performed by Merkel and the Wiltschkos, the results have been consistent.

Wiltschko and Wiltschko (1972) conducted additional experiments to explore the way in which their orienting birds might use the magnetic cues by manipulating independently the horizontal and vertical components of the artifical magnetic field surrounding the cages. This led to the surprising discovery that the robins apparently did not use the polarity of the field to determine the north direction. Rather, the horizontal field component provides the birds with a north–south axis, the north end of which is defined by the smaller angle between the resultant field vector and gravity. This model and the experimental data are shown in Fig. 7. When the vertical component of the field was reversed (from $+66°$ to $-66°$), the robins' orientation was reversed (Fig. 7c). Likewise, if the horizontal component pointed southward rather than northward but the vertical component was unaltered ($+66°$), the birds also reversed directions (Fig. 7b). If horizontal direction and polarity were changed together (Fig. 7d), no shift in orientation occurred, and the birds were unable to orient in a field with a zero vertical component (Fig. 7e). The inclination of the earth's magnetic field is thus apparently used to assign directionality to the north–south axis defined by the horizontal component.

Despite the volume of negative data on magnetic effects in homing ex-

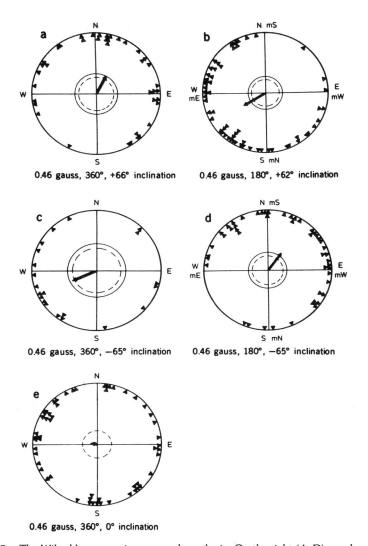

Fig. 7. The Wiltschko magnetic compass hypothesis. On the right (A–D) are shown four magnetic field configurations used in the tests. (*g*) force of gravity; (*H*$_e$) local earth's magnetic field vector; (*H*) experimental magnetic field vector; (*H*$_h$) horizontal field component; (*H*$_v$) vertical field component; (γ) angle between magnetic field vector and gravity. Robins (*Erithacus rubecula*) tested in spring in a closed room under the natural magnetic field at Frankfurt (A) oriented northeast (a). In an artificial field of equal intensity but with the horizontal component pointing southward (B), the birds also oriented magnetic northeast, but celestial southwest (b). Inverting only the vertical component of the field (C) resulted in orientation toward magnetic and celestial southwest (c). The robins oriented toward magnetic northeast (d) when

308

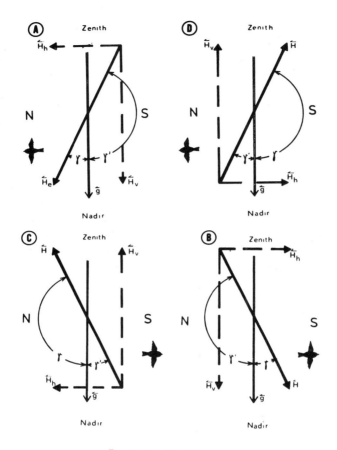

Fig. 7. (*Continued*)

both horizontal and vertical components were reversed (D). In a field with no vertical component the bird's hopping was random (e). In the orientation diagrams each triangle represents the mean direction of activity of a single bird on one night; the arrows represent the mean of these nightly means. Their significance (Rayleigh test) is indicated by the circles (dashed = $p < .05$; solid = $p < .01$) in the center of each diagram. (From Wiltschko and Wiltschko, 1972; copyright 1972 by the American Association for the Advancement of Science.)

periments, Keeton was moved to reanalyze the situation as a result of his discovery that experienced homing pigeons flew off in the homeward direction under solid overcast skies even when clock-shifted (Keeton, 1969). By attaching bar magnets to the backs of pigeons (brass bars of equal weight were glued to controls) he found no effect on vanishing bearings when the sun was visible. But when similar releases were performed under solid overcast skies, the experimental birds usually gave random bearings. Young, inexperienced pigeons were disoriented when carrying magnets even under sunny skies (Keeton, 1971). Southern (1977) has obtained similar results with juvenile ring-billed gulls *(Larus delawarensis)* homing to the colony for the first time from 30 km.

Walcott (1972, 1977) has observed similar magnetic effects in pigeons fitted with miniature Helmholtz coils around their heads. In addition, he sees some disruption of homeward orientation even under sunny conditions, perhaps suggesting interaction between sun and magnetic compasses rather than a simple switching between systems. In a very interesting experiment Walcott and Green (1974) released two groups of homing pigeons carrying identical Helmholtz coils except that the direction of current flow was opposite in the two groups. The two types of coils produced fields of similar strength (ca. 0.6 G) but one coil type (NUP) produced a field pointing up through the bird's head whereas the other type (SUP) produced a downward pointing field. When released under solid overcast skies there was a tendency for the two groups to fly off in opposite directions with the SUP's being homeward directed. Replications of these experiments have not been performed and are badly needed, but the preliminary results are interpretable by the Wiltschkos' model of the magnetic compass in the robin. Pigeons wearing SUP coils would experience the smallest angle between the magnetic field vector and gravity when flying northward, whereas a bird wearing a NUP coil would experience the minimal angle when flying southward. There is thus some evidence suggesting a similar mechanism in homing pigeons and a migratory bird.

The earth's magnetic field varies regularly in both time and space and is also subject to irregular variation (so-called magnetic storms) as a function of sun spot and solar flare activity and the presence of deposits of magnetic material. If birds can be shown to respond systematically to such variables it would indicate a sensitivity of very small changes in the magnetic field (on the order of gammas where 1 gamma $= 10^{-5}$ G).

Recent evidence of sensitivity to such small changes in the magnetic field has come from several sources. Southern (1969, 1972) was the first to report an effect of magnetic storm activity on the orientation of very young ring-billed gull chicks walking in a circular arena from which they can see only the sky. Chicks as young as 2 days of age show a southeastward orientation that is believed to be a precursor of the migration direction of their population. In a recent reanalysis of data gathered over a 15-year period, Southern (1978) has found that under overcast skies gull chicks were not significantly oriented at levels of magnetic disturbance greater than $K = 1^*$. Under clear skies the chicks were well oriented at all $K \leq 4$, although there was a suggestion of a directional shift with increasing disturbance. Although these results were somewhat variable and inconsistent (e.g., with different arena designs) they suggest a sensitivity on the part of the birds to changes in the magnetic field on the order of perhaps 20 gamma or so.

A somewhat different effect of natural magnetic fluctuations has been found in homing pigeons. Keeton et al. (1974) found that the vanishing bearings of experienced homers shifted leftward with increasing levels of magnetic disturbance when released under sunny skies. That this shift in compass direction is caused by magnetic events per se is strongly suggested by the results of subsequent experiments (Larkin and Keeton, 1976) in which attaching magnets to the pigeons eliminated the effect. Walcott (1977) has found a slight counterclockwise shift under sun in pigeons wearing Helmholtz coils and in the series of releases by Larkin and Keeton (1976) the birds' bearings were shifted leftward by a few degrees when they wore magnets compared to their flights with brasses. However, Moore (1975) found a slight clockwise shift in the southward orientation of herring gull (Larus argentatus) chicks carrying ceramic disc magnets on their heads.

Magnetic effects in birds engaged in migration have been difficult to document. Several studies have failed to reveal any relationship between the degree of orientation of migrants and K values (Able, 1974a; Richardson, 1974, 1976). However, the effect, if any, is likely to be very subtle and a large data set with many cases under intense magnetic disturbances may be necessary to reveal it. The only report of a positive correlation between spread in flight directions and magnetic storms is Moore's (1977) analysis of nocturnal migration in the southeastern United States. In both spring and fall, the flight directions of passerine migrants observed passing through the beam of a portable ceilometer (Gauthreaux, 1969) became increasingly spread with increasing K value. Assuming that magnetic storm intensity is not correlated with some relevant aspect of weather (e.g., cloud cover or wind velocity), these results are consistent with those of Southern and some of the pigeon data. Larkin and Sutherland (1977) have reported a slight influence on birds migrating over the underground antenna of Project Seafarer in Wisconsin: birds tracked by radar were more likely to turn or change altitude when the ac field was being turned on or off. Gull chicks

---

* Magnetic disturbances are ranked on a scale of K values, ranging from 0 to 9, where $K = 0$–3 indicates little or no disturbance, $K = 4$ weak storm activity, $K = 5$ moderate disturbance, and $K \geq 6$ indicates moderately severe storms (see Keeton, 1974a).

tested in circular arenas over this antenna oriented in the customary south-easterly direction when the antenna was not energized (255 trials), but were not significantly oriented when it was turned on (642 trials) (Southern, 1975).

Of a similar nature may be the apparent effects on homing pigeon flight of anomalies in the earth's magnetic field. Wagner (1976) observed a tendency for his pigeons to orient leftward of the home direction and to follow the magnetic gradient of a weak anomaly. Over stronger anomalies (variations in field strength to 3000 gamma or so over distances of less than 1 km) Walcott (1978) found a signficant decrement in the homeward orientation of radio-tracked pigeons correlated with the magnitude of variation in the earth's field in the homeward direction.

These apparent effects of temporal and spatial variations in the earth's magnetic field, subtle as they are, suggest that the behavior of birds is sensitive to very small changes in the magnetic field (probably less than 100 gamma or a very few percent of the local field strength). These levels of sensitivity are comparable to those of bees and would allow a bird to obtain useful orientation information from the earth's magnetic field. It is important to note, however, that most of the magnetic influences I have described could be classified as disturbances of one kind or another imposed on the orientation behavior. Decreases in the accuracy of orientation or slight directional shifts do not provide unambiguous evidence that a magnetic compass is involved in orientation. Results from most studies have been highly variable and affected by minor details of equipment and experimental design. The most persuasive data supporting the use of magnetic cues for orientation by birds are clearly those of the Wiltschkos in which orientation has been predictably altered by shifting the surrounding field. Whereas a healthy degree of skepticism still exists among workers in the field (e.g., Griffin, 1978), I believe the present weight of evidence supports a real, albeit subtle, influence of terrestrial magnetism on bird orientation, although based on present knowledge it is not easy to see what role it might play in natural migration (but see Section IV).

There does exist a large body of negative data, but some of it may be attributable to methodological subtleties or failure to ask questions in a way meaningful to and answerable by the birds. Because we know nothing about the possible magnetic receptor or the manner in which magnetic information impinges on the animal's behavior, experiments must be designed largely in the dark. The recent exciting success of Bookman (1977) in conditioning pigeons to distinguish between the presence or virtual absence of an earth-strength vertical magnetic field may open the door to further advances.

In some other groups of animals the data regarding magnetic influences on orientation seem less ambiguous, although this is perhaps because fewer experiments have been performed. Responses to magnetism have been reported in a variety of insects: Isoptera (Becker, 1963a), Coleoptera (Schneider, 1963, 1975), Diptera (Becker, 1963b, 1964; Picton, 1966; Wehner and Labhart, 1970), and Hymenoptera (Lindauer and Martin, 1968; Martin and Lindauer, 1977; Kisliuk and Ishay, 1977). Unfortunately, none of these responses is directly related to migratory orientation.

Von Frisch (1967) had found that the waggle dances of honeybees (*Apis mellifera*) performed on the vertical comb surface did not exactly compensate for the changing position of the sun during the day. There were errors in the dances of the bees (up to about 15%) and this *Missweisung* followed a regular diurnal pattern and was influenced by whether the vertical comb was oriented east–west or north–south (Lindauer and Martin, 1968, 1972). Lindauer and Martin showed that these small errors in the dance disappeared when the earth's magnetic field was cancelled with Helmholtz coils. As with many of the magnetic influences on bird orientation, this effect involves a disturbance. Because the bees' dances within the dark hive are oriented with respect to gravity, these findings suggest an interaction between gravity orientation and terrestrial magnetism. A similar result has been found with the negative geotactic orientation in *Drosophila* (Wehner and Labhart, 1970). Interestingly, Larkin and Keeton (1978) have found a possible parallel in homing pigeons. The day-to-day variations in the mean vanishing bearings of pigeons released repeatedly at a single site were found to be significantly related to the day of the lunar month. A relationship existed in all six experimental series, but in some seasons the changes in bearings cycled with the day of the lunar month beginning at new moon whereas in others they cycled beginning with full moon. Thus there appeared to be a relationship that was stable in one of two states with opposite phase. The interpretation of these data must be considered preliminary because no direct relationship to any gravitational variable could be found and the possible involvement of natural magnetic fluctuations is unclear and problematic.

The only direct use of magnetic information for orientation in insects appears to come from studies of nest or comb construction. Lindauer and Martin (1972) presented preliminary results showing that bees enclosed in cylinders built comb oriented in the same direction as in their home hive. The orientation of comb construction could be manipulated with Helmholtz coils surrounding the cylinders (but cf. Gould *et al.*, 1978, footnote 4). In the hornet (*Vespa orientalis*), Kisliuk and Ishay (1977) found effects on survival and comb-building in imposed magnetic fields of several to many times the strength of the earth's.

Phillips (1977) reported limited success in training cave salamanders

(Eurycea lucifuga) to select one of the two directional axes inside an enclosed corridor. Although some methodological problems existed (the animals were tested in groups and the same apparatus was used for both training and tests, raising the specter of olfactory cues), in six of 13 tests the salamanders moved non-randomly and along the expected corridors relative to the magnetic field. As in so many other cases, the relevance of this ability to the life history of this sedentary species is not obvious. However, preliminary data suggest that adult newts *(Notophthalmus viridescens)* may respond at least to the strength of a nonuniform field produced by placing a bar magnet at one end of an elongate tank (Phillips and Alder, 1978). These are the only studies performed to date on possible magnetic sensitivity in amphibians and they are sufficiently suggestive to warrant further study.

Because of the vastly greater electrical conductivity of water, the use of magnetic information in this medium, at least by means of induced voltage gradients or currents, presents fewer theoretical problems than in air. Indeed, the extreme sensitivity of marine elasmobranchs to weak electric fields has been known for many years (recently reviewed by Kalmijn, 1978). With regard to migratory orientation, these abilities are of interest because information about currents could be obtained, at least in salt water.

Ocean currents flowing through the earth's static magnetic field result in an induced electric field that is perpendicular to the direction of water flow. In the Northern Hemisphere this field is also directed to the left with respect to the current. Measurements with towed electrodes have shown that voltage gradients ranging from 0.05 to 0.5 $\mu V/cm$ occur naturally in waters of the North Atlantic (maximal values were recorded in Gulf Stream waters) (Rommel and McCleave, 1972). For a fish oriented either upstream or downstream with respect to the current a potential gradient perpendicular to the body axis would be generated. The gradient would be parallel to the body axis if the fish oriented across the current. If the animal could also sense the polarity of the field, it could discriminate upstream from downstream.

By means of conditioned cardiac deceleration, Rommel and McCleave (1972) were able to demonstrate sensitivity in the migratory American eel *(Anguilla rostrata)* to perpendicular electric fields as weak as 0.668 $\mu V/cm$ (8 of 8 fish) and 0.067 $\mu V/cm$ (4 of 8 fish). None appeared to perceive fields oriented parallel to the body axis (McCleave *et al.*, 1971). The eel's perception of perpendicular gradients appears to fall within the range of naturally occurring oceanic values. Marine elasmobranchs have similar sensitivity (Kalmijn, 1978). Other reports indicate possible magnetic detection and orientation in European eels *(Anguilla)* (Branover *et al.*, 1971; Ovchinnikov *et al.*, 1973), but attempts to repeat these experiments with American eels were not successful (Zimmerman and McCleave, 1975).

In a similar manner, a fish swimming through the earth's magnetic field induces voltage gradients that are a function of the orientation of the fish. These gradients appear to be of sufficient magnitude to be detectable by a fish swimming at a speed of 2 cm/sec or so. Kalmijn (1978) has presented data on sharks and rays indicating the possibility that they possess an electromagnetic compass sense. Such an ability could provide at least one means of orienting with respect to currents during oceanic migration and might be especially important in regions of the world where other information (e.g., visual cues) is often unavailable. Unfortunately, except for the eel, no data are available on the electrosensitivity of any migratory fish species.

One of the most frustrating aspects of studies of possible magnetic orientation is that for no organism, with the possible exception of the bacteria of marine mud studies by Blakemore (1975), do we know in detail how magnetic information is sensed and processed. Various possibilities have been discussed by Keeton (1972, 1974a), Leask (1977, 1978), Wallraff (1978a), and Griffin (1978). General models of magnetic detection usually involve (1) induction of a measurable current when an organism moves through the earth's magnetic field, (2) some sort of paramagnetic material which responds in a magnetic field, or (3) deposits of magnetic material in the animal's body which respond in a detectable way to an external field.

Gould *et al.* (1978) have reported the exciting discovery of deposits of transversely oriented magnetic material (apparently magnetite) in the anterior part of the abdomen of honeybees. The magnetic moment apparently develops during the pupal stage and dead bees lacked a natural field although remanence could be induced by application of an external field. Live bees obtained from hives with the customary north–south comb orientation had their north poles pointing either left or right. During the several days of motionless pupal development the earth's field would be transverse to the bee's bodies and if the orientation of the magnetic deposits occurs at that time one would predict that bees from opposite sides of the comb would have opposite horizontal polarity. These data demonstrated an orderly array of magnetic material in the bee that could serve as a detector. Further experiments will be necessary to determine whether these crystals are actually used in orientation. The application of similar technology to other animals could prove extremely rewarding.˙

˙Note added in proof: Walcott *et al.* (1979) have found deposits of magnetite in several places in the heads of homing pigeons. Their relevance to magnetic orientation has not yet been established.

B. Orientation by Currents and Waves in Water

To the land-locked observer, the oceans seem virtually trackless. However, while they do not independently yield compass directions, currents provide relatively stable directional information that could be calibrated by or used in concert with compass information. In the absence of other cues, orientation by currents could at least permit an animal to maintain directed movement.

Many kinds of fish are known to be able to orient with respect to current directions in streams, probably using visual, tactile, or lateral line senses to maintain orientation (see Royce *et al.*, 1968; Leggett, 1977). Such simple mechanisms are not so feasible in the sea, but there is nonetheless considerable evidence that currents may be important sources of orientation information for migrating fish. In the northern Pacific Ocean, for example, much descriptive data suggests that most migrating salmon follow the general pattern of current flow. The fish are not passively drifting with the water and after summarizing a large body of data, Royce *et al.* (1968, p.457) concluded that "the migrations of salmon on the high seas are actively directional in a way which somehow relates to the ocean currents."

The direction of currents could be assessed by a fish immersed in the medium in several ways. Voltage gradients induced by the movements of fish and/or water through the earth's magnetic field could provide such information as discussed in Section III,A. Also, as in air masses, the interface zones between water masses or currents have distinct physical characteristics such as turbulence, chemical composition, and marked temperature and salinity gradients. If fish could detect such interfaces the tracking of currents would be facilitated. There are virtually no data, aside from mark–recapture studies, directly testing the importance of currents in the orientation of migratory fish. The difficulties in obtaining such data are obvious, but it is an important question that relates intimately to the question of homing in anadromous fish (see Section V,A).

Currents are not the only potentially important structural features of large bodies of water. The passage of waves or swells on the ocean surface produces horizontal oscillations on the bottom called wave surge. The direction of wave surge is often independent of currents, but nonetheless constant over long periods of time. It may be detected in the absence of current or when superimposed on it. The migratory spiny lobsters *(Panulirus argus)* on the Bahama Bank have been studied by Herrnkind and colleagues. In their natural autumn migrations, performed by walking along the sea floor in long queues (Herrnkind and Kanciruk, 1978, and references cited therein), the lobsters move at angles to both current and wave surge. In local movements, however, wave surge may provide an important cue. At Bimini, surge is produced primarily by swells approaching from deep water to the west or northwest. Lobsters often move from shallow inshore waters to the shelter of reefs in deeper water. Lobsters whose eyes were covered with opaque tape showed little inclination to move and wandered in random directions when displaced and tested in a calm lagoon with no detectable current or surge (Fig. 8A). In an area with surge displacements of 10–50 cm the lobsters consistently oriented into the surge (the offshore direction), and this behavior was replicated in a wave tank (Fig. 8B) (Walton and Herrnkind, 1978). Orientation with respect to currents in these tests was much more variable.

Currents and waves, where they provide stable information and can be related to some external frame of reference, may be important to aquatic animals both in compass orientation and as "landmarks" used in navigation.

C. Wind Direction

Bats, birds, and insects are the only migratory animals to have successfully included the air as a major component of their living space. As a

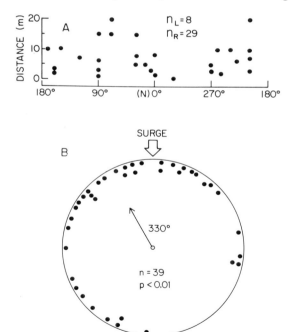

Fig. 8. Orientation of spiny lobsters (*Panulirus argus*) blinded with opaque eyecaps. (A) In a quiet area in Bimini Lagoon lacking both current and surge lobsters were disoriented and moved only short distances. (B) Orientation into wave surge in a tank in which waves were generated with motor-driven paddles. Lobsters that moved on straight paths traveled into the wave surge (V test, $p < .01$). (Redrawn from Walton and Herrnkind, 1978.)

result of this evolutionary breakthrough, they are capable of undertaking migrations and dispersal of very long distances. In species for which flight is extremely important, we should expect natural selection to mold highly sophisticated responses to wind and other related aspects of atmospheric structure; there is much evidence that this is the case. Failure to respond appropriately must often mean the difference between successful migration and death.

The problem of avoiding undesirable wind displacement is particularly severe for insects because of their low air speeds. Virtually no insects are capable of flying at air speeds greater than 10 m/sec and most fly at considerably less than 5 m/sec during sustained movements (Johnson, 1969; Weis-Fogh, 1976). As a result, one would expect insects to be frequently displaced by winds to areas beyond their normal range. On the other hand, the evolution of day-to-day or seasonal behavioral selection could result in most individuals departing under conditions favorable for the directed movement.

Many species such as aphids are essentially at the mercy of winds and convection currents. Flying at air speeds of less than 1 m/sec, they are carried almost passively by the wind. The altitudes at which they fly are determined by the turbulent-convective process and they may be carried to high altitudes during the day, only to descend after dark (Dixon, 1971). Even in larger, faster-flying species the effects of wind on flight are often obvious. Although the seasonal timing of the flights of most species is thought to have evolved to correspond with the most favorable winds (Johnson, 1960, 1969), large groups are frequently caught in unfavorable air flow. The moth, Laphygma (= Spodoptera) exigna, occasionally irrupts from central Asia and reaches northern Europe and Great Britain. Large flights are correlated with winds blowing from source areas and can be mapped on an almost day by day basis according to the direction and speed of winds aloft (Hurst, 1969; Mikkola, 1970). Arrivals in Britain of North American insect species (e.g., monarch butterflies, Danaus plexippus; the moth, Phymatra biloba) can be predicted on the basis of broad wind patterns and often coincide with the occurrence of vagrant bird migrants from the same source areas (Hurst, 1969). Whereas it is often possible to exercise some control over the direction of movement by selecting appropriate winds in which to initiate migration or dispersal, once aloft many, perhaps most, insects appear to be largely at the mercy of the wind (Roer, 1974). In some species this potentially dangerous situation has been elaborated into a highly adaptive dispersal system.

The intensively studied desert locust (Schistocerca gregaria) engages in extensive movements in arid parts of Africa. Their direction of movement is now known to be very largely or wholly determined by wind and the altitudes they reach controlled by the height of thermals (Rainey, 1976). Airplane tracking of locust swarms in East Africa showed that on 42 occasions there was only a root-mean-square difference of 15° between track and wind direction, a value within observational error (Rainey, 1976), and a similar value was found in a whole year's data (Rainey and Aspliden, 1963). Timed as they are, however, these downwind flights are highly adaptive. In tropical regions, trade winds and monsoons originating from the Northern and Southern Hemispheres, respectively, meet in the so-called Intertropical Convergence Zones. Downwind flight in these regions leads ultimately to the areas of convergence. This is extremely important for the locusts because it is in these areas that rain will fall, providing the temporary flush of vegetation that is necessary for successful reproduction. Figure 9 shows the typical pattern of winds in the area of an intertropical front explored by airplane flights, and a radar presentation showing the concentrating effect such systems have on airborne insects (in this case, moths in Africa). Similar wind-induced concentrations of insects have also been observed frequently along both stable and moving fronts in temperate regions (Rainey, 1976; Schaefer, 1976).

Detailed studies of swarms of desert locusts have shown that not all aspects of their flight are determined by atmospheric features. The cohesion of moving swarms is maintained by the orientation of the individual insects. Locusts at the edges of the swarm consistently head inward toward its center. One group monitored as it moved 370 km over 70 hr varied in size by only 10%, whereas convection currents encountered during the flight should have lead to a 10-fold increase in swarm size if the locusts were entirely passive (Rainey, 1976). In light winds it is possible for the swarm to stay over the same area for long periods of time by heading into the wind with an air speed equal to the wind speed (Scheepers and Gunn, 1958; Rainey, 1973, 1976).

Although locusts seem to consistently move downwind, the radar studies of Schaefer have shown that this is not invariably the case with other species, especially in daytime when visual cues might be used to gauge wind drift. Spruce budworms (Choristoneura fumiferana) observed with 3-cm radar in New Brunswick had flight tracks up to 60° from downwind during the day, but were obviously downwind at night, showing changes in directions with altitude (Fig. 10) (Schaefer, 1976). Pierid butterflies were observed heading into the Sahara Desert with a crosswind during the day and in the Sudan grasshoppers (Aiolopus) oriented SSW at night in all but strongly opposed winds. Furthermore, grasshoppers in the Sudan were not passively carried to flight altitudes by convection, but climbed actively in flapping flight (Rainey, 1976). Thus the strategies of species appear to vary considerably. For example, Schaefer (1976) pointed out that the Aiolopus

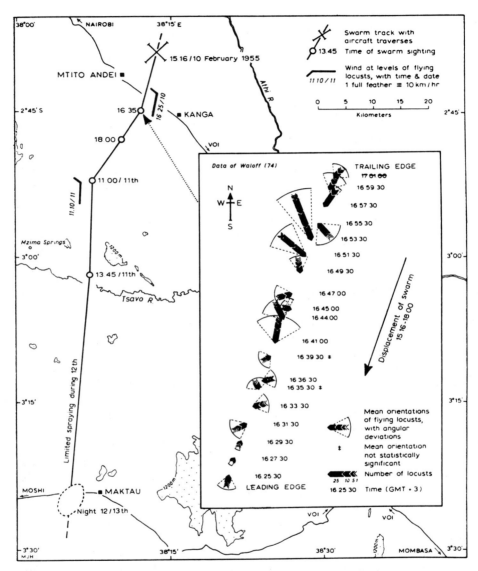

Fig. 9. Swarm track and orientation of flying locusts *(Schistocerca gregaria).* The map shows the downwind flight track of the swarm, estimated at 10⁹ individuals, over 2 days. Inset shows the orientation of locusts within the swarm as shown by photgraphs taken at 2-min intervals as the swarm passed overhead. Note the into-swarm headings of the individual locusts. (From Rainey, 1976.)

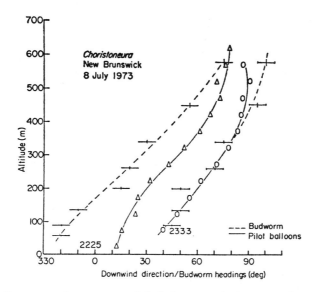

Fig. 10. Orientation during nocturnal flight by spruce budworm moths (*Choristoneura fumiferana*) determined by radar. Earlier in the night (2224), just after takeoff, moths headed about 25° to the downwind direction, especially at low altitudes. Later in the night (2333) all moths showed downwind orientation which closely followed the wind changes with altitude. (From Schaefer, 1976.)

probably could not achieve their goal area in the few nights of flight they perform without actively orienting at angles to the wind. Riley (1975) has also observed a nocturnal flight of insects (probably grasshoppers) at nearly right angles to the wind in Mali (under nearly full moon).

Even in terrestrial insects wind direction may serve as an important orienting cue. The desert ants, *Cataglyphis bicolor*, apparently integrate orientation information from several sources, including wind direction. Thus when orienting on the basis of polarized light alone, they are bimodal unless the wind is blowing. If wind cues are available, they orient in a single direction, that indicated by the wind at the time of training. Finally, Linsenmair (1972) reported menotactic orientation in scorpions and tenebrionid beetles in wind speeds from 0.02 (scorpions) to 4.5 m/sec.

In one manner or another, wind has been recognized as an important factor in the orientation of migratory birds for a very long time. The occurrences of migrant birds far out of their normal ranges has often been correlated with winds during the preceding few days (references cited in Able, 1980). Furthermore, it is well documented that weather factors, especially wind direction, seem to be very important determinants of whether a migrant bird will initiate flight on a particular day (see Richardson, 1978a, for a comprehensive review). Finally, the wind provides a directional cue that might be used by itself or in combination with compass information to establish flight orientation. Being variable in time and space, the wind cannot, of course, provide absolute compass information. However, the directions of winds aloft are quite predictable on the basis of information available on the ground. If birds are as good at meterology as evidence leads us to believe, they could obtain at least crude compass information even in the absence of the more reliable systems. In the northeastern United States, for example, the passage of a cold front in autumn is usually signaled by falling temperature and relative humidity, clearing skies, and rising barometric pressure. A bird sensing these or other signs of a frontal passage could gamble, with a very high probability of being correct, that the winds are blowing from the northwest. In addition, knowing the direction of winds on the surface often enables one to predict wind direction in the lower thousand meters of atmosphere where most birds migrate. It is, therefore, not unreasonable to expect natural selection to have molded response packages based on the covariance between favorable winds and other meterological factors. In the absence of all other compass information, a bird might well be able to select a night for migration on the basis of ambient weather, orient solely with respect to wind direction, and have a high probability of flying in the correct seasonal direction.

Many years ago, Vleugel (1954, 1959, 1962) proposed the hypothesis that nocturnal migrants used the position of sunset to select their flight direction and then maintained that direction after dark by flying at a fixed angle to the wind. For such a system to work, reference to close-range visual landmarks would seem to be required and the model assumes that wind direction remains constant over relatively large areas. It now appears that both sunset (Section II,A,3) and wind direction are important directional cues for nocturnal migrants, but not in exactly the sense that Vleugel envisioned.

Two unambiguous orientation responses to wind direction are flight with or into the wind. Virtually all field studies of bird migration, especially night migration, have shown that birds generally fly with a following wind component. The energetic advantages of such behavior are obvious and it may often mean the difference between the successful completion of a flight and death when long overwater movements are involved (Tucker, 1971, 1974; Able, 1974a; Williams and Williams, 1978). However, the observation that migrants usually fly with a tail wind does not demonstrate the use of wind as a cue in orientation. A stronger case may be made if birds fly downwind even when winds are not blowing in the seasonal migration direction. On this basis, the data from nocturnal songbird migrants in the inland southeastern United States reported by Gauthreaux and Able (1970) and Able (1974a) suggested that wind direction was being used as a primary directional cue. Figure 11A shows data from fall migrants observed in Louisiana and Georgia. The high correlation between wind direction and the directions of the birds showed that their flight tracks were statistically indistinguishable from downwind (i.e., the difference between tracks and wind were within the range of observational error of the two variables). It is most important to note that many of these flights were oriented downwind in directions at large angles or even opposed to the normal seasonal direction of migration at these localities (approximately SW). This fact, coupled with significant changes in flight orientation accompanying wind shifts within a single night (see Able, 1974a, Fig. 3), lead us to conclude that nocturnal passerine migrants in that area selectively oriented downwind, regardless of wind direction and even under clear skies. Flocked migrants (waterfowl and shorebirds), on the other hand, flew in appropriate directions regardless of the wind.

Night-to-night variability in the flight directions of migrating birds correlated with wind direction has been observed in many radar studies. Several workers (Evans, 1966; Nisbet and Drury, 1967; Alerstam, 1976) have proposed that this variability is due largely to differential migration of groups of birds with different preferred migration directions. The result of this behavior would be changes in flight direction from night to night with a consistent downwind tendency. Some of the changes in orientation observed in the field are undoubtedly due to this factor. However, when one consistently observes large numbers of birds moving almost precisely downwind in directions that cannot be toward the migratory goal of any of the species, I find the wind selection hypothesis untenable. It is hard to imagine that at all times there exist substantial numbers of migrants waiting around for winds to blow in any given direction.

On the other hand, it is very difficult to explain convincingly why birds should fly downwind in seasonally inappropriate directions when they might better remain on the ground and wait for more favorable winds. In spring, a downwind reversed flight could often remove a bird from an area that is going to become colder and perhaps temporarily uninhabitable. Northward autumn flights, however, cannot be explained in the same way. If the downwind flights of passerine migrants in inexplicable directions in the inland southeastern United States form part of an adaptive response, how they do so is not apparent to me. All that can be said with certainty is that in this region the average winds during migration seasons blow roughly in the migratory directions of the birds; even if a bird flew nightly, taking numerous detours, it would undergo a net movement toward its destination. Furthermore, flight in other directions in this inland region is not likely to place a bird in an overly hazardous situation.

In other regions, winds are generally much less favorable for migration

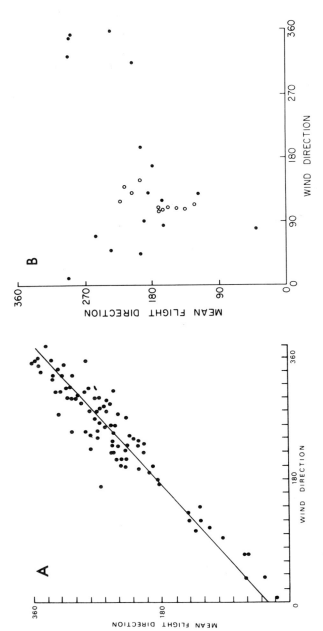

Fig. 11. Flight directions of passerine nocturnal migrants observed with portable ceilometers as a function of wind direction 300 m above ground. (A) Fall migration at Lake Charles, Louisiana, and Athens, Georgia. Birds flew downwind regardless of wind direction or speed. (B) Fall migration near Albany, New York. The birds flew in appropriate southerly direction on most nights regardless of wind direction. Each point in the figures represents the hourly mean flight direction of the birds and wind directions are those toward which the winds were blowing. (From Able, 1974a,b.)

or pose distinct hazards to a bird flying downwind. In these areas downwind flight in seasonally inappropriate directions is not generally observed: in the Florida peninsula in autumn (Williams *et al.*, 1977), the northeastern United States (Drury and Nisbet, 1964; Nisbet and Drury, 1967; Richardson, 1971, 1972; Able 1974b, 1978), or in the Caribbean area (Richardson, 1974, 1976). It is perhaps not too surprising that migrants flying in coastal areas or embarking on overwater flights might carefully avoid downwind flight in some situations; the success of their migration might often depend on such behavior. But even in inland eastern New York forward migration often occurs at large angles to the wind or even into headwinds (e.g., Fig. 11B; Able 1974b, 1978). Because the species and populations migrating through the northeastern and southeastern United States are largely the same, the striking differences in their behavior imply geographic variability in the responses of birds to directional information. In the Southeast wind direction seems to predominate over other cues whereas in the Northeast other input (apparently sun and stars) is primary (see Section IV). The adaptive significance of this difference may lie in the relative dangers of downwind flight in the two regions. In eastern New York in both spring and fall the most frequent and strongest winds are from the northwest, posing the threat of carrying migrants offshore into the Atlantic Ocean.

How birds might accomplish orientation by wind direction has generated considerable discussion (reviewed by Griffin, 1969; Emlen, 1975). If a flying bird can maintain visual contact with some nearby visual landmark such as an object on the ground or a distinctive cloud surface, the angle of orientation relative to the wind could be determined by gauging lateral drift. There is at present no evidence which demands a more complicated explanation. It is, however, theoretically possible for a flying bird to determine wind direction from consistent patterns of turbulence structure in the lower atmosphere (Nisbet, 1955; Bellrose, 1967). Such patterned atmospheric structure has been observed on both large and small scales, but its occurrence may be so limited in both space and time as to render it of little use to migrating birds (see Able, 1977b).

D. ACOUSTIC CUES

As with wind and currents, sounds do not provide animals with compass information per se. However, although their possible role in orientation has been studied very little, acoustic cues might provide at least useful back-up input to migrating animals. Considerable speculation and a small amount of data are available with regard to the roles sounds might play in the migratory orientation of birds and whales.

In several recent papers, Griffin and his colleagues (Griffin, 1969, 1976;

D'Arms and Griffin, 1972; Griffin and Hopkins, 1974) have discussed the kinds of sounds that may reach migrating birds from the ground. It turns out that a variety of sounds emanating from ground sources should be audible to birds at normal altitudes of migration. Particular attention has been given to breeding choruses of frogs which can probably be heard by migrants flying 1 km above the ground (Griffin and Hopkins, 1974). These and other terrestrial sound sources (e.g., ocean whitecaps) might, under special conditions, serve as acoustic guideposts that might aid a bird in maintaining orientation, for example, when flying out of visual contact with the ground.

The flight calls emitted by birds migrating at night are mysterious. Anecdotal evidence, at least, suggests that calling increases under poor visibility or other conditions when migrants might become confused. Certainly flying birds must often hear calls emitted by other migrants aloft and there exists at least the potential that this information might aid in orientation. Individuals flying under unfavorable conditions (e.g., inside a cloud) might use calls of other birds to maintain straight and level flight, as suggested by Griffin (1969), or some sort of pooling of orientation information could theoretically take place (Hamilton, 1967). All that can be said at present is that migrating birds call frequently at night and that some information potentially relevant to orientation is available.

Another way in which flight calls might provide useful input to a migrating bird is through their echoes. Griffin and Buchler (1978) have shown that ground echoes of sounds similar in some ways to nocturnal flight calls might be audible to flying birds up to altitudes of 1 km. While there is as yet no evidence that birds perceive or use such echoes in any way, they could give information about flight altitude and perhaps about the nature of the terrain over which the bird was flying.

Finally, the recent fascinating discovery that pigeons can hear sounds of less than 0.1 Hz (infrasound) (Yodlowski et al., 1977; Kreithen, 1978) opens up yet another possibility. Infrasound travels thousands of kilometers with very little attenuation, creating a loud and complex world of sounds we cannot hear. Among other sources of infrasound are numerous stationary and continuous emitters such as wind flow over certain mountain ranges and breaking ocean waves. Formidable problems confront a bird attempting to sort out infrasound signals from other pressure variations because of local wind and turbulence (pseudosound) and there is as yet no evidence that birds make any use of the sounds. However, spatially fixed sources could act as infrasound beacons if they could be directionally localized, an extremely difficult problem in itself at the wavelengths involved.

The remarkable low frequency sounds made by baleen whales may be available to other whales over distances of hundreds of kilometers (Payne and Webb, 1971). Altough their function is unknown, they are apparently sufficiently localizable that it is not unreasonable to suggest that they might serve some role in the orientation or navigation of the migratory large whales.

IV. Integration of Orientation Cues

Perhaps the most important advance in the study of animal orientation during the past decade has been the demonstration of considerable redundancy in orientation abilities. The relevant data have come primarily, but by no means exclusively, from work on birds and they have had profound effects on our understanding of the behavior as well as on the approaches taken in research. It is now clear that demonstrating a particular capability in an animal is only the first step in understanding its orientation behavior. The ability under consideration may be only one of an array upon which the animal may rely, depending on environmental conditions or other factors. This makes the interpretation of so-called negative data very difficult. Depriving an animal of a given set of compass information, for example, may have no discernible effect on its orientation. But that does not imply an inability to utilize that cue when it is available. In its absence the animal may simply have switched to an alternate reference system to accomplish the task [see Emlen (1975) for a discussion of the ways in which this problem has hampered research].

Once we know that natural selection has provided animals with back-up orientation systems, research questions become more difficult to formulate and experiments harder to design. Only by observing its behavior under natural conditions can we be certain that the entire repertoire of an individual might be revealed. It becomes extremely important to know not only what cues an animal uses, but also how those cues are related to one another. Although the task has clearly become more difficult, the analysis of complex orientation systems and the discovery of how redundant capabilities are integrated is an exciting prospect for the next decade.

Because of the relative newness of this area of research, few cases have been explored in detail. I will describe three examples from diverse taxa as exemplifying the kinds of relationships and complexity we may expect to find in other groups.

Adult fiddler crabs *(Uca)* exhibit well-developed escape responses oriented toward their home burrows. Two discrete types of behavior appear to be involved: (1) short-range (to ca. 1 m) homing to individual burrows based on kinesthetic orientation, and (2) movement usually directed landward regardless of the position of the burrow and oriented by visual cues (Herrnkind, 1972). This latter far-orientation is of interest here.

Fiddler crabs foraging on the lower beach do not hold individual burrows and may wander as much as 50 m from the area of burrows. If disturbed, they move rapidly away from the water, entering any unoccupied burrow encountered. The landward orientation of these movements appears to be based on celestial cues and grass landmarks near the horizon. If tested in circular arenas from which only the sky is visible, the crabs move in the landward direction (Fig. 12) and this orientation can be manipulated by presenting a mirror image of the sun (Altevogt and von Hagen, 1964; Herrnkind, 1968).

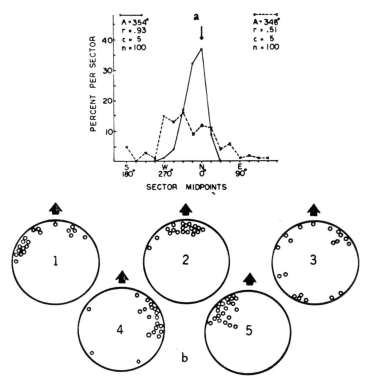

Fig. 12. Landward orientation in fiddler crabs (*Uca pugilator*) in a test chamber affording a view of the sky. (a) Orientation of groups versus single crabs. The solid line shows the distribution for groups of five crabs, the dashed line the distribution of the same crabs tested individually. *A*, mean angle; *r*, mean vector length; *c*, number of test crabs; *n*, number of positions recorded. Landward direction is shown by the arrows. (b) Scatter diagrams showing the direction and distance of movement of the five crabs tested individually. (From Herrnkind, 1972.)

Most crabs in familiar surroundings appear to orient primarily using celestial cues, switching to landmarks if they are persistently pursued, become desiccated, or are transported to a novel beach with a different orientation. There exists in the populations a proportion of crabs which orient to landmarks even under clear skies. As Fig. 12 shows, groups of crabs show better landward directionality than do single individuals, perhaps as a result of crabs with weak directional preferences following well-oriented individuals (Herrnkind, 1972). As in the case of y-axis orientation in amphibians (Section II,A), new shoreline orientations can be learned in both natural and laboratory situations.

Insight into the relationships among orientation capabilities in the sand fiddler crab *(Uca pugilator)* has been obtained by studying the behavior during ontogeny (Herrnkind, 1972). Crabs of various ages were collected from a Florida beach with a landward direction of 11.5°. Also, eggs from some of these individuals were reared in the laboratory. These larvae as well as the wild-caught early instar individuals were raised indoors without exposure to natural celestial objects or a shoreline.

No crabs, from 1 day to 8 months of age, showed significant orientation when tested in an arena under uniform, diffuse white light. If given a contrasting landmark (a black screen covering a 45° sector of the otherwise white chamber), all groups moved toward the screen. This response was equally strong in wild and laboratory-reared individuals, but tended to wane with age. In nature, the first few instars do not dig burrows, but rather take shelter under objects on the beach; thus their orientation in laboratory conditions appeared to correlate with what is known of their behavior in the field.

Orientation with respect to celestial cues varied with age, landward movement increasing as the crabs became older. By 5 to 8 months of age, the wild crabs oriented generally landward, although the spread in directions was greater than that in adults. Laboratory-reared crabs showed no consistent celestial orientation. During ontogeny celestial orientation (apparently using either the sun or polarized light) gradually comes to dominate landmark orientation and the simple positive phototaxis shown by younger instars. These latter mechanisms are probably sufficient to orient the rather haphazard movements of young crabs, whereas more complex mechanisms are required by adults returning to burrows. Considerable plasticity is built into the adult system because the landward directions can be modified through experience.

Many insects appear to possess a host of orientation mechanisms although major adaptive differences occur between species. Both celestial cues and odors are important in the foraging flights of honeybees, for example (reviewed by Gould, 1976). Harvester ants *(Pogonomyrmex badius)* normally forage for seeds within 10–20 m of their nest entrance. Because of intercolony aggression in this species, it is highly adaptive for foragers to return to their own nests. The ants appear to rely primarily on scent trails laid down during the outward journey. But the Dufour's gland secretion in this species is not colony-specific so additional information is probably often important. Hölldobler (1971) found that reversing the position of an artificial sun caused ants foraging in a circular arena to move in the opposite direction in spite of the presence of an odor trail. Landmarks appeared to be of at least tertiary importance in this species.

The situation appears to be quite different in the desert ant, *Cataglyphis bicolor*. This species does not use scent trails and relies mainly on the sun, switching to use of landmarks when the sun is unavailable (Wehner and Menzel, 1969; Wehner, 1972). Experiments designed to place the directions indicated by sun and landmarks in competition showed that the ants moved in the direction toward food indicated by the sun, ignoring a black screen landmark moved 60° from the training position. When the sun is not directly visible, the ants use the plane of polarization coupled with wind direction. If the wind remains from the same direction, the ants are well oriented in one direction. If wind reverses direction, the ants move 180° to the home direction; if wind information is not available the ants are bimodal before sunrise or after sunset (Wehner, 1972).

In general, the orientation system in *Cataglyphis bicolor* appears to be hierarchical, the animals switching from one cue to another. Similar sorts of alternative orientation have been found in bees (sun versus landmarks; von Frisch and Lindauer, 1954), spiders (sun versus kinesthetic memory; Moller, 1970), beetles (sun versus wind direction; Birukow, 1958), and scorpions (sun versus wind direction; Linsenmair, 1968). There appears to be simultaneous integration of wind and polarization information in *Cataglyphis*, with wind direction being employed to resolve the inherent ambiguity in the *e*-vector of plane polarized light. Some sort of integration is probably also involved in cases in which compromise directions are observed when cues are experimentally opposed (Jander, 1957; Moller, 1970; Görner, 1966).

Investigations of compass cue interaction in birds are considerably more difficult to design and have proceeded along two rather independent lines. On the one hand, migrants confined in orientation cages have been used to examine the role of stars, magnetic field, and sun in nocturnal direction finding, while on the other, techniques of visual and radar observation have been used to evaluate sun, stars, landmarks, and meterological factors in free-flying birds. Bearing in mind the caveat that a complete understanding of the interactions of cues can probably never be obtained under controlled conditions, I shall discuss first the recent results from orientation cage studies.

An important question left unanswered by Emlen's (1970) experiments on celestial rotation in the development of stellar orientation in the indigo bunting *(Passerina cyanea)* (see Section II,C) was, what frame of reference determines the way in which the birds respond to the rotation? Wiltschko and Wiltschko (1975a,b) have explored this question and may have provided at least a partial answer. Three species of transequatorial migrants (the warblers, *Sylvia borin, communis,* and *cantillans)* were captured on migration and tested outdoors at night in standard octagonal, radial-perch cages. Control tests were performed under clear skies in the earth's magnetic field. Experimental tests were conducted under similar conditions, but Helmholtz coils around the cages produced an earth-strength field with magnetic north corresponding to celestial east–southeast (120°). The same birds tested in this situation oriented with respect to magnetic directions, seemingly ignoring the stars. The best of their results are shown in Fig. 13 and the data certainly suggest that the magnetic compass takes precedence over the star compass. In fact, when placed in cages in which the magnetic field strength was reduced to 0.32 G and the horizontal component eliminated, the warblers did not yield significant directionality although stars were visible.

Spring experiments on the short-distance European robin *(Erithacus rubecula)* yielded basically similar results, with two notable exceptions: the robins did not respond to the altered magnetic directions immediately, and the birds persisted in northward orientation in the weak field lacking a horizontal component.

The interpretation of these experiments is somewhat difficult because of a variety of factors: (1) The same possibilities for cage artifacts existed and the same special statistical procedures were required as in the earlier magnetic orientation experiments (see Section III,A). (2) In the experiments with warblers, the control groups were not always signficantly oriented even when data were pooled, although the directional trends are those that would be expected. (3) Individuals of some of the same species studied here, as well as numerous others, have oriented in appropriate celestial directions in a planetarium when stellar and magnetic directions did not conform. (4) Reversal of planetarium skies has consistently resulted in a similar shift in the orientation of birds, at least over the short term. (5) Magnetic compass cues change abruptly when a bird crosses the equator as two of the warbler species do. (6) Data obtained in the Wiltschko cages are not strictly comparable to others because the birds' view of the sky is restricted to a 95° sector centered on the zenith. (7) In the experiments with robins,

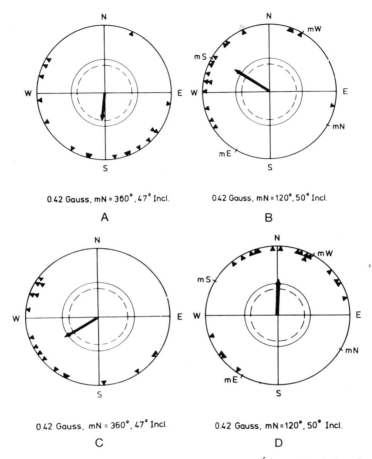

0.42 Gauss, mN = 360°, 47° Incl.   0.42 Gauss, mN = 120°, 50° Incl.

A   B

0.42 Gauss, mN = 360°, 47° Incl.   0.42 Gauss, mN = 120°, 50° Incl.

C   D

Fig. 13. Migratory orientation under clear night skies with normal and altered magnetic fields. (A) Directional preferences of whitethroats *(Sylvia communis)* under stars in the earth's magnetic field. (B) Orientation under stars and an induced magnetic field with magnetic north turned to celestial ESE. (C) Orientation of garden warblers *(S. borin)* under control conditions (as in A). (D) Directional preference of garden warblers in test conditions (as in C). Symbols as in Fig. 7. (From Wiltschko and Wiltschko, 1975a.)

the directional shift under the altered magnetic field appeared only after the data were subjected to an arbitrary a posteriori subdivision into two groups comprising earlier and later experiments. Even so, the shifted orientation is statistically significant in only one of the two seasons.

These possible problems do not, however, detract from the importance of these experiments. Further studies of this type may well confirm the basic framework of the scheme the Wiltschkos have put forth to explain these data. While there are many inconsistencies between their results and those of earlier experiments, they may not be insurmountable. At the same time, the Wiltschkos' hypothesis accounts for a number of previously inexplicable observations.

Wiltschko and Wiltschko (1975a) proposed that the magnetic compass is the primary orientation system for these migratory birds. Stars provide a secondary source of information that must be calibrated periodically by reference to the magnetic compass. Contrary to Emlen (1970), the star compass does not become permanently set by the time of the first migrations, but can be repeatedly reset by magnetic cues. By this view, then, the star and magnetic compasses do not represent two alternative compass systems but rather a single independent compass (the magnetic field) from which information may be transferred onto other cues. The obvious difference between the orientation system of the indigo bunting and the Wiltschko scheme may be a product of phylogenetic differences, or indigo buntings may respond more slowly to changed magnetic conditions, as suggested for robins. In the latter case, Emlen's experiments might not have been of sufficient duration to reveal the effects. In any event it is not clear why rotation should be used at all if magnetic information can be used to set the star compass. If magnetic orientation is primary in the sense of the Wiltschkos' model, I find it incongruous that the experimental evidence to date that birds are sensitive to the earth's field is so subtle as to be barely detectable above noise. Why should a bird use what seems to be an incredibly crude compass to set what appears to be a quite accurate one? One possible reason is that magnetic information is more reliable in time and space. Especially for a long-distance migrant, visible star patterns change as the journey progresses and the seemingly important circumpolar stars disappear below the horizon if the bird moves far south. On the other hand, magnetic cues also change (e.g., the apparently important dip angle reverses upon crossing the equator), so neither system can be used in the same way for the entire migration. Perhaps we have just not yet presented birds with an optimal experimental situation in which to reveal magnetic sensitivity.

The Wiltschkos' data and some other studies (see Keeton, 1974a) suggest that the magnetic sense has a long time constant, perhaps requiring the transfer of magnetic information to another cue system that can be used on short notice. Support for this notion comes from experiments in which European robins apparently calibrated an artificial pattern of 16 stars using an artificial magnetic field (Fig. 14) (Wiltschko and Wiltschko, 1976). If the selection and maintenance of directions are relatively independent processes, and the basic magnetic compass is consulted only occasionally, the results of Wallraff and Gelderloos (1978) may be explained. Under an arti-

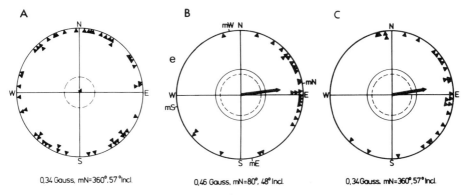

Fig. 14. The orientation of robins *(Erithacus rubecula)* under an artificial pattern of 16 "stars." (A) Random hopping under the star pattern before a magnetic field of sufficient intensity was turned on. (B) Orientation in the presence of a magnetic field of sufficient intensity to be used in orientation. (C) Orientation on subsequent nights under the same star pattern, but with the magnetic field strength again reduced to a low level. (From Wiltschko and Wiltschko, 1976.)

ficial star pattern coupled with a slowly rotating magnetic field, their birds (European robins, blackcaps, *Sylvia atricapilla*, and a redstart, *P. phoenicurus*) relied on the visual cues, seemingly ignoring the magnetic field. Under rotating stars and a stationary magnetic field the birds shifted with the artifical star patterns. Taken at face value these data suggest primacy of stellar cues, but other interpretations are possible. The Wiltschkos (1978) have found that in a magnetic field of 0.34 G, birds were unable to determine the correct migratory direction. However, if artificial star patterns were present the birds took up a direction apparently at random, and then maintained it with great accuracy. Once such a direction is selected, and this process may occur only a very few times per night, stars or lights appear to be used to maintain the orientation.

On the other hand, there is evidence that birds may respond rapidly to magnetic information. This certainly seems to be the case with effects on pigeon homing (e.g., Keeton, 1971, 1972; Walcott and Green, 1974) and in Bookman's (1977) training experiments. However, it must be remembered that the influence of magnets on pigeon homing are often unspecific disturbances and in Bookman's experiments the pigeons had only to discriminate whether the field was on or off. Extracting compass information from magnetic cues may involve qualitatively or quantitatively different processes.

It seems likely that age and species differences may add another layer of complexity to the problem. Keeton (1971, 1974a) found that first flight homing pigeons were usually unable to orient in the homeward direction both under overcast and sunny skies when carrying bar magnets. The idea that less experienced individuals require more information seems reasonable and this factor may explain some of the results of Southern (1977) and Wiltschko and Wiltschko (1975b).

The number and kinds of orientation cues and the possibilities for interaction among them are greatly enhanced for a bird in nature. Studies in orientation cages described earlier (Section II,A,3) have shown that the sun may be an important, although not necessary, source of information in nocturnal migratory orientation. Recently, data pointing to the same conclusion have come from field studies of free-flying migrants. Tracking radar studies and visual observations in upstate New York have shown that passerine nocturnal migrants generally flew in appropriate migration directions, even in opposing winds, if either stars or the sun near the time of sunset was visible (Able, 1978). When solid, thick overcast began several hours before sunset and continued after dark, birds migrating on nights with opposing winds often appeared unable to determine the correct migration directions and headed downwind in seasonally inappropriate directions. Figure 15 compares the flight orientation of passerine migrants in opposed winds under clear versus solid overcast skies that prevented a view of both stars and the sun late in the day. In comparably opposed winds, the two distributions are significantly different, the birds heading downwind when prevented from seeing either sun or stars and flying into the wind in an appropriate direction under clear skies. On the rare occasions when solid overcast commenced near dusk, affording the birds a view of the sun near sunset, but preventing them from seeing stars, headings were indis-

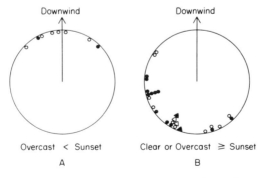

Fig. 15. Headings of passerine nocturnal migrants tracked by radar or observed with portable ceilometers near Albany, New York. The headings are plotted with respect to the downwind direction, at the top of each figure. Solid figures represent spring data, open figures refer to fall. (A) Data obtained in opposed winds under solid overcast which began at least 8 hr before sunset. (B) Data obtained in opposed winds under clear skies (circles) or when overcast began near the time of sunset (triangles).

tinguishable from those under clear skies (Fig. 15B). For these nocturnal migrants a view of either stars or the sun near sunset appears to be sufficient for the performance of correct orientation. In the absence of both of these inputs the birds were unable to determine the seasonal migration direction and relied on an apparently tertiary cue, wind direction. If magnetic information is used at all on such a short-term basis, it does not seem to prevail over wind direction under overcast conditions.

Other recent evidence also supports the idea that information obtained from the sun is integrated with other cues by nocturnal migrants. Emlen and Demong (1980) released white-throated sparrows (*Zonotrichia albicollis*) carried aloft by balloon-borne boxes and tracked by radar after release (for details of the technique, see Demong and Emlen, 1978). Among other experiments, they released a small number of birds during the brief interval between sunset and the time the first stars became visible. Birds initiating apparent migratory flight at this time were able to select the appropriate direction. Other data, summarized by Able (1980), reveals a deterioration in the orientation of nocturnal songbird migrants during periods of solid overcast lasting several days. Thus a number of independent lines of evidence point to the sun as an important source of orientation information in at least some species. It should be noted that all the data at hand are from relatively short-distance migrants.

V. Homing and Navigation

The apparent facility with which many kinds of animals return to nests, burrows, or other familiar areas has intrigued us for centuries. Whereas the plodding journey of a tiny salamander returning from an artificial displacement of several kilometers may be a physical task of no less magnitude than a bird homing several hundred kilometers, we must be very cautious when considering mechanisms or invoking true navigation (type III orientation, see Section I,B). Many animals move over considerable distances during the course of normal daily activities. Homing within these familiar areas may require nothing more than piloting by learned features, retracing the outward journey, or sensory contact with the home site. Experimental analyses of homing require that the animal be artifically transported to a locality outside its experience and from which it cannot establish sensory contact with the goal. Uncertainty about these criteria, based on inadequate knowledge of natural movements or sensory capabilities, obscures the interpretation of many homing studies.

All experimental investigations of homing suffer from an uncontrollable bias: they require that animals be displaced. Although the trip is not made under their own power, this in no way prevents the animals from accumulating information during the outward journey that may be used in homing. Indeed, recent work with homing pigeons, discussed below, points to a number of influences of information perceived during the displacement, although we do not know that this information is necessary.

Various models of true navigation have been suggested or implied in a number of inappropriate cases. Examples in which there exists uncertainty concerning the home range area of an individual or the possibility of direct sensory contact with home should be assumed explicable without invoking navigation until proved otherwise. Furthermore, at the present time there exists no evidence that any animal can navigate to a place it has never been. The recent discovery in Mexico of very localized winter roosts inhabited by staggering numbers of monarch butterflies *(Danaus plexippus)* (Urquhart and Urquhart, 1976, 1978; Brower, 1977) leads one to wonder how the butterflies find these isolated, but traditionally used, sites at the end of a flight that may well exceed 2500 km. Although some overwintering adults may make a return flight northward in the spring (Urquhart, 1960), there is no reason to think that any individuals return to the winter roost a second time. For the autumn migration of these butterflies to be goal-directed in the absence of any opportunity for learning requires a seemingly impossible level of complexity for genetic programming. Similar problems surround the first autumn migrations of young birds, discussed below.

A. Homing without Navigation

Numerous cases of undisputed homing fail to fulfill all or some of the criteria discussed above. The best documented cases are listed in Table I along with information on homing distances and possible cues involved. For none of these cases does it seem necessary to postulate navigation to explain the existing data, although future work may reveal more sophisticated behavior than is now demonstrated. Homing in some of the groups tabulated has been extensively studied and deserves more detailed comment.

For their relatively limited locomotor abilities, many species of salamanders engage in remarkable migrations to and from breeding ponds and several have been found to be adept at homing. The pioneering studies of Twitty and colleagues set the stage for much subsequent work with homing in urodeles. In 1959 he showed that newts *(Taricha granulosa)* could home from distances of at least 8 km, even if displaced over a high ridge into an adjacent drainage system. Furthermore, they showed predicted orientation upon leaving the release site even at distances as great as 12.8 km (Twitty *et al.*, 1967). These observations and the finding that blinded newts could

TABLE I

Cases of Homing Which Can Probably Be Explained without Invoking True Navigation

Taxa	Max. distance	Presumed cues used	Reference
Mollusca			
Limpets (*Patella* spp., *Siphonaria normailis, Cellana tramoserica*)	10 cm ±	Chemical trails	Cook *et. al.*, 1969; Cook, 1969; Mackay and Underwood, 1977
Arthropoda			
Crustacea	200 m	Familiar landmarks or spatial features of terrain; blinded animal returned to within 10 m of capture site	Herrnkind and McLean, 1971
Spiny lobster (*Panulirus argus*)			
Insecta			
Ants			
Pogonomyrmex badius	10–20 m	Odor trail laid down on outward Journey	Hölldobler, 1971
Cataglyphis bicolor	< 10 m	Sun, landmarks, wind direction	Wehner, 1972
Vertebrata			
Pisces			
Various species within or near home streams or lakes	?	Odors and odor gradients	Hasler *et al.*, 1978
Salmon	Hundreds of km	Directionally biased movement and odors near goal	Leggett, 1977
Amphibia			
Salamanders (*Plethodon jordani, Ambystoma maculatum, Desmognathus fuscus*)	100–500 m	Odors and odor gradients	Madison, 1969, 1972; Shoop, 1965; Barthalmus and Bellis, 1969, 1972
Newts (*Taricha* spp.)	Few km	Odors and odor graidents	Twitty, 1959; Grant *et al.*, 1968
Toad (*Bufo boreas*)	200 m	Odors and odor gradients	Tracy and Dole, 1969

TABLE I (Continued)

Taxa	Max. distance	Presumed cues used	Reference
Reptilia			
Turtles (Terrapene caro- lina, Chrysemys picta, Clemmys insculpta)	3–5 km ±	Odors and odor gradients; vis- ual landmarks	Gould, 1957, 1959; Emlen, 1969c; Ernst, 1970; Lemkau, 1970; Carroll and Ehrenfeld, 1978
Sea Turtle (Chelonia)	Hundreds of km	Currents, odors, and odor gra- dients	Koch, et al., 1969; Carr, 1972
Aves			
Swallows (Riparia riparia, Progne subis)	To 800 km	Visual land- marks and random search	Sargent, 1962; Downhower and Windsor, 1971; South- ern, 1968
Mammalia			
Bats (various species)	To 500 km	Visual land- marks and random search	Davis, 1966; Williams and Williams, 1967, 1970; Wilson and Findley, 1972
Rodents (various species)	600 m ±	Familiar odors or visual landmarks	Joslin, 1977

sometimes home successfully, whereas those with olfactory nerves severed did not, led to the hypothesis that odors might be important in the homing process.

Grant et al. (1968) tested the olfactory hypothesis of homing in newts in experiments involving very large numbers of animals. Newts were rendered anosmic by perfusing the nasal cavity with formaldehyde (shams were per- fused only in the oral cavity). All animals were collected from the same stream segment and 502 anosmic animals, 451 shams, and 617 untreated animals were displaced 1.2 km upstream. The newts moved slowly, with the first trap recoveries 228.5 m from the release site about 2 weeks after release. Of 89 recoveries of anosmic newts, 58 were downstream from the release area, whereas 31 were upstream. The sham controls and normal newts were captured downstream significantly more frequently (78 of 82 and 117 of 136, respectively). Whereas the anosmic newts were not recaptured at random, the results clearly imply an important role for olfaction, a sensible cue for this species to use because its migrations often take place on cloudy, rainy nights (the same is true for many other salamanders as well; Shoop, 1965).

Experiments performed on several other species of salamanders imply some generality for the olfactory hypothesis. Madison (1969) and Madison and Shoop (1970) studied homing in the terrestrial *Plethodon jordani.* Animals with olfactory nerves severed (33 individuals; 26 recaptured) dispersed randomly from release sites beyond their home ranges, whereas both blinded and normal animals homed. Observations on the behavior of the displaced salamanders also supported the olfactory hypothesis: they climbed on vegetation significantly more frequently than animals on their home ranges, and also did so more frequently on the night of release than later, perhaps suggesting use of airborne odors.

Similar results have also been obtained with the more aquatic *Desmognathus fuscus.* Barthalmus and Bellis (1969, 1972) displaced a large number of individuals between 3–30 m downstream from their small (<3 m along stream) home ranges. The only significant difference in their results was the low homing success rate in the anosmic group (16.4% versus 59.6% in normal animals, 48.6% in the sham-operated group, and 47.9% in blind salamanders). Those anosmic animals that did home were all released 3 m or less from home. Anosmic California toads *(Bufo boreas)* also moved at random when displaced from breeding ponds (Tracy and Dole, 1969). Most of these studies failed to control for trauma and other behavioral effects of blocking olfactory input.

As in so many cases, there is evidence that several cues may be involved in salamander homing orientation. Barthalmus and Bellis (1969) found that significantly more dusky salamanders returned from downstream versus upstream displacements. This result is explicable via the olfactory hypothesis, but may also suggest a rheotropic response. Finneran (1951) proposed down-current movement as a mechanism to bring spotted salamanders *(Ambystoma maculatum)* to breeding ponds, but Shoop's (1968) data cannot be explained in this way. Endler (1971) presented some evidence supporting the existence of kinesthetic memory in the newt *Taricha torosa,* and Madison (1972) noted that existing data do not exclude the possibility of some kind of inertial guidance at least over short distances.

Homing ability in several species of turtles has been known for many years (Romanes, 1883; Nichols, 1939). Most if not all of the existing data may be explained by what Carroll and Ehrenfeld (1978) termed "intermediate-range homing": the animals are released outside an immediately familiar area, but close enough that some local cues (e.g., airborne odors, large visual landmarks) may be available. In general, the homing speeds are very slow, often requiring months or even years (fastest documented speeds are about 0.5–1.0 km/day at displacement distances < 2 km in the wood turtle, *Clemmys insculpta*; Carroll and Ehrenfeld, 1978). Homing success diminishes drastically at distances beyond 2 km or less in most species studied: *Clemmys insculpta* (Carroll and Ehrenfeld, 1978); painted turtle (*Chrysemys picta*) (Gould, 1959; Emlen, 1969c; Ernst, 1970); and box turtle (*Terrapene carolina*) (Gould, 1957, 1959; Lemkau, 1970). A sufficient number of controlled experiments have not been performed and species differences may be significant. Blindfolded painted turtles dispersed at random, suggesting the importance of visual landmarks (Emlen, 1969a) and box turtles transported to the same release site several times tended to depart along the same paths (Lemkau, 1970). However, one naturally blind, 18-year-old wood turtle homed from 1.35 km by 3 years later (Carroll and Ehrenfeld, 1978).

The only data that are difficult to explain under the hypothesis of piloting by visual or olfactory landmarks are those of Gould (1957, 1959). An evaluation of the results of these experiments is slightly complicated by the fact that individuals were selected for experiments on the basis of homeward orientation in earlier trials, the same individuals were tested repeatedly and initial bearings appeared to be influenced by previous releases. Box turtles, which are reputed to have a very small home range (cf. Carroll and Ehrenfeld, 1978), showed significant homeward movement upon release to distances of at least 2.7 km. Release site bias was evaluated by simultaneously releasing turtles with opposite homeward directions: the animals moved in opposite directions. Initial departure directions became random under overcast. The animals were transported in opaque bags and the data seem to imply some role of the sun. However, the displacement distances are too small for any navigation system involving celestial coordinates to be used and I feel further experiments are required to fully eliminate the possibility that some sort of landmarks are being used. In many ways, turtles seem to provide an ideal experimental system in which to study homing.

The final well-studied group in which most if not all cases of homing can be explained without invoking true navigation is the bats. Interpretation of the available data is somewhat complicated by the fact that many temperate zone species migrate relatively long distances and individuals may therefore be familiar with large areas. Davis (1966) summarized the homing experiments to that date, and as noted by Wilson and Findley (1972), all cases of tropical homing can be explained by a model based on random selection of a flight direction at the release point and maintenance of that direction until a familiar area is encountered. The results of Williams and

Williams (1967, 1970) with *Phyllostomus hasitatus* in Trinidad are typical. They found good homing success at displacement distances to about 30 km, but a marked deterioration beyond that range. Blindfolded bats flew erratically, and coupled with the rapid decrement in homing ability with distance their results strongly suggested piloting by familiar landmarks. At very short ranges piloting may be possible using only echolocation. Other experiments with blinded bats have also yielded very low percentage returns (maximum about 11% when released with normal bats) (Davis, 1966).

The only data which seem inexplicable by a hypothesis of random search and piloting once a familiar area has been reached are those of Cope *et al.* (1961). They released 36 female *Eptesicus fuscus* at a distance of 402 km from home. An extraordinarily high 85% return was recorded at a high average speed of 10.1 km/hr, assuming that the bats flew 10 hr per night. The region of familiarity in this species is probably large (perhaps a radius of 225 km; Davis, 1966), but the percentage and speed of homing in this experiment vastly exceeded random predictions. Thus in spite of the generally negative evidence of the existence of type III homing in bats, the possibility cannot be certainly rejected.

The return to natal streams of many species of anadromous fish may at least be discussed as two processes: the open ocean phase during which the mouth of the home stream must be localized, and recognition of and movement in the home stream system. The mechanisms of home stream recognition have been elucidated by the extensive studies of Hasler and his students. Hasler and Wisby (1951) expanded the idea of Craigie (1926) into a hypothesis of home stream recognition based on the odors of the stream water. The hypothesis has been tested using a variety of species and the results have been reviewed by Hasler (1966) and Hasler *et al.* (1978).

Recent field experiments performed by Hasler's group provided data typical of the sort of convincing support they have consistently marshaled in favor of the olfactory hypothesis. Cooper *et al.* (1976) exposed a large number of young coho salmon (*Oncorhynchus kisutch*) to the soluble odorant morpholine at a concentration of 5×10^{-5} mg/liter for 34–36 days. The fish were then released into Lake Michigan at two localities, near the mouth of Oak Creek and 13 km north of Oak Creek. During the spawning migration period for these fish some 19 months later, morpholine was dripped into Oak Creek and fish moving into the stream were sampled. Over a 2-year period during which four such experiments were performed, 1739 imprinted versus 197 nonimprinted fish were caught. In the third year no morpholine was dripped into the creek and only 51 imprinted fish (versus 55 nonimprinted ones) were taken. Similar results have been obtained with trout, *Salmo gairdneri* (Cooper and Scholz, 1976). Furthermore, the effect is not specific to morpholine, although it is the odorous substance most commonly used in imprinting experiments. Scholz et al. (1976) used two groups of fish, one imprinted to morpholine (5×10^{-5} mg/liter) and the other to phenethyl alcohol (5×10^{-3} mg/liter). After release the fish were successfully decoyed to different streams by introducing one of the two substances into the water and the results were highly significant.

The results of these naturalistic experiments have received support from a technique developed by Hara et al. (1965). They were able to show that perfusing the nasal cavity of homing salmon with water from their home stream produced a characteristic EEG pattern as recorded from the olfactory bulb that could be distinguished from responses to other water. In subsequent experiments by Ueda et al. (1967) possible artifacts because of concentration effects were eliminated by using salmon from three different spawning grounds. The EEG response appeared definitely specific to "home" water. Interestingly, weaker, but detectable responses were produced by water traversed by the fish en route to the spawning site, water from a bypassed sidestream, and water from a point upstream from the spawning site. These data suggest that anadromous fish may create an "olfactory map" on the downstream migration that is remembered and subsequently used on the return trip.

More recent EEG studies have somewhat clouded the seemingly straightforward picture that emerged earlier. Cooper and Hasler (1974) noted that the EEG response is not entirely specific to home stream water, that it is quite variable within a population, and that it is complicated by pH and other nonspecific stimulatory products in the water. Nonetheless, Cooper and Hasler (1974, 1976) found the characteristic EEG response in morpholine-imprinted fish captured in decoy streams on their first spawning migration. In spite of criticisms by Hara and MacDonald (1975) that there is no direct evidence that morpholine is detected by olfaction, the clear difference between experimentals and controls seems to eliminate a nonspecific effect (Cooper and Hasler, 1975). Morpholine imprinting was found to last for at least 1½ years (Cooper and Hasler, 1974), but very little is known about the odors that presumably guide fish in natural situations. The available data (see Hara, 1970) suggest that they are volatile, water-soluble, neutral, dialyzable, heat-labile substances and it is possible that they derive from the fish themselves.

The idea that the stream odorant used by fish is a pheromone was suggested long ago by White (1934). Nordeng (1971) has revived this hypothesis, suggesting that the population in a stream conditions the water, perhaps with substances in the mucous covering their bodies. No direct test of this idea has been carried out, but Oshima et al. (1969) showed on the basis of EEG response that chinook salmon (Oncorhynchus tshawytscha)

could distinguish between water that had contained conspecifics overnight
and that which had held coho salmon. At least in theory the orientation
mechanism involving odors could be relatively simple. Harden Jones (1968)
and Hara (1970) noted that the presence of the imprinted odor might
simply induce the fish to swim upstream. Data supporting this idea have
been obtained in elvers of *Anguilla* spp. (Creutzberg, 1961; Miles, 1968)
and chinook salmon (Delacy *et al.*, 1969).

The major unsolved dilemma in long-distance fish migration is, of
course, the means by which the fish find their way to the vicinity of the
home stream where olfactory guidance may take over. At present there is
no direct evidence that true navigational ability is involved. Indeed, several
models involving movements with only slight degrees of bias in the homeward direction can account for most if not all of the observed cases (Saila,
1961; Saila and Shappy, 1963; Patten, 1964; Leggett, 1977; Westin and
Nyman, 1977). The small orientational bias necessary could be accounted
for by any of several cues (sun, currents, magnetism, water temperature,
etc.) discussed earlier.

The success with which such a pseudo-random search strategy could be
employed will depend in part on the extent to which cues indicating the
home stream (e.g., odors) penetrate the lake or ocean from which the fish
must return. Unfortunately, this is not known precisely in any case although there is reason to hypothesize that a river with large discharge
might be detectable at a distance of a few hundred kilometers (Harden
Jones, 1968). Nordeng (1977) has in fact proposed that odors emanating
from the home stream in the form of population-specific pheromones released by smolts migrating downstream provide the cues by which adults
pilot to the home streams. In most, but not all, species of salmon the homeward migration of adults overlaps partially with the downstream migration
of smolts. Furthermore, water masses within the ocean differ in chemical
and biological composition and such information might be used in homing
although the mechanism is by no means apparent (see Leggett, 1977).
Before more elaborate hypotheses are advanced and tested, it should be
firmly established that random or simple searching movements coupled
with short-range piloting are insufficient to explain the behavior of the fish.

B. Homing Probably Involving Navigation

1. Evidence of Navigation

The homing accomplishments of migratory birds present a considerable
contradiction. It is well documented that adult birds return repeatedly to
precise breeding and wintering localities occupied in previous years.

Furthermore, unequivocal navigation has been demonstrated by rapid homing from unfamiliar distant localities in several kinds of birds: Manx shearwater *(Puffinus puffinus)* (see Matthews, 1964), Laysan albatross *(Diomedia immutabilis)* (Kenyon and Rice, 1958), Leach's storm-petrel *(Oceanodroma leucorhoa)* (Griffin, 1940; Billings, 1968), homing pigeons *(Columba livia)*, and perhaps swallows (Rüppell, 1934, 1936, 1937; Southern, 1968) (for details of these cases, see Matthews, 1968; Keeton, 1974a; Emlen, 1975; Schmidt-Koenig, 1979b). On the other hand, there are no convincing data to show when or even if true navigation is employed during the normal process of migration. Several questions arise: Do migrants returning to a previously occupied nesting or wintering area perform behavior analogous to that of a homing pigeon? If so, does navigation take place throughout the migratory journey or only when nearing the goal? Finally, if navigation occurs, are the mechanisms similar to those in homing pigeons? Homing navigation with emphasis on pigeons was recently reviewed by Keeton (1974a). This discussion will emphasize migrant species and only the most recent developments from research on homing pigeons will be described.

The difficulty involved in observing homing in wild birds or inducing orientation clearly related to navigation in an experimental situation has retarded advances in this area. While many instances of homing in migratory birds have been documented, very few individuals have actually been followed from release to home and in many cases simple search until familiar landmarks are encountered cannot be ruled out. At present we need to know some very basic things about homing in migrant species: over what distances and at what speeds it occurs and details of the routes followed by the birds. Until these aspects of the behavior are well documented, it is premature to attempt to formulate hypotheses concerning mechanisms.

Given sufficient time, even small passerine migrants can return to previously occupied areas. Mewaldt's (1964a,b) classic displacement experiments with *Zonotrichia* sparrows yielded returns from as far as 3860 km. However, the birds were displaced during winter and none was known to have returned until the following winter. Their whereabouts in the interim remains unknown; presumably they went to northern breeding areas. More recently, Ralph and Mewaldt (1976) displaced 905 wintering individuals of two species of *Zonotrichia* over distances ranging from 5 to 160 km. Return rates were much longer than one would expect if the birds performed a direct flight. Adult birds returned with similar success from all distances, although there was a slight suggestion that homing performance might have been lower at intermediate ranges (60–100 km). Subadults did not return from distances greater than 60 km.

There now exists a consistent body of evidence that migratory goals are
in no way predetermined. Rather, they appear to be established through
experience in an imprinting-like process that appears to depend upon the
birds' ability to move about in the vicinity (Berndt and Winkel, 1980). This
has been demonstrated for both nesting (Löhrl, 1959; Perdeck, 1958) and
overwintering sites (Schwartz, 1963; Ralph and Mewaldt, 1975). It is
presumed that until these goal areas are established through experience, no
navigation can occur. The young bird on its first autumn migration, then,
might be somehow programmed to move in a given direction for a distance
that would place it within the species' wintering range, but it should not
show evidence of navigation toward a specific goal. Most available data
support this hypothesis, including the classic displacement experiments of
Perdeck (1958, 1967, 1974) with starlings (*Sturnus vulgaris*), Schüz (1949)
with storks (*Ciconia ciconia*), and Rüppell (1944) and Rüppell and Schüz
(1948) with hooded crows (*Corvus corone cornix*).

Exactly how the course of migration might be programmed in a young
bird has been the subject of considerable experimental scrutiny. Gwinner,
Berthold, and their colleagues have examined the role of endogenous
rhythmic factors in the control of migratory aspects of the annual cycle of
several species of sylviid warblers (e.g., Gwinner, 1968a,b, 1972, 1977;
Berthold, 1973, 1975, 1978). They propose that molt, migratory fat deposition, and nocturnal restlessness are under the control of an endogenous circannual rhythm. For purposes of this discussion it is not necessary to
evaluate these results in detail, but only to point out that comparative
studies on five species of *Sylvia* and two *Phylloscopus* have been done. In
birds held under constant conditions of photoperiod and temperature in the
laboratory, differences in the timing, duration, and intensity of *Zugunruhe*
among the species and between populations of the same species were correlated with the timing and distance of their migrations in the field. These
results certainly suggest that at least on their first fall migration, the
behavior of the birds is to a large degree under endogenous control. Indeed,
Gwinner (1972, 1977) has simulated the first fall migration of species with
quite different winter ranges in Africa. Based on quantitative data on
Zugunruhe in caged birds converted to flight distance by use of banding
recovery data, he was able to show that the birds could reach their species'
wintering area by so-called vector navigation (Schmidt-Koenig, 1973), i.e.,
flying in a given direction for a programmed length of time.

Several attempts have been made to elicit navigational responses from
artificially displaced birds tested in orientation cages (Dol'nik and
Shumakov, 1967; Potapov, 1966; Sauer and Sauer, 1959; Hamilton,
1962a,b; Rabøl, 1969, 1970, 1972, 1973; see Emlen, 1975). In no case have
the data so obtained shown unequivocally that the birds recognized the

geographical displacement, although some were suggestive. Yet during the course of migration, birds are frequently displaced by winds or otherwise find themselves off the normal route. Especially in coastal areas, impressive numbers of birds are often involved and individuals finding themselves offshore at dawn have often been observed to reorient and fly toward shore (Myres, 1964; Baird and Nisbet, 1960; DeSante, 1973; Richardson, 1978b). Such drifted migrants, including many immature birds on their first migration, have also been tested in orientation cages shortly after displacement (Evans, 1968; Able, 1977a). The orientation of some of these birds corresponded to the reorientation direction and could therefore suggest a navigational response. However, as discussed by Emlen (1975) and Able (1977a), a more or less automatic landward reorientation response accomplished by simple compass orientation could well have evolved in those migrant populations that move near coastal areas. The occurrence of reorientation in both free-flying and caged immature migrants strengthens an explanation involving only a compass-based response and on the basis of currently existing data I feel it is premature to attribute navigational ability to immature birds on their first autumn migration.

In inland South Carolina, Gauthreaux (1978) has studied oriented, low altitude movements of nocturnal migrants in the early morning hours. These flights are similar in many ways to the reoriented movements seen in offshore and coastal areas and Gauthreaux believes the birds involved are performing compensatory flights to counteract wind drift suffered during nocturnal flight. The predominant direction of the movement is toward the northwest, the expected direction if birds are correcting for drift by the prevalent westerly wind component. Rare morning flights toward the southeast appeared to be correlated with the occurrence of easterly winds at night within a few days before the observed morning movement. These "redetermined migrations" seemed to be better oriented when the sun was visible and to involve a disproportionate number of adult birds (S.A. Gauthreaux, personal communication). As in the case of the coastal reoriented morning flights, it is not clear that true navigation is involved. Birds could perceive in a number of ways that they are being drifted during nighttime flight. Flight oriented generally into those winds the following morning would likely effect at least partial compensation for the drift accrued during the night. Nor is it clear how general this phenomenon is in inland areas. Morning flights of nocturnal migrants are also often observed in eastern New York, but in that area the birds fly primarily in the same direction as the previous night's migration (Bingman, 1980).

The Gwinner hypothesis assumed that the direction as well as the timing of the migration was under the control of an endogenous program, but this aspect has been tested only recently. The garden warbler *(Sylvia borin)*

departs from northern Europe on a southwestward track passing through
Spain. Somewhere near the latitude of the Mediterranean, the orientation
of the birds apparently shifts to south or southeast, carrying them to the
winter range in southern Africa (see Fig. 16). Naive birds held under constant conditions during their first autumn migration period should show
such a shift in orientation during the course of the season if the hypothesis

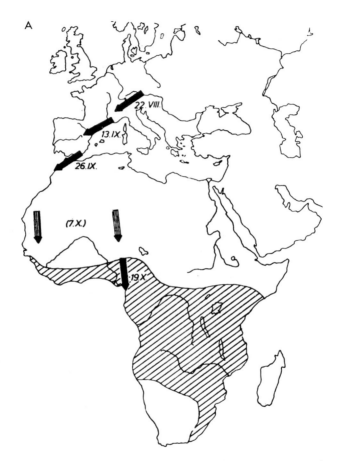

Fig. 16. Autumn migratory orientation of hand raised garden warblers *(Sylvia borin)*
tested indoors. (A) Map showing the African winter range and fall migration route of the
garden warbler. (B) Orientation of warblers during August and September (left) and during
October and November (right). Each triangle on the periphery represents the mean heading of
a single bird-night. Arrows indicate the direction of the mean vectors. The arrows on the map
(A) are the directional preferences of the experimental birds during periods at which garden
warblers normally pass through the indicated areas. (From Gwinner and Wiltschko, 1978.)

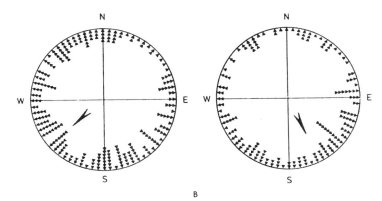

Fig. 16 (Continued)

of endogenous control is correct. Gwinner and Wiltschko (1978) hand-raised 59 garden warblers taken from nests before their eyes were open. The birds were kept under the natural photoperiod until late July–early August, at which time they were moved indoors and held on a constant L:D 12:12 photoperiod at constant temperature. Orientation was tested in a closed room in the standard Wiltschko cage which was covered with transluscent glass and surrounded by an opaque black curtain. Thus the orientation of the birds was recorded under so-called visually cueless conditions in which magnetic cues are presumably used. As would be expected with naive birds, the directions of individual bird-nights were highly variable and many were probably not significantly oriented. Figure 16 shows that the pooled nightly means for 404 tests yielded a significant vector toward 233° during the early part of the season (August and September) and toward 154° later (October, November, and December). This result supports the hypothesis of endogenous control and the two means of pooled data are significantly different by the parametric Watson–Williams test.

The data contained in Fig. 16 were actually obtained over four different years (1972–1975). In three years there was a counterclockwise shift in orientation between early and late in the season, but the difference was significant only in two of the years (largely because of the very great spread in directions). In the fourth year there was a nonsignificant clockwise shift of 20° in the mean directions. This degree of variability is unfortunately to be expected when naive, hand-raised birds are tested in orientation cages, especially without a view of the sky.

Several mechanisms are possible to explain orientation in species in which the normal migration involves a large change in direction and in which young birds do not migrate in social groups with adults (Wallraff,

1977). There might exist (1) a time-dependent program based on a seasonal rhythm or the duration of flight; (2) a distance-dependent program; or (3) a position-dependent program based on topographic cues or navigational coordinates. The important results of Gwinner and Wiltschko and the earlier data of Perdeck (1974) from Chaffinches (*Fringilla coelebs*) support the hypothesis of a time-dependent program with an endogenous rhythmic basis (actual migratory flight is not necessary). Other field data, primarily from displacement experiments, cannot be interpreted as unambiguously, but they are not inconsistent with that idea. Companion orientation cage experiments with a species such as the lesser whitethroat (*S. curruca*) that passes around the eastern end of the Mediterranean would be most important.

A similar, but somewhat more complex model for the programming of the migratory journey was proposed by Rabøl (1969). He envisioned a goal area of unknown size which moves through time along the migratory route of the species. If displaced from the route of migration a bird would employ navigation, presumably using the stars, not toward the ultimate goal of its migration, but toward the current position of the moving goal area. Thus a bird captured, displaced, and held in the displacement site for a sufficient length of time should show gradual shifts in orientation as the coordinates of the moving goal change. Rabøl's (1969) initial displacement experiments with immature and adult garden warblers suggested, as had Sauer's early work with the same species, that the birds might have perceived the geographic displacement. However, later experiments with garden warblers, willow warblers (*Phylloscopus trochilus*), and redstarts (*P. phoenicurus*) gave variable and equivocal results on both points (Rabøl, 1970, 1972). The arbitrariness of the size, position and rate of movement of the goal area makes rigorous tests of the hypothesis very difficult and alternate explanations of the data are possible (see Rabøl, 1972; Wiltschko and Wiltschko, 1978).

Although some possible exceptions exist, I believe the current weight of evidence supports the notion that migrant birds employ some type of direction and distance orientation (vector navigation) on their first migratory flight. They may employ directional corrections to counteract displacements by the wind, but there is no compelling evidence that first trip migrants are capable of bicoordinate navigation or goal directed homing to the wintering ground. Experience on wintering or breeding areas appears to be necessary to permit homing to those localities in subsequent seasons. Experienced individuals of at least some species (e.g., some sea birds and homing pigeons) have the ability to navigate home from unfamiliar release points at great distances from the home site. Although the return of a migrant to a previously occupied nesting or wintering locale is in many ways analogous, there is at present no evidence that it is performed in the same way or involves the same degree of sophistication. As Emlen (1975) has suggested, by prudently selecting nights to migrate based on favorable winds and in some way compensating for drift, a bird might well reach the general vicinity of its goal without performing navigation. We know nothing of how long it takes a migratory bird to reach the precise goal once it has arrived in the area or how much wandering or searching it does. The recent work of Perdeck (1974) even suggests that the migrant need not depart the natal area under its own power in order to return there the following spring. Nearly 3000 juvenile starlings were displaced from Holland to Switzerland. No differences were found between the number of these birds that returned to the capture site and the number of banded, nondisplaced juveniles. These young starlings had, of course, become familiar with the position of the capture site. Although the courses taken by these birds and the speed with which they made the trip are not known, the data cannot be explained by route reversal mechanisms involving landmarks or compass bearings.

2. Models of Navigation

Homing experiments on several species of birds leave no doubt that at least some can return rapidly from unfamiliar release points at great distances from home. Various models, based primarily on data from homing pigeons, have been proposed. No specific model has withstood experimental testing and at present there exists no detailed hypothesis that can account for the available observations. Below I will summarize models based on single cues and hybrid systems.

One means by which a bird might home would involve determination of latitude and longitude at the displacement site and comparison of these values with the coordinates of home. In theory, there are several ways in which this might be done, as proposed long ago, by using the sun (Matthews, 1951, 1953a, 1968), stars (Sauer and Sauer, 1960; Sauer, 1961), and earth magnetism coupled with the strength of the Coriolis force (Yeagley, 1947, 1951). These older hypotheses of bicoordinate navigation have been soundly rejected on the basis of numerous experimental tests recently reviewed by Keeton (1974a).

Some form of inertial navigation in which a bird detected and integrated all the accelerations experienced during the displacement journey could provide a unitary model of homing. In its simplest form this would involve an exact reversal of the outward journey, something clearly not observed in many experiments with homing pigeons. Barlow (1964, 1966, 1971) proposed a more elaborate model in which the bird must compute the most direct homeward route from accelerations logged during the outward journey. Experiments testing this idea were reviewed by Keeton (1974a), and although Merkel and Fischer-Klein (1973; Merkel, 1978) have recently shown that quail (*Excalfactoria chinensis*) running or passively transported through a runway containing a single turn can compensate for that angular deviation, the evidence with regard to inertial navigation is overwhelmingly negative.

The idea that a navigating bird might perform two tasks analogous to consulting a map and a compass was first explicitly stated by Kramer (1953). Whereas the cue or cues that provide the map component of this hybrid system remain elusive, the map and compass model has been of great heuristic value. Extensive analyses of the role of various compass cues in pigeon homing have been done. There now exists much evidence, based primarily on clock shifting experiments, that pigeons employ the sun compass in determining the homeward direction (see Keeton, 1974a). However, the sun is clearly not necessary for homeward orientation to be established and data discussed in Section III,A suggest that a magnetic compass may also serve in this capacity. There is as yet no evidence that a star compass is employed in any case of avian homing although there are instances of nocturnal homing (see, e.g., Keeton 1974a; Billings, 1968; Southern, 1968). The map component of navigation has remained frustratingly elusive. One of the most exciting recent developments in the study of bird navigation has been the discovery by Papi and his colleagues that olfaction plays some role in pigeon homing. The hypothesis is based on the pigeons becoming familiar with odors in the vicinity of their home loft and odors from more distant areas carried to the loft vicinity by winds (Papi *et al.*, 1972; Papi, 1976). The birds may associate the odors with the compass directions from which the wind carries them and with sufficient experience it could establish an olfactory map of the surrounding region. This, coupled with a compass, could allow the pigeon to determine the approximate homeward direction from unfamiliar sites. Obviously there should be problems in using such a system at very great distances from the loft. To overcome this problem, Papi *et al.* (1978) has proposed that pigeons also use odors encountered during the first part of the displacement journey as a clue to which direction they are being transported.

A variety of experiments have been performed by Papi's group to test the olfactory hypothesis of pigeon homing and a large number of papers have been published. It will not be possible to describe all the experiments in detail here, but I will attempt to outline the most important.

a. Olfactory Nerve Section.   Severing the olfactory nerves of homing pigeons would seem a straightforward way of testing the hypothesis that odors are vital to homing behavior. In fact, this was one of the first techniques employed by Papi's group (Papi *et al.*, 1971, 1972). These early experiments revealed a marked effect of bilateral nerve section or unilateral nerve section with the contralateral nostril plugged at release in experimental birds. The vanishing bearings of the experimental birds tended to be more spread, but the most noticeable effect was on homing success. Experimental birds tended to land shortly after release and their homing times were much longer than controls. Later experiments by Benvenuti et al. (1973a) also showed little effect on initial orientation, but a marked reduction in homing success from unfamiliar sites. Performance improved when the birds were released at familiar sites and similar results have been obtained by Snyder and Cheney (1975) using nostril plugs. Baldaccini et al. (1975a) reported on some well-controlled experiments involving larger numbers of birds. Twenty-eight pigeons were divided into two groups matched for experience. All had one olfactory nerve severed. A preliminary release of both groups about 40 days after surgery showed both well oriented in the homeward direction with similar homing speeds. The test release was made 3 days later. Control birds had the nostril ipsilateral to nerve section plugged with cement whereas the experimentals had the contralateral nostril plugged. The controls showed much more spread than in the initial release, but were still homeward oriented. The anosmic experimentals were not oriented in the homeward direction and had much poorer homing success. Sectioning the olfactory nerves of pigeons is a traumatic operation and although no other noticeable behavioral deficits have been reported, the effects on homing behavior are largely a result of the birds' reluctance to fly.

   b. Application of Masking Odors.   If pigeons use odors as Papi proposed, preventing the birds from smelling at the release site might render them unable to home. The early experiments by Benvenuti et al. (1973b) yielded somewhat variable results, but a variety of different procedures were employed. Later experiments (Benvenuti et al., 1977) controlled for the possible acquisition of olfactory information on the outward journey by plugging nostrils during transport. At release, α-pinene (experimentals) or Vaseline (controls) was painted thinly on the beaks and nostrils of the birds. The experimental birds vanished at random in all 10 releases whereas the controls were significantly oriented in every case, and were oriented in the homeward direction in 7 of 10 cases. The experimental birds also had significantly longer homing times. Similar results were obtained in two releases in Switzerland (Fiaschi and Wagner, 1976). Other attempts to repeat these experiments (Keeton and Brown, 1976; Hartwick et al., 1978) were unsuccessful.

   c. Detour Experiments and Other Manipulations during the Outward Journey. For releases made at considerable distances from the home loft, Papi's hypothesis proposes that odors detected during the early part of the displacement trip inform the pigeons about the direction they are being carried. If correct, transporting two or more groups of pigeons to the same release site via widely divergent routes should cause large differences in initial orientation as the birds attempt to retrace the portion of the route recognized by its odor. A similar technique was used long ago to test ideas of inertial navigation and it is important to note that simple detour experiments do not unambiguously test the olfaction hypothesis. For example, recent work (Wiltschko et al., 1978; Kiepenheuer, 1978; Papi et al., 1978) has at least suggested that magnetic influences may occur during the outward journey, although the effects documented so far are slight and inconsistent.

Papi et al. (1973a) reported the results of five detour experiments in which experienced pigeons were released at unfamiliar sites between 87.9 and 130.8 km from the home loft. In each case, two groups of pigeons were driven to a single release site over different routes selected such that their initial segments carried the pigeons in nearly opposite directions. The prediction is that both groups will be deflected from the homeward bearing, but in opposite directions. In all five releases the birds were deflected in the predicted directions and in each case the clockwise and counterclockwise birds were significantly different in mean vanishing bearing by the Watson–Williams F test. Later experiments (Papi, 1976; Fiaschi and Wagner, 1976) and attempts by other groups to repeat the results of the earlier tests (Keeton, 1974b; Hartwich et al., 1978) yielded highly variable results that provided no support for the hypothesis. Recently, Papi et al. (1978) summarized the quite variable results of 27 detour experiments. One of the two groups of pigeons was random in 13 releases and both were random in one. If all the releases are considered, the clockwise and counterclockwise groups are shifted relative to each other and the clockwise group is significantly shifted relative to the home direction, but the counterclockwise group is not different from homeward. If one considers only those releases in which both groups were significantly oriented and in which the two groups were significantly different from each other, only seven cases remain. In all these the deflections are in the predicted directions, but the counterclockwise group is still homeward oriented. Many factors, including wind patterns in the loft region, the intensity of odors along the two displacement routes, and release site biases could introduce a great deal of variability.

In an attempt to more directly test the involvement of odors in the detour effect, Papi included some birds with plugged nostrils in nine of these recent experiments. In only three releases were both groups oriented and in no case did they differ from one another. However, both groups tended to be deflected considerably clockwise of the home direction. Birds subjected to anosmic transport did have significantly more scattered vanishing bearings, but this could be an effect of the trauma involved in plugging their nostrils. While acknowledging the variability in the results of detour experiments, Papi *et al.* (1978) reiterated his belief that the effect is primarily a function of differences in olfactory information perceived.

d. Nasal Tubes. In an effort to avoid the traumatic side effects of surgical sectioning of pigeons' olfactory nerves or plugging of nostrils, Keeton *et al.* (1977) developed a technique for rendering pigeons anosmic by inserting thin plastic tubes into the nostrils. The birds should then be able to breathe freely, but the air should not come into contact with the olfactory epithelium. Releases at unfamiliar sites from 18 to 113 km revealed no differences in vanishing bearings or vanishing intervals. However, experimental birds had much poorer homing success from long distances, an effect the authors attributed to trauma or interference with respiration produced by the tubes. Using a similar technique, Hartwick *et al.* (1977) performed two releases with Italian pigeons, one at a familiar site (54.7 km west of the loft) and the other at an unfamiliar site (48.4 km south of the loft). At the familiar site, both experimentals and controls were significantly oriented and in the homeward direction. The nasal tube birds had significantly more scattered vanishing bearings and more of them landed close to the release point. At the unfamiliar site the control birds were homeward oriented; the experimentals were oriented, but not in the homeward direction. Again, many more birds with tubes landed before vanishing from sight and homing times were much longer in the experimentals at both release sites.

Hartwick *et al.* (1977) used the ability of the birds with tubes to home successfully from the familiar site as an argument against Keeton's proposal that the effect results from a disruption of the birds' motivation to home rather than their ability to do so. The only real difference between the two sets of experiments is the fact that the Italian pigeons were not homeward oriented at the familiar site. Their vanishing bearings were well oriented, however, and it seems dangerous to draw strong conclusions from this single release. The poor homing success and large number of experimentals that landed are very similar to Keeton's results.

e. Deflector Cages. Baldaccini *et al.* (1975b) devised an ingenious test of the olfactory navigation hypothesis. If an "olfactory map" is developed through experience at the home loft as they propose, then one ought to be able to induce the birds to learn an incorrect map if the winds carrying odors are systematically rotated so that they appear to come from a different direction. In the initial experiments three small cages (2.1-m cubes) were used, one with solid, but transparent wind deflectors oriented to deflect winds about 70° clockwise, one deflecting winds about 70° counterclockwise, and the control cage with no deflectors and sides that allowed air to flow through. Young, inexperienced pigeons were confined to these cages except for three 2-day exercise periods and three training releases close to the loft. During these forays all birds wore masks over their nostrils.

The pigeons were transported to release sites from 9.0 to 105.3 km from the loft in closed baskets. They were not allowed to see any landmarks during the journey, but were ventilated with air piped in from outside the van. In three releases, all groups were oriented, the mean vectors of the experimental groups were all deflected in the predicted directions, and in each case the experimental groups differed significantly from each other and from the controls.

More recent experiments (Baldaccini *et al.*, 1978) gave somewhat more variable results. In tests with naive birds, the expected results were obtained in nine of 13 releases. Other more equivocal data suggested that the deflector cage effect can be reversed by switching experimental groups and can be induced even in experienced birds that have presumably already established their map with respect to familiar release sites.

The results of the deflector cage experiments as well as the decrements in navigational ability observed in pigeons housed in aviaries impervious to wind (Papi *et al.*, 1973b; Baldaccini *et al.*, 1974) are suggestive of the "atmospheric factor" postulated by Wallraff (1966, 1970; see Keeton, 1974a). However, it is important to note again that this paradigm does not prove that odors per se are involved in the effect. Olfaction may be the explanation, but other potentially relevant orientation information (e.g., sun position and polarization patterns) is also altered by the deflector cages (W.T. Keeton, personal communication).

f. Fan Experiments. In an attempt to more directly demonstrate the involvement of odors in the pigeons' homing navigation, Papi *et al.* (1974) raised pigeons from fledging in elongate corridors in which they were shielded from natural winds and odors. Instead, artificial odorous winds were created by fans at the ends of the corridors. One group of birds was exposed no more than once each day to an odorous wind of olive oil from the south or an odorous wind of turpentine from the north; the other group received the opposite treatment. The birds were transported to two release sites 21.0 and 26.5 km east of the loft in closed containers ventilated with bottled air. Upon release, either olive oil or turpentine was applied to the birds' beaks and nostrils and the two groups within each treatment were predicted to vanish in opposite directions. The pooled results of two releases supported the hypothesis.

More recently, Ioalé *et al.* (1979) used similar corridors aligned approximately parallel to the prevailing wind near Pisa. When winds blew from one of these two quadrants, one group of pigeons in a closed corridor was exposed to fan-generated winds from that direction whereas another group was presented with air blown from the opposite end of the corridor. When winds blew from other directions, the fans were inactive. A third group in a middle corridor with screen ends experienced natural winds from the two experimental quadrants while a fourth group was never exposed to natural or artificial winds. In releases made along the axis of prevailing winds it was predicted that the two experimental groups (fans) would initially fly in opposite directions with the first group homeward oriented. Birds exposed to natural winds along this axis should orient homeward and birds never exposed to wind of any kind should fly off at random. Releases made off the axis of orientation of the corridors should result in random orientation in all groups since none had experienced winds from those directions.

In general, the results were consistent with the predictions. Experimental birds exposed to artificial winds from opposite the ambient direction did tend to fly more or less away from home (five of eight releases); in other cases they were random (three). The experimentals exposed to artificial winds from the ambient direction were homeward oriented in every case, as were the birds from the corridor with open ends. Birds shielded from all winds were statistically random in all four releases using these birds, although they showed a strong tendency toward bimodality along the homeward axis. Homing success was markedly impaired in the reversed wind birds and in those shielded from wind.

These experiments are perhaps the clearest in implicating some wind-related factor in the observed effects on initial orientation. The results reported by Ioalé *et al.* do not definitely implicate odors as the cause of the observed differences, but that seems the most likely possibility.

As dramatic as the results from the Pisa group have obviously been, attempts by other groups to repeat some of the experiments have been frustratingly unsuccessful. In hopes of resolving some of the apparent differences between experiments performed by Keeton's group at Cornell and Papi's in Italy, a series of collaborative studies was performed in Ithaca during the summer of 1977. The following is a brief outline of the results of this work published in two papers (Papi *et al.*, 1979; Waldvogel *et al.*, 1979).

1. *Olfactory nerve section.* Two releases were performed using pigeons in which one nerve had been sectioned and the same (controls) or opposite (experimentals) nostril plugged prior to release. No significant differences in initial orientation were found. In one of the two releases experimentals had significantly slower homing speed and more failed to return.

2. *Nasal tubes.* So as to minimize trauma, the tubes were inserted the night before the releases were performed. Two releases from familiar sites

and two from unfamiliar sites yielded rather variable results. The only significant difference was that experimentals had much lower homing performance from the unfamiliar site.

3. *α-Pinene.* Unfamiliar release sites were used and the birds' nostrils were plugged during transport. Seven releases were made and the results were again quite variable (controls were oriented in only two releases and experimentals in only two). No significant differences were found.

4. *Anosmic transport.* Two types of experiments were performed: pigeons were transported with their nostrils plugged or in aluminum containers supplied with bottled air. No clear directional trends were apparent, but once again experimentals showed significant decrements in homing speed.

5. *Detour experiments.* In two of four releases a significiant difference in the predicted direction was observed between the two groups. There were no differences in homing speed or success.

6. *Deflector cage experiments* (Waldvogel *et al.*, 1979). At distances between 16 and 20 km inexperienced pigeons that had been housed in deflector cages for 6 weeks beginning at 4–7 weeks of age were significantly oriented in four releases. The predicted geometric relationship between clockwise and counterclockwise-expected groups occurred in each release and the two experimental groups were significantly different in three of the four. The deflections were not as great as that usually seen in the Italian experiments and the deflector cage groups often did not differ in directionality from the controls.

Overall, the results of these experiments must be regarded as less than overwhelming replications of the work of Papi and his colleagues with Italian pigeons. The obvious question arises whether the differences were due to the different strains of pigeons being used or to environmental differences between New York and Italy. These questions cannot be answered by the results of experiments performed to date and, in fact, the Pisa and Ithaca groups were unable to reach complete agreement as to the interpretation of the collaborative experiments: the paper (Papi *et al.*, 1979) was published with separate discussions by the two groups. Keeton and Brown concluded "that there must be major differences in the homing behavior of the Pisa and Ithaca pigeons," whereas Papi and Benvenuti state "that no major differences in the homing mechanism of American and Italian pigeons are apparent."

The major difference cited by Keeton and Brown is that the Pisa workers have consistently found an effect of olfactory manipulations on initial orientation, while the Ithaca group has not. This appears to be true inasmuch as Papi and his colleagues have obtained significant differences in initial orientation in at least some experiments with nerve section, masking odors, nasal tubes, detours, and deflector cages. In the collaborative experiments, detours produced an occasional effect (on one of six routes examined) and deflector cages produced a clear effect. Note, however, that neither of these techniques necessarily implicates odors although that is certainly a likely possibility. Many of the treatments employed in the collaborative experiments produced marked effects on homing success. Keeton and Brown believe that because the birds in these experiments usually departed from the release site on the proper course, the poorer performance of the experimentals may have nothing to do with navigation per se but rather may result from a diminished motivation to home because of trauma or interference with respiration. In support of this, they cite the large number of experimental birds that landed near the release points.

In light of all the available evidence I believe it is clear that olfactory cues sometimes play a role in homing navigation in pigeons. On the other hand, although dramatic effects are sometimes achieved (especially in the detour, deflector cage, and fan experiments), it is not clear that odors are a necessary or primary component in the system. At best, they appear to be only one of the host of cues involved in what was already known to be an extremely flexible and redundant system. After all, like virtually every other experimental manipulation to which pigeons have been subjected, the vast majority of the birds ultimately homed, most at speeds that did not suggest random searching.

Most other models of homing navigation involve a map consisting of some system of coordinates, although what the axes might be is unclear. The interpretation of map information might be quite precise as required by the early models of sun or magnetic navigation. However, this is probably not necessary to explain many and perhaps all of the available data and Wallraff (1974, 1978b, 1979) has proposed a less restrictive hypothesis. First he noted that homing pigeons very often have a preferred compass direction or directional bias that is reflected in their vanishing bearings regardless of the homeward direction from a given release site (see also Windsor, 1975). Second, he postulated a grid of two coordinates or gradient fields that intersect at an angle that must not be too small, although they need not be orthogonal. This model and some release data are shown in Fig. 17. He then assumed that the bird can detect at least the sign of displacement along and the direction of one of the two gradient (gradient y in this case). It can therefore fly toward y_0, the axis of gradient y that passes through home. On the other hand, assume that there is a broad zone on both sides of home wherein movement along the x-gradient cannot be detected (bounded by Tx_1 and Tx_2). The bird would not know whether to fly up or down the x-gradient. Its realized flight direction will be a function of its perceived displacement along the y-gradient and its preferred com-

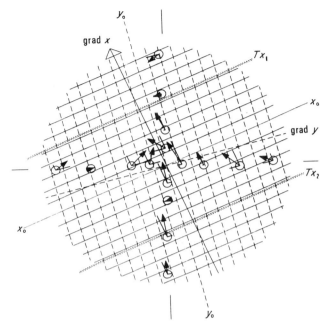

Fig. 17. Wallraff's null-axis hypothesis of pigeon navigation arbitrarily fitted to release data from Würzburg. The loft is in the center of the 200 km circle, north is at the top. The hypothetical x- and y-gradients and Tx_1 and Tx_2 have been positioned to provide an optimal fit to the release data. The arrows show the mean vector of 18–27 bearings. The mean vector was also split into two components aligned with the two gradients and these are shown as vector lines bounding the arrows. Note that with three exceptions, the vector components follow the general rule of paralleling the x-gradient with polarization in the preferred compass direction and paralleling the x-gradient toward y_0. See text for explanation of the model. (From Wallraff, 1979.)

pass direction. Flight in this direction will lead it eventually to the y_0 axis, which it will approach at some acute angle due to the influence of the preferred compass direction. Upon reaching the null axis of gradient y the bird will turn left or right and proceed until it (1) reaches home, or (2) crosses Tx_1 or Tx_2, at which point it should perceive the value of its displacement on the x-gradient and reverse its direction of movement along y_0. As noted by Wallraff (1978b) and shown in Fig. 17, by using information on the directional biases of pigeons, one can arbitrarily align the hypothetical gradients in such a way as to account for the observed vanishing bearings of a number of pigeon releases.

In Wallraff's scheme, the directional biases usually shown by homing pigeons are functionally related to one of the axes of the navigation system.

In numerous studies (Wallraff, 1978b) he investigated various aspects of these biases. The preferred compass directions appear to be loft specific, although at 10 of 11 lofts the bias was between NNW and WSW. They develop without flight experience in the vicintiy of the loft. Homing experience tends to change the relationship between the preferred compass direction and the homeward direction such that initial orientation becomes more homeward. However, no experimental manipulation at the loft (including long-term clock shifts and olfactory deprivation) produced a consistent change in the bias except housing the pigeons in a corridor open on only two opposing sides. The orientation of the axis formed by the open sides was related to the preferred direction of German pigeons, although an explanation for this effect was not obvious (Wallraff, 1978b).

Windsor (1975) examined the pattern of directional biases at a large number of release sites around the Ithaca loft. In general, his results are in agreement with Wallraff's predictions. At sites east of a line oriented NNW–SSE through the loft, the Cornell pigeons chose initial bearings that deviated clockwise from the homeward direction; west of that line their bias was counterclockwise. Although Wallraff (1978b) sees little exceptional in the pattern of preferred compass directions exhibited by Cornell pigeons, Keeton (1973, 1974a) has been impressed by several sites at which vanishing bearings consistently deviated greatly from homeward. He noted (1974a) that two sites quite close together may have very different biases although each site remains consistent over many releases. Thus Keeton and Wallraff approach the matter from fundamentally different viewpoints. Wallraff believes that whatever produces the directional preferences or biases acts at the home loft. Keeton, on the other hand, following Kramer (1959), proposed that some peculiarity at the release sites (he therefore calls them release-site biases) rather than loft sites was responsible for the biases. He believes that some unknown component of the navigational map may be rotated or otherwise distorted at certain release sites and that the observed biases in flight direction might provide a key to the elusive map.

Although there exists a general overall pattern to the release-site biases around Ithaca that supports Wallraff's hypothesis, there are some important facts that are very difficult to explain by a model depending upon the action of influences at the loft. For example, at Castor Hill, New York, pigeons usually deviate by 50°–90° clockwise from the homeward direction. Virtually identical biases were observed in pigeons from the Cornell loft (143 km SSW), from Schenectady, New York (182 km SE), and from Fredonia, New York (267 km SW), as well as in bank swallows (Riparia riparia) homing to Ithaca. It is possible that Castor Hill and the few other sites with very large deviations from homeward directions may be exceptional, as Wallraff (1978b) suggests. Only further investigation will resolve the important differences in interpretation, but both Keeton and Wallraff may be correct in believing that the consistent directional biases by homing pigeons have something important to tell us about homing navigation.

In attempting to study bird navigation we currently find ourselves in the rather embarassing position of not knowing even what questions to ask. On the one hand, recent developments in this field warn us not to search for unitary explanations, whereas on the other, we have no comprehensive, testable hypothesis to suggest experimental approaches. One step has been made by the important studies of Schlichte and Schmidt-Koenig (1971; Schmidt-Koenig and Schlichte, 1972; Schlichte, 1973; Schmidt-Koenig and Walcott, 1973) which showed that pigeons wearing frosted contact lenses could home successfully even though their vision was so drastically reduced that they could not recognize landmarks at 6-m distance. Although such pigeons utilize the sun compass in the determination of initial flight bearings (Schmidt-Koenig and Keeton, 1977), the navigational component of the homing process may well be based largely or entirely on nonvisual information.

VI. Conclusions

Research on animal orientation, navigation, and homing is currently in a state of turmoil. On the one hand, the past decade has seen rapid advances on many fronts, but on the other, each new discovery has in a real sense made the next questions and experiments harder to formulate. In nearly all animal groups an understanding of the behavior under natural conditions lags behind the documentation of sensory and behavioral capabilities. In most cases, unitary mechanisms will probably not prove viable, if one may project the results obtained with the most extensively studied species. Therefore we are faced with the difficult task of designing critical tests of hypotheses that will not unnaturally constrain the plasticity or mask the redundancy in the system. I believe these problems can best be addressed through field-experimental studies carefully guided by the results of both experimental and descriptive research.

A second problem is that even in the most extensively studied taxa, only a handful of species have been examined in detail. There is a justifiable tendency to generalize from those models and yet we should bear in mind that for most animals migration may greatly affect fitness and will therefore be subject to intense selection pressure. We might thus expect considerable divergence in the behavior of even reasonably closely related forms. The solution to this problem calls for a comparative approach which attempts to elucidate species, as well as age, sex, and motivational differences.

A small number of studies has revealed the power of a developmental approach to understanding orientation. Although new variables such as age and experience may obscure patterns, ontogenetic studies provide an excellent window through which to examine the plasticity characteristic of orientation mechanisms. Comparative developmental studies on carefully selected species are likely to provide important insights in the foreseeable future.

Persistent indications of magnetic sensitivity and even the use of magnetic information in orientation come from a variety of vertebrates and invertebrates. Yet in no case do we have a paradigm for revealing magnetic sensitivity that yields repeatable results of low variability. Behaviorists need to come up with new, more effective ways of asking animals to respond to magnetic stimuli. This task might be greatly facilitated if more was known about the sensory mechanisms of magnetic perception, an area in which interaction with neurophysiologists might prove profitable.

Finally, the ability of true navigation found in birds remains one of the most enduring mysteries in biology. There seems no doubt that at least a handful of species can goal orient from unfamiliar localities and we lack a viable hypothesis to entirely account for that phenomenon. On the other hand, while it is no less important to discover how pigeons, for example, perform homing navigation, we badly need to determine the extent of this ability among other species. This will call for basic descriptive studies on a spectrum of carefully selected typical species.

Despite these problems, we should take some encouragement from the rapid progress being made at this time. As this review points out, there are established and emerging commonalities in the orientation mechanisms of animals. It is the very difficulty of prying loose the answers to questions about orientation and navigation that makes the research so challenging, at times frustrating, and always exciting.

Acknowledgments

I express my sincere thanks to William Gergits and Verner P. Bingman for critically reviewing a draft of this chapter, and to William Brown for providing some references. It is also appropriate to acknowledge those who have aided my work in various ways over the years: Sidney A. Gauthreaux, Carl W. Helms, Donald R. Griffin, and my wife, Mary. The preparation of this review was facilitated by the opportunity to attend the Symposium on Animal Migration, Navigation, and Homing in Tübingen in 1977 and the International Ornithological Congress in Berlin in 1978. Thanks are due Klaus Schmidt-Koenig, William T. Keeton, Stephen T. Emlen, Wolfgang Wiltschko, the Deutsche Forschungsgemeinschaft, the National Research Council, and the Research Foundation of the State University of New York. I thank the National Science Foundation for research support during recent years.

References

Able, K. P. (1974a). *Anim. Behav.* **22**, 224–234.

Able, K. P. (1974b). *In* "Biological Aspects of the Bird/Aircraft Collision Problem" (S. A. Gauthreaux, ed.), pp. 331–357. Clemson, South Carolina.

Able, K. P. (1977a). *Auk* **94**, 320–329.

Able, K. P. (1977b). *Anim. Behav.* **25**, 924–935.

Able, K. P. (1978). *In* "Animal Migration, Navigation, and Homing" (K. Schmidt-Koenig and W. T. Keeton, eds.), pp. 228–238. Springer-Verlag, Berlin and New York.

Able, K. P. (1980). *Proc. Int. Ornithol. Congr., 17th, 1978* (in press).

Able, K. P., and Dillon, P. M. (1977). *Condor* **79**, 393–395.

Adler, K. (1970), *J. Herpetol.* **4**, 99–112.

Adler, K., and Taylor, D. H. (1973). *J. Comp. Physiol.* **87**, 203–212.

Alerstam, T. (1976). Ph.D. Thesis. Univ. of Lund, Sweden.

Alerstam, T., and Pettersson, S. G. (1976). *Nature (London)* **259**, 205–207.

Altevogt, R., and von Hagen, H. O. (1964). *Z. Morphol. Oekol. Tiere* **53**, 636–656.

Auburn, J. S., and Taylor, D. H. (1979). *Anim. Behav.* **27**, 658–668.

Baird, J., and Nisbet, I. C. T. (1960). *Auk* **77**, 119–149.

Baker, R. R. (1968a). *Proc. R. Entomol. Soc. London, Ser. A* **43**, 89–95.

Baker, R. R. (1968b). *Philos. Trans, R. Soc. London, Ser. B* **253**, 309–341.

Baker, R. R. (1969). *J. Anim. Ecol.* **38**, 703–746.

Baker, R. R. (1978). "The Evolutionary Ecology of Animal Migration." Holmes & Meier, New York.

Baldaccini, N. E., Benvenuti, S., Fiaschi, V., Ioalé, P., and Papi, F. (1974). *J. Comp. Physiol.* **94**, 85–96.

Baldaccini, N. E., Benvenuti, S., Fiaschi, V., and Papi, F. (1975a). *In* "Olfaction and Taste V" (D. A. Denton and J. P. Coglan, eds.), pp. 351–353. Academic Press, New York.

Baldaccini, N. E., Benvenuti, S., Fiaschi, V., and Papi, F. (1975b). *J. Comp. Physiol.* **99**, 177–186.

Baldaccini, N. E., Benvenuti, S., Fiaschi, V., Ioalé, P., and Papi, F. (1978). *In* "Animal Orientation, Navigation, and Homing" (K. Schmidt-Koenig and W. T. Keeton, eds.), pp. 78–91. Springer-Verlag, Berlin and New York.

Barlow, J. S. (1964). *J. Theor. Biol.* **6**, 76–117.

Barlow, J. S. (1966). *J. Inst. Navigation* **19**, 302–316.

Barlow, J. S. (1971). *Ann. N. Y. Acad. Sci.* **188**, 333–335.

Barthalmus, G. T., and Bellis, E. D. (1969). *Copeia* pp. 148–153.

Barthalmus, G. T., and Bellis, E. D. (1972). *Copeia* pp. 632–642.

Becker, G. (1963a). *Naturwissenschaften* **50**, 455.

Becker, G. (1963b). *Naturwissenschaften* **50**, 664.

Becker, G. (1964). *Z. Angew. Entomol.* **54**, 75.

Bellrose, F. C. (1958). *Bird-Banding* **29**, 75–90.

Bellrose, F. C. (1963). *Auk* **80**, 257–289.

Bellrose, F. C. (1967). *Proc. Int. Ornithol. Congr., 14th, 1966* pp. 281–310.

Benvenuti, S., Fiaschi, V., Fiore, L., and Papi, F. (1973a). *J. Comp. Physiol.* **83**, 81–92.

Benvenuti, S., Fiaschi, V., Fiore, L., and Papi, F. (1973b). *Monit. Zool. Ital.* [N.S.] **7**, 117–128.

Benvenuti, S., Fiaschi, V., and Foá, A. (1977). *J. Comp. Physiol.* **120**, 173–179.

Berndt, R., and Winkel, W. (1980). *Proc. Int. Ornithol. Congr., 17th 1978* (in press).

Berthold, P. (1973). *Ibis* **155**, 594–599.

Berthold, P. (1975). *In* "Avian Biology" (D. S. Farner and J. R. King, eds.), Vol. 5, pp. 77–128. Academic Press, New York.

Berthold, P. (1978). *In* "Animal Migration, Navigation, and Homing" (K. Schmidt-Koenig and W. T. Keeton, eds.), pp. 273-282. Springer-Verlag, Berlin and New York.

Billings, S. M. (1968). *Auk* **85**, 36-43.

Bingman, V. P. (1980). *Auk* **97**, 465-472.

Bingman, V. P., and Able, K. P. (1979). *Anim. Behav.* **27**, 621-622.

Birukow, G. (1958). *Z. Tierpsychol.* **15**, 265-276.

Birukow, G. (1960). *Cold Spring Harbor Symp. Quant. Biol.* **25**, 403-412.

Blakemore, R. (1975). *Science* **190**, 377-379.

Blunck, H. (1954). *Beitr. Entomol.* **4**, 485-528.

Bookman, M. A. (1977). *Nature (London)* **267**, 340-342.

Branover, G. G., Vasil'yer, A. S., Gleizer, S. I., and Tsinober, A. B. (1971). *J. Ichthyol.* **11**, 608-614.

Brower, L. P. (1977). *Nat. Hist.* **86**, 40-52.

Brown, F. A., Jr. (1971). *Ann. N.Y. Acad. Sci.* **188**, 224-241.

Carr, A. F. (1972). *NASA Spec. Publ.* **262**, 469-483.

Carroll, T. E., and Ehrenfeld, D. W. (1978). *Copeia* pp. 117-126.

Cook, S. B. (1969). *Anim. Behav.* **17**, 679-682.

Cook, A., Bramford, O. S., Freeman, J. D. B., and Teideman, D. J. (1969). *Anim. Behav.* **17**, 330-339.

Cooper, J. C., and Hasler, A. D. (1974). *Science* **183**, 336-338.

Cooper, J. C., and Hasler, A. D. (1975). *Science* **187**, 81-82.

Cooper, J. C., and Hasler, A. D. (1976). *J. Fish Res. Board Can.* **33**, 668-694.

Cooper, J. C., and Scholz, A. T. (1976). *J. Fish Res. Board Can.* **33**, 826-829.

Cooper, J. C., Scholz, A. T., Horrall, R. M., Hasler, A. D., and Madison, D. M. (1976). *J. Fish Res. Board Can.* **33**, 703-710.

Cope, J. B., Koontz, K., and Churchwell, E. (1961). *Proc. Indiana Acad. Sci.* **67**, 316-321.

Craigie, E. H. (1926). *Trans. R. Soc. Can.* **20**, 215-224.

Creutzbuerg, F. (1961). *Neth. J. Sea Res.* **1**, 257-338.

D'Arms, E., and Griffin, D. R. (1972). *Auk* **89**, 269-279.

Davis, R. (1966). *Ecol. Monogr.* **36**, 201-237.

DeLacy, A. C., Donaldson, L. R., and Brannon, E. L. (1969). *Res. Fish., Fish. Res. Inst., Univ. Wash.* **300**, 59-60.

Demong, N. J., and Emlen, S. T. (1978). *Bird-Banding* **49**, 342-359.

De Rosa, C. T., and Taylor, D. H. (1978). *J. Herpetol.* **12**, 25-28.

DeSante, D. F. (1973). Ph.D. Dissertation, Stanford University, Stanford, California.

Dietz, M. (1972). *Naturwissenschaften* **59**, 316.

Dingle, H. (1972). *Science* **175**, 1327-1335.

Dixon, A. F. G. (1971). *Sci. Prog. Oxford* **59**, 41-53.

Dolñik, V. R., and Shumakov, M. E. (1967). *Bionika* pp. 500-507.

Downhower, J. F., and Windsor, D. M. (1971). *BioScience* **21**, 570-572.

Drury, W. H., Jr., and Nisbet, I. C. T. (1964). *Bird-Banding* **35**, 69-119.

Duelli, P. (1975). *J. Comp. Physiol.* **102**, 43-56.

Duelli, P., and Wehner, R. (1973). *J. Comp. Physiol.* **86**, 37-53.

Edrich, W., and von Helversen, O. (1976). *J. Comp. Physiol.* **109**, 309-314.

Emlen, S. T. (1967a). *Auk* **84**, 309-342.

Emlen, S. T. (1967b). *Auk* **84**, 463-489.

Emlen, S. T. (1969a). *Living Bird* pp. 113-126.

Emlen, S. T. (1969b). *Science* **165**, 716-718.

Emlen, S. T. (1969c). *Behaviour* **33**, 58-76.

Emlen, S. T. (1970). *Science* **170**, 1198-1201.

Emlen, S. T. (1971). *Science* **173**, 460-461.

Emlen, S. T. (1975). *In* "Avian Biology" (D. S. Farner and J. R. King, eds.), Vol. 5, pp. 129–219. Academic Press, New York.

Emlen, S. T., and Demong, N. J. (1978). *In* "Animal Migration, Navigation, and Homing" (K. Schmidt-Koenig and W. T. Keeton, eds.), pp. 283–293. Springer-Verlag, Berlin and New York.

Emlen, S. T., and Demong, N. J. (1980). *Proc. Int. Ornithol. Congr., 17th, 1978* (in press).

Emlen, S. T., and Emlen, J. T. (1966). *Auk* **83**, 361–367.

Emlen, S. T., Wiltschko, W., Demong, N. J., Wiltschko, R., and Bergman, S. (1976). *Science* **193**, 505–508.

Endler, J. (1971). *Behavior* **37**, 15–23.

Enright, J. T. (1961). *Biol. Bull. (Woods Hole, Mass.)* **120**, 148–156.

Enright, J. T. (1972). *NASA Spec. Publ.* **262**, 523–555.

Ercolini, A., and Badino, G. (1961). *Boll. Zool.* **28**, 421–432.

Ernst, C. H. (1970). *Herpetologica* **26**, 399–403.

Evans, P. R. (1966). *J. Zool.* **150**, 319–369.

Evans, P. R. (1968). *Br. Birds* **61**, 281–303.

Ferguson, D. E. (1967). *In* "Animal Orientation and Navigation" (R.M. Storm, ed.), pp. 21–34. Oregon State Univ. Press, Corvallis.

Ferguson, D. E. (1971). *Ann. N.Y. Acad. Sci.* **188**, 30–36.

Ferguson, D. E., Landreth, H. F., and Turnipseed, M. R. (1965). *Copeia* pp. 58–66.

Fiaschi, V., and Wagner, G. (1976). *Experientia* **32**, 991–993.

Finneran, L. C. (1951). *Copeia* p. 81.

Forward, R. B., Horch, K. W., and Waterman, T. H. (1972). *Biol. Bull. (Woods Hole, Mass.)* **143**, 112–126.

Gauthreaux, S. A. (1969). *Bird-Banding* **40**, 309–320.

Gauthreaux, S. A. (1978). *In* "Animal Migration, Navigation, and Homing" (K. Schmidt-Koenig and W. T. Keeton, eds.), pp. 219–227. Springer-Verlag, Berlin and New York.

Gauthreaux, S. A., and Able, K. P. (1970). *Nature (London)* **228**, 476–477.

Goodyear, C. P. (1970). *Science* **168**, 603–605.

Görner, P. (1966). *Z. Vergl. Physiol.* **53**, 253–276.

Gould, E. (1957). *Biol. Bull. (Wood Hole, Mass.)* **112**, 336–348.

Gould, E. (1959). *Copeia* pp. 174–176.

Gould, J. L. (1975). *J. Comp. Physiol.* **104**, 161–174.

Gould, J. L. (1976). *Q. Rev. Biol.* **51**, 211–244.

Gould, J. L., Kirschvink, J. L., and Deffeyes, K. S. (1978). *Science* **201**, 1026–1028.

Grant, D., Anderson, O., and Twitty, V. (1968). *Science* **160**, 1354–1356.

Griffin, D. R. (1940). *Auk* **57**, 61–74.

Griffin, D. R. (1955). *In* "Recent Studies in Avian Biology" (A. Wolfson, ed.), pp. 154–197. Univ. of Illinois Press, Urbana.

Griffin, D. R. (1964). "Bird Migration." Doubleday, Garden City, New York.

Griffin, D. R. (1969). *Q. Rev. Biol.* **44**, 255–276.

Griffin, D. R. (1976). *Anim. Behav.* **24**, 421–427.

Griffin, D. R. (1978). *Harvey Lect.* **71**, 133–172.

Griffin, D. R., and Buchler, E. R. (1978). *In* "Animal Migration, Navigation, and Homing" (K. Schmidt-Koenig and W. T. Keeton, eds.). pp. 201–208. Springer-Verlag, Berlin and New York.

Griffin, D. R., and Hock, R. J. (1949). *Ecology* **30**, 176–198.

Griffin, D. R., and Hopkins, C. D. (1974). *Anim. Behav.* **22**, 672–678.

Gwinner, E. (1968a). *Z. Tierpsychol.* **25**, 843–853.

Gwinner, E. (1968b). *J. Ornithol.* **109**, 70–95.

Gwinner, E. (1972). *NASA Spec. Publ.* **262**, 321–338.

Gwinner, E. (1977). *Annu. Rev. Ecol. Syst.* **8**, 381–405.

Gwinner, E., and Wiltschko, W. (1978) *J. Comp. Physiol.* **125**, 267-273.

Hamilton, W. J., III (1962a). *Auk* **79**, 208-233.

Hamilton, W. J., III (1962b). *Wilson Bull.* **74**, 357-366.

Hamilton, W. J., III (1967). *In* "Animal Orientation and Navigation" (R. M. Storm, ed.), pp. 57-71. Oregon State Univ. Press, Corvallis.

Hamner, W. M., Smyth, M., and Mulford, E. D. (1968). *Anim. Behav.* **16**, 405-409.

Hara, T. J. (1970). *J. Fish. Res. Board Can.* **27**, 565-586.

Hara, T. J., and MacDonald, S. (1975). *Science* **187**, 81-82.

Hara, T. J., Ueda, K., and Gorbman, A. (1965). *Science* **149**, 884-885.

Harden Jones, F. R. (1968). "Fish Migration." Arnold, London.

Hartwick, R. F., Foá, A., and Papi, F. (1977). *Behav. Ecol. Sociobiol.* **2**, 81-89.

Hartwick, R., Kiepenheuer, J., and Schmidt-Koenig, K. (1978). *In* "Animal Orientation, Navigation, and Homing" (K. Schmidt-Koenig and W. T. Keeton, eds.), pp. 107-118. Springer-Verlag, Berlin and New York.

Hasler, A. D. (1966) "Underwater Guideposts." Univ. of Wisconsin Press, Madison.

Hasler, A. D., and Wisby, W. J. (1951). *Am. Nat.* **85**, 223-238.

Hasler, A. D., Scholz, A. T., and Horrall, R. M. (1978). *Am. Sci.* **66**, 347-355.

Hebrard, J. (1971). *Condor* **74**, 106-107.

Heran, H. (1962). *Z. Vergl. Physiol.* **46**, 129-149.

Herrnkind, W. F. (1968). *Am. Zool.* **8**, 585-598.

Herrnkind, W. F. (1972). *In* "Behavior of Marine Animals" (H. E. Winn and B. L. Olla, eds.), Vol. I, pp. 1-59. Plenum, New York.

Herrnkind, W. F., and McLean, R. (1971). *Ann. N.Y. Acad. Sci.* **188**, 359-377.

Herrnkind, W. F., and Kanciruk, P. (1978). *In* "Animal Migration, Navigation, and Homing" (K. Schmidt-Koenig and W. T. Keeton, eds.), pp. 430-439. Springer-Verlag, Berlin and New York.

Hisada, M. (1972). *NASA Spec. Publ.* **262**, 511-522.

Hoffmann, K. (1954). *Z. Tierpsychol.* **11**, 453-475.

Hölldobler, B. (1971). *Science* **171**, 1149-1151.

Hurst, G. W. (1969). *Endeavour* **28**, 77-81.

Huth, H., and Burkhardt, D. (1972). *Naturwissenschaften* **59**, 650.

Ioalé, P., Papi, F., Fiaschi, V., and Baldaccini, N. E. (1979). *J. Comp. Physiol.* **128**, 285-295.

Jander, R. (1957). *Z. Vergl. Physiol.* **40**, 162-238.

Jander, R. (1975). *Annu. Rev. Ecol. Syst.* **6**, 171-188.

Johnson, C. G. (1960). *Nature (London)* **186**, 348-350.

Johnson, C. G. (1969). "Migration and Dispersal of Insects by Flight." Methuen, London.

Joslin, J. K. (1977). *Adv. Ecol. Res.* **10**, 63-89.

Kalmijn, A. J. (1978). *In* "Animal Migration, Navigation, and Homing" (K. Schmidt-Koenig and W. T. Keeton, eds.), pp. 345-353. Springer-Verlag, Berlin and New York.

Keeton, W. T. (1969). *Science* **165**, 922-928.

Keeton, W. T. (1971). *Proc. Natl. Acad. Sci. U.S.A.* **68**, 102-106.

Keeton, W. T. (1972). *NASA Spec. Publ.* **262**, 579-594.

Keeton, W. T. (1973). *J. Comp. Physiol.* **86**, 1-16.

Keeton, W. T. (1974a). *Adv. Study Behav.* **5**, 47-132.

Keeton, W. T. (1974b). *Monit. Zool. Ital.* [N.S.] **8**, 227-234.

Keeton, W. T., and Brown, A. I. (1976). *J. Comp. Physiol.* **105**, 259-266.

Keeton, W. T., Larkin, T. S., and Windsor, D. M. (1974). *J. Comp. Physiol.* **95**, 95-103.

Keeton, W. T., Kreithen, M. L., and Hermayer, K. L. (1977). *J. Comp. Physiol.* **114**, 289-300.

Kennedy, J. S. (1961). *Nature (London)* **189**, 785-791.

Kenyon, K. W., and Rice, D. W. (1958). *Condor* **60**, 3-6.

Kiepenheuer, J. (1978). *Naturwissenschaften* **65**, 113.

Kimeldorf, D. J., and Fontanini, D. F. (1974). *Environ. Physiol. & Biochem.* **4**, 40-44.

Kirschfeld, K., Lindauer, M., and Martin, H. (1975). *Z. Naturforsch., Teil C* **30**, 88–90.

Kisliuk, M., and Ishay, J. (1977). *Experientia* **33**, 885–887.

Kleerekoper, H., Matis, J. H., Timms, A. M., and Gensler, P. (1973). *J. Comp. Physiol.* **86**, 27–36.

Koch, A. L., Carr, A., and Ehrenfeld, D. W. (1969). *J. Theor. Biol.* **22**, 163–179.

Kramer, G. (1949). *In* "Ornithologie als biologische Wissenschaft, Festschrift Erwin Stresemann" (E. Mayr and E. Schüz, eds.), pp. 269–283. Carl Winter, Heidelberg.

Kramer, G. (1950). *Naturwissenschaften* **37**, 377–378.

Kramer, G. (1951). *Proc. Int. Ornithol. Congr., 10th, 1950*, pp. 269–280.

Kramer, G. (1952). *Ibis* **94**, 265–285.

Kramer, G. (1953). *Verh. Dsch. Zool. Ges. Frieburg, 1952* pp. 72–84.

Kramer, G. (1959). *Ibis* **101**, 399–416.

Kramer, G., and von St. Paul, U. (1950). *Z. Tierpsychol.* **7**, 620–631.

Kreithen, M. L. (1978). *In* "Animal Migration, Navigation, and Homing" (K. Schmidt-Koenig and W. T. Keeton, eds.), pp. 25–34. Springer-Verlag, Berlin and New York.

Kreithen, M. L., and Eisner, T. (1978). *Nature (London)* **272**, 347–348.

Kreithen, M. L., and Keeton, W. T. (1974). *J. Comp. Physiol.* **89**, 83–92.

Landreth, H. F., and Ferguson, D. E. (1967). *Science* **158**, 1459–1461.

Larkin, R. P., and Sutherland, P. J. (1977). *Science* **195**, 777–779.

Larkin, T. S., and Keeton, W. T. (1976). *J. Comp. Physiol.* **110**, 227–231.

Larkin, T. and Keeton, W. T. (1978). *In* "Animal Migration, Navigation, and Homing" (K. Schmidt-Koenig and W. T. Keeton, eds.), pp. 92–106. Springer-Verlag, Berlin and New York.

Leask, M. J. M. (1977). *Nature (London)* **267**, 144–145.

Leask, M. J. M. (1978). *In* "Animal Migration, Navigation, and Homing" (K. Schmidt-Koenig and W. T. Keeton, eds.), pp. 318–322. Springer-Verlag, Berlin and New York.

Leggett, W. C. (1977). *Annu. Rev. Ecol. Syst.* **8**, 285–308.

Lemkau, P. J. (1970). *Copeia* pp. 781–783.

Lindauer, M., and Martin, H. (1968). *Z. Vergl. Physiol.* **60**, 219–243.

Lindauer, M., and Martin, H. (1972). *NASA Spec. Publ.* **262**, 559–567.

Linsenmair, K. E. (1968). *Z. Vergl. Physiol.* **60**, 445–449.

Linsenmair, K. E. (1972). *NASA Spec. Publ.* **262**, 501–510.

Löhrl, H. (1959). *J. Ornithol.* **100**, 132–140.

Lowery, G. H., Jr., and Newman, R. J. (1966). *Auk* **83**, 547–586.

Lüters, W., and Birukow, G. (1963). *Naturwissenschaften* **50**, 737–738.

McCleave, J. D., Rommel, S. A., and Cathcart, C. L. (1971). *Ann. N.Y. Acad. Sci.* **188**, 270–282.

Mackay, D. A., and Underwood, A. J. (1977). *Oecologia* **30**, 215–237.

Madison, D. M. (1969). *Anim. Behav.* **17**, 25–99.

Madison, D. M. (1972). *NASA Spec. Publ.* **262**, 485–498.

Madison, D. M., and Shoop, C. R. (1970). *Science* **168**, 1484–1487.

Martin, D. D., and Meier, A. H. (1973). *Condor* **75**, 472–474.

Martin, H., and Lindauer, M. (1977). *J. Comp. Physiol.* **122**, 145–188.

Matthews, G. V. T. (1951). *J. Inst. Navigation* **4**, 260–275.

Matthews, G. V. T. (1953a). *J. Exp. Biol.* **30**, 243–267.

Matthews, G. V. T. (1953b). *J. Exp. Biol.* **30**, 370–396.

Matthews, G. V. T. (1961). *Ibis* **103**, 211–230.

Matthews, G. V. T. (1963). *Proc. Int. Ornithol. Congr., 13th, 1962* pp. 415–429.

Matthews, G. V. T. (1964). *Auk* **81**, 132–146.

Matthews, G. V. T. (1968). "Bird Navigation," 2nd ed. Cambridge Univ. Press, London and New York.

Merkel, F. W. (1971). *Ann. N.Y. Acad. Sci.* **188**, 283–294.

Merkel, F. W. (1978). In "Animal Migration, Navigation, and Homing" (K. Schmidt-Koenig and W. T. Keeton, eds.), pp. 269–274. Springer-Verlag, Berlin and New York.
Merkel, F. W., and Fischer-Klein, K. (1973). Vogelwarte 27, 39–50.
Merkel, F. W., and Fromme, H. G. (1958). Naturwissenschaften 45, 499–500.
Merkel, F. W., and Wiltschko, W. (1965). Vogelwarte 23, 71–77.
Merkel, F. W., Fromme, H. G., and Wiltschko, W. (1964). Vogelwarte 22, 168–173.
Mewaldt, L. R. (1964a). Science 146, 941–942.
Mewaldt, L. R. (1964b). West. Bird-Bander 39, 1–2.
Michener, M. C., and Walcott, C. (1967). Science 155, 1136.
Mikkola, K. (1970). J. Anim. Ecol. 39, 593–598.
Miles, S. G. (1968). J. Fish. Res. Board Can. 25, 1591–1602.
Miller, L. J., and Weise, C. M. (1978). Condor 80, 94–96.
Moehn, L. (1974). J. Herpetol. 8, 175–183.
Moller, P. (1970). Z. Vergl. Physiol. 66, 78–106.
Moore, F. R. (1975). Auk 92, 655–664.
Moore, F. R. (1977). Science 196, 682–684.
Moore, F. R. (1978). Nature (London) 274, 154–156.
Myres, M. T. (1964). Ibis 106, 7–51.
Newcomer, R. T., Taylor, D. H., and Guttman, S. I. (1974). Herpetologica 30, 194–200.
Nichols, J. T. (1939). Copeia pp. 125–127.
Nielsen, E. T. (1961). Biol. Medd. 23, 1–81.
Nisbet, I. C. T. (1955). Br. Birds 48, 557–559.
Nisbet, I. C. T., and Drury, W. H. (1967). Bird-Banding 38, 173–186.
Nordeng, H. (1971). Nature (London) 233, 411–413.
Nordeng, H. (1977). Oikos 28, 155–159.
Oshima, K., Hahn, W. E., and Gorbman, A. (1969). J. Fish. Res. Board Can. 26, 2111–2121.
Ovchinnikov, V. V., Gleizer, S. I., and Galaktionov, G. Z. (1973). J. Ichthyol. 13, 455–463.
Papi, F. (1960). Cold Spring Harbor Symp. Quant. Biol. 25, 475–480.
Papi, F. (1976). Verh. Dtsch. Zool. Ges. 69, 184–205.
Papi, F., and Pardi, L. (1953). Z. Vergl. Physiol. 35, 490–518.
Papi, F., and Pardi, L. (1959). Z. Vergl. Physiol. 41, 583–596.
Papi, F., and Pardi, L. (1963). Biol. Bull. (Woods Hole, Mass.) 124, 97–105.
Papi, F., and Tongiorgi, P. (1963). Ergeb. Biol. 26, 259–280.
Papi, F., Fiore, L., Fiaschi, V., and Benvenuti, S. (1971). Monit. Zool. Ital. [N.S.] 5, 265–267.
Papi, F., Fiore, L., Fiaschi, V., and Benvenuti, S. (1972). Monit. Zool. Ital. [N.S.] 6, 85–95.
Papi, F., Fiaschi, V., Benvenuti, S., and Baldaccini, N. E. (1973a). Monit. Zool. Ital. [N.S.] 7, 129–133.
Papi, F., Fiore, L., Fiaschi, V., and Benvenuti, S. (1973b). J. Comp. Physiol. 83, 93–102.
Papi, F., Ioalé, P., Fiaschi, V., Benvenuti, S., and Baldaccini, N. E. (1974). J. Comp. Physiol. 94, 187–193.
Papi F., Ioalé, P., Fiaschi, V., Benvenuti, S., and Baldaccini, N. E. (1978). In "Animal Migration, Navigation and Homing" (K. Schmidt-Koenig and W. T. Keeton, eds.), pp. 65–77. Springer-Verlag, Berlin and New York.
Papi, F., Keeton, W. T., Brown, A. I., and Benvenuti, S. (1979). J. Comp. Physiol. 128, 303–317.
Pardi, L. (1954). Z. Tierpsychol. 11, 175–181.
Pardi, L. (1960). Cold Spring Harbor Symp. Quant. Biol. 25, 395–401.
Pardi, L., and Papi, F. (1952). Naturwissenschaften 39, 262–263.
Patten, B. C. (1964). J. Cons., Cons. Int. Explor. Mer. 28, 410–417.
Payne, R., and Webb, D. (1971). Ann. N.Y. Acad. Sci. 188, 110–141.
Perdeck, A. C. (1958). Ardea 46, 1–37.
Perdeck, A. C. (1967). Ardea 55, 194–202.

Perdeck, A. C. (1974). *Ardea* **62**, 190–195.

Phillips, J. B. (1977). *J. Comp. Physiol.* **121**, 273–288.

Phillips, J. B., and Adler, K. (1978). In "Animal Migration, Navigation, and Homing" (K. Schmidt-Koenig and W. T. Keeton, eds.), pp. 325–333. Springer-Verlag, Berlin and New York.

Picton, H. D. (1966). *Nature (London)* **211**, 303–304.

Potapov, R. L. (1966). *Dokl. Akad. Nauk SSSR* **171**, 226–228.

Rabøl, J. (1969). *Dan. Ornithol. Forens. Tidsskr.* **63**, 93–104.

Rabøl, J. (1970). *Ornis Scand.* **1**, 27–43.

Rabøl, J. (1972). *Z. Tierpsychol.* **30**, 14–25.

Rabøl, J. (1973). *Dan. Ornithol. Foren. Tidsskr.* **67**, 85–94.

Rainey, R. C. (1973). *Weather* **28**, 224–239.

Rainey, R. C. (1976). *Symp. R. Entomol. Soc. London* **7**, 75–112.

Rainey, R. C., and Aspliden, C. I. H. (1963). *Anti-Locust Mem.* **7**, 54–103.

Ralph, C. J. (1971). *Condor* **73**, 243–246.

Ralph, C. J., and Mewaldt, L. R. (1975). *Auk* **92**, 698–705.

Ralph, C. J., and Mewaldt, L. R. (1976). *Auk* **93**, 1–14.

Richardson, W. J. (1971). *Am. Birds* **25**, 684–690.

Richardson, W. J. (1972). *Am. Birds* **26**, 10–17.

Richardson, W. J. (1974). *Ibis* **116**, 172–193.

Richardson, W. J. (1976). *Ibis* **118**, 309–332.

Richardson, W. J. (1978a). *Oikos* **30**, 224–272.

Richardson, W. J. (1978b). *Auk* **95**, 717–732.

Riley, J. R. (1975). *Nature (London)* **253**, 113–114.

Roer, H. (1974). *Folia Entomol. Hung.* **27**(Suppl.), 49–59.

Romanes, G. J. (1883). "Animal Intelligence." Appleton, New York.

Rommel, S. A., and McCleave, J. D. (1972). *Science* **176**, 1233–1235.

Royce, W. F., Smith, L. S., and Hartt, A. (1968). *Fish. Bull.* **66**, 441–462.

Rüppel, W. (1934). *Vogelzug* **5**, 161–166.

Rüppel, W. (1936). *J. Ornithol.* **84**, 180–198.

Rüppel, W. (1937). *J. Ornithol.* **85**, 120–135.

Rüppel, W. (1944). *J. Ornithol.* **92**, 106–132.

Rüppel, W., and Shüz, E. (1948). *Vogelwarte* **15**, 30–36.

Saila, S. B. (1961). *Limnol. Oceanogr.* **6**, 292–298.

Saila, S. B., and Shappy, R. A. (1963). *J. Cons., Cons. Int. Explor. Mer.* **28**, 153–166.

Santschi, F. (1911). *Rev. Suisse Zool.* **19**, 303–338.

Sargent, T. D. (1962). *Auk* **79**, 234–246.

Sauer, E. G. F. (1957). *Z. Tierpsychol.* **14**, 29–70.

Sauer, E. G. F. (1961). *Psychol. Forsch.* **26**, 224–244.

Sauer, E. G. F. (1971). *Science* **173**, 459–460.

Sauer, E. G. F., and Sauer, E. M. (1959). *Vogelwarte* **20**, 4–31.

Sauer, E. G. F., and Sauer, E. M. (1960). *Cold Spring Harbor Symp. Quant. Biol.* **25**, 463–473.

Schaefer, G. W. (1976). *Symp. R. Entomol. Soc. London* **7**, 157–197.

Scheepers, C. C., and Gunn, D. L. (1958). *Bull. Entomol. Res.* **49**, 273–285.

Schlichte, H. J. (1973). *Z. Tierpsychol.* **32**, 257–280.

Schlichte, H. J., and Schmidt-Koenig, K. (1971). *Naturwissenschaften* **58**, 329–330.

Schmidt-Koenig, K. (1965). *Adv. Study Behav.* **1**, 217–278.

Schmidt-Koenig, K. (1973). *Naturwissenschaften* **60**, 88–94.

Schmidt-Koenig, K. (1975). "Migration and Homing in Animals." Springer-Verlag, Berlin and New York.

Schmidt-Koenig, K. (1979a). *Behav. Processes* **4**, 73–78.

Schmidt-Koenig, K. (1979b). "Avian Orientation and Navigation." Academic Press, New York.
Schmidt-Koenig, K., and Keeton, W. T. (1977). *Auk* **94**, 143–145.
Schmidt-Koenig, K., and Schlicte, H. J. (1972). *Proc. Natl. Acad. Sci. U.S.A.* **69**, 2446–2447.
Schmidt-Koenig, K., and Walcott, C. (1973). *Naturwissenschaften* **60**, 108–109.
Schneider, F. (1963). *Ergeb. Biol.* **40**, 252–279.
Schneider, F. (1975). *Vierteljahresschr. Naturforsch. Ges. Zuerich* **120**, 33–79.
Scholz, A. T., Horrall, R. M., Cooper, J. C., and Hasler, A. D. (1976). *Science* **192**, 1247–1249.
Schöne, H. (1963). *Z. Vergl. Physiol.* **46**, 496–514.
Schüz, E. (1949). *Vogelwarte* **15**, 63–78.
Schwartz, P. (1963). *Proc. Int. Ornithol. Congr., 13th, 1962* pp. 481–484.
Shoop, C. R. (1965). *Science* **149**, 558–559.
Shoop, C. R. (1968). *Biol. Bull. (Woods Hole, Mass.)* **135**, 230–238.
Snyder, R. L., and Cheney, C. D. (1975). *Bull. Psychon. Soc.* **6**, 592–594.
Southern, W. E. (1968). *Living Bird* **7**, 71–84.
Southern, W. E. (1969). *Condor* **71**, 418–425.
Southern, W. E. (1972). *Condor* **74**, 102–105.
Southern, W. E. (1975). *Science* **189**, 143–145.
Southern, W. E. (1977). *Mar. Sci. Res. Lab. Tech. Rep. No. 20*, 101–139.
Southern, W. E. (1978). *In* "Animal Migration, Navigation, and Homing" (K. Schmidt-Koenig and W. T. Keeton, eds.), pp. 311–317. Springer-Verlag, Berlin and New York.
Taylor, D. H., and Adler, K. (1973). *Science* **181**, 285–287.
Taylor, D. H., and Adler, K. (1978). *J. Comp. Physiol.* **124**, 357–361.
Taylor, D. H., and Ferguson, D. E. (1970). *Science* **168**, 390–392.
Tongiorgi, P. (1962). *Boll. Zool.* **28**, 683–689.
Tongiorgi, P. (1969). *Redia* **51**, 1–19.
Tracy, C. R., and Dole, J. W. (1969). *Copeia* pp. 693–700.
Tucker, V. A. (1971). *Am. Zool.* **11**, 115–124.
Tucker, V. A. (1974). *Publ. Nuttall Ornithol. Club* **15**, 289–333.
Twitty, V. C. (1959). *Science* **130**, 1735–1743.
Twitty, V. C., Grant, D., and Anderson, O. (1967). *Science* **155**, 352–353.
Ueda, K., Hara, T. J., and Gorbman, A. (1967). *Comp. Biochem. Physiol.* **21**, 133–143.
Urquhart, F. A. (1958). *Contrib. R. Ont. Mus. Div. Zool. Palaeontol.* **50**, 1–11.
Urquhart, F. A. (1960). "The Monarch Butterfly." Univ. of Toronto Press, Toronto.
Urquhart, F. A., and Urquhart, N. R. (1976). "Ecological Studies of the Monarch Butterfly," Natl. Geog. Res. Rep. 1968 Proj., pp. 437–443.
Urquhart, F. A., and Urquhart, N. R. (1978). *Can. J. Zool.* **56**, 1756–1764.
Viehmann, W. (1979). *Behaviour* **68**, 24–30.
Viehmann, W., and Wiltschko, W. (1977). *J. Ornithol.* **118**, 439–440.
Vleugel, D. A. (1954). *Gerfaut* **44**, 259–277.
Vleugel, D. A. (1959). *Ornis Fenn.* **36**, 78–88.
Vleugel, D. A. (1962). *Vogelwarte* **21**, 307–313.
von Frisch, K. (1949). *Experientia* **5**, 142–148.
von Frisch, K. (1950). *Experientia* **6**, 210–221.
von Frisch, K. (1967). "The Dance Language and Orientation of Bees." Harvard Univ. Press, Cambridge Massachusetts.
von Frisch, K., and Lindauer, M. (1954). *Naturwissenschaften* **41**, 245–253.
von Helversen, O., and Edrich, W. (1974). *J. Comp. Physiol.* **94**, 33–47.
von St. Paul, U. (1953). *Behaviour* **6**, 1–7.
von St. Paul, U. (1956). *Auk* **73**, 203–210.
Wagner, G. (1972). *NASA Spec. Publ.* **262**, 259–273.

Wagner, G. (1976). *Rev. Suisse Zool.* **83**, 883–890.

Walcott, C. (1972). *NASA Spec. Publ.* **262**, 283–292.

Walcott, C. (1977). *J. Exp. Biol.* **70**, 105–123.

Walcott, C. (1978). *In* "Animal Migration, Navigation, and Homing" (K. Schmidt-Koenig and W. T. Keeton, eds.), pp. 143–151. Springer-Verlag, Berlin and New York.

Walcott, C., Gould, J. L., and Kirschvink, J. L. (1979). *Science* **205**, 1027–1029.

Walcott, C., and Green, R. P. (1974). *Science* **184**, 180–182.

Walcott, C., and Michener, M. (1967). *Proc. Int. Ornithol. Congr., 14th, 1966* pp. 311–329.

Waldvogel, J. A., Benvenuti, S., Keeton, W. T., and Papi, F. (1979). *J. Comp. Physiol.* **128**, 297–301.

Wallraff, H. G. (1966). *Z. Vergl. Physiol.* **52**, 215–259.

Wallraff, H. G. (1970). *Z. Vergl. Physiol.* **68**, 182–201.

Wallraff, H. G. (1972). *Z. Tierpsychol.* **30**, 374–382.

Wallraff, H. G. (1974). "Das Navigationssystem der Vogel." Oldenbourg, Munich and Vienna.

Wallraff, H. G. (1977). *Vogelwarte* **29**, 64–76.

Wallraff, H. G. (1978a). *Oikos* **30**, 188–194.

Wallraff, H. G. (1978b). *In* "Animal Orientation, Navigation, and Homing" (K. Schmidt-Koenig and W. T. Keeton, eds.), pp. 171–183. Springer-Verlag, Berlin and New York.

Wallraff, H. G. (1980). *Proc. Ornithol. Congr., 17th, 1978* (in press).

Wallraff, H. G., and Gelderloos, O. G. (1978). *Oikos* **30**, 207–215.

Walton, A. S., and Herrnkind, W. F. (1978). *Mar. Sci. Res. Lab., Tech. Rep.* No. 20.

Waterman, T. H. (1966). *Am. Sci.* **54**, 15–45.

Waterman, T. H. (1972). *NASA Spec. Publ.* **262**, 437–456.

Waterman, T. H., and Forward, R. B. (1970). *Nature (London)* **228**, 85–87.

Waterman, T. H., and Forward, R. B. (1972). *J. Exp. Zool.* **180**, 33–54.

Waterman, T. H., and Haskimoto, H. (1974). *J. Comp. Physiol.* **95**, 1 –12.

Waterman, T. H. and Westell, W. E. (1956). *J. Mar. Res.* **15**, 149–169.

Wehner, R. (1972). *NASA Spec Publ.* **262**, 421–436.

Wehner, R., and Labhart, T. (1970). *Experientia* **26**, 967–968.

Wehner, R., and Menzel, R. (1969). *Science* **164**, 192–194.

Weis–Fogh, T. (1976). *Symp. R. Entomol. Soc. London* **7**, 48–72.

Wellington, W. G. (1974). *Science* **183**, 550–551.

Westin, L., and Nyman, L. (1977). *Contrib. Asko Lab., Univ. Stockholm*, No. 17.

White, H. C. (1934). *Trans. Am. Fish. Soc.* **64**, 360–362.

Williams, C. B. (1958). "Insect Migration." Collins, London.

Williams, T. C., and Williams, J. M. (1967). *Science* **155**, 1435–1436.

Williams, T. C., and Williams, J. M. (1970). *Anim. Behav.* **18**, 302–309.

Williams, T. C., and Williams, J. M. (1978). *In* "Animal Migration, Navigation and Homing" (K. Schmidt-Koenig and W. T. Keeton, eds.), pp. 239–251. Springer-Verlag, Berlin and New York.

Williams, T. C., Berkeley, P., and Victor, H. (1977). *Bird-Banding* **48**, 1–10.

Wilson, D. E., and Findley, J. S. (1972). *Am. Nat.* **106**, 418–423.

Wiltschko, R., and Wiltschko, W. (1978). *Oikos* **30**, 195–206.

Wiltschko, R., Wiltschko, W., and Keeton, W. T. (1978). *In* "Animal Orientation, Navigation, and Homing" (K. Schmidt-Koenig and W. T. Keeton, eds.), pp. 152–161. Springer-Verlag, Berlin and New York.

Wiltschko, W. (1968). *Z. Tierpsychol.* **25**, 537–558.

Wiltschko, W. (1974). *J. Ornithol.* **115**, 1–7.

Wiltschko, W. (1978). *In* "Animal Migration, Navigation, and Homing" (K. Schmidt-Koenig and W. T. Keeton, eds.), pp. 302–310. Springer-Verlag, Berlin and New York.

Wiltschko, W., and Merkel, F. W. (1965). *Verh. Dtsch. Zool. Ges. Jena* **32**, 362–367.

Wiltschko, W., and Merkel, F. W. (1971). *Vogelwarte* **26**, 245–249.
Wiltschko, W., and Wiltschko, R. (1972). *Science* **176**, 62–64.
Wiltschko, W., and Wiltschko, R. (1975a). *Z. Tierpsychol.* **37**, 337–355.
Wiltschko, W., and Wiltschko, R. (1975b). *Z. Tierpsychol.* **39**, 265–282.
Wiltschko, W., and Wiltschko, R. (1976). *J. Comp. Physiol.* **109**, 91–100.
Wiltschko, W., and Wiltschko, R. (1978). *Oikos* **30**, 177–187.
Windsor, D. M. (1975). *Anim. Behav.* **23**, 335–343.
Winn, H. E., Salmon, M., and Roberts, N. (1964). *Z. Tierpsychol.* **21**, 798–812.
Yeagley, H. L. (1947). *J. Appl. Phys.* **18**, 1035–1063.
Yeagley, H. L. (1951). *J. Appl. Phys.* **22**, 746–760.
Yodlowski, M. L., Kreithen, M. L., and Keeton, W. T. (1977). *Nature (London)* **265**, 725–726.
Zimmerman, M. A., and McCleave, J. D. (1975). *Helgol. Wiss. Teeresunters.* **27**, 175–189.
Zolotov, V., and Frantsevich, L. (1973). *J. Comp. Physiol.* **85**, 25–36.

Subject Index

Species Index

A

Abies (Fir), 120, 124
Acanthephyra, 7
Acer (Maple), 121
Acheta domesticus, 265
Acipenser transmontanus (White sturgeon), 148
Acris gryllus (Southern cricket frog), 259, 292, 294, 302
Acrocephalus arundinaceus (Great reed warbler), 182
Acroperus harpae, 129
Acyrthosiphon pisum, 86
Adansonia (Baobab), 26, 27
Aedes, 22
 aegypti, 21
Aelia, 33
Agkistrodon contortrix, 132
Agrotis ipsilon, 265
Aiolopus, 319
Airaphilus elongatus, 129
Alca torda (Razorbill), 134
Alectoris graeca (Rock partridge), 135
Alnus, (Alder). 122, 124
Alopex lagopus (Arctic fox), 137
Alosa
 pseudoharengus (Alewife), 243
 sapidissima (American shad), 44, 49, 148
Amazilia fimbriata (Glittering-throated emerald), 204
Ambystoma, 52
 maculatum (Spotted salamander), 338, 340
 tigrinum (Tiger salamander), 293, 294, 295, 299
Amphibolurus ornatus, 52–53
Anas

gibberifrons (Grey teal), 62
platyrhynchos (Mallard), 289
Anolis carolinensis (Green anole), 259
Anguilla
 anguilla (European eel), 41, 241, 242, 244, 247, 249, 314, 344
 japonica, 249
 rostrata (American eel), 41, 314
Antilocapra americana (Pronghorn), 138, 139
Aphelonema simplex, 28, 29
Aphis fabae (Black bean aphid), 3, 20, 29, 36, 37, 178, 186–187, 203
Apis mellifera (Honeybee), 284, 313
Apodemus
 aqrarius, 289
 sylvaticus, 76
Archilochus colubris (Ruby-throated hummingbird), 189, 191, 194
Arctosa variana, 289, 291
Ascia, 288
 monuste, 205, 209
Astyanax mexicanus (Mexican tetra), 39

B

Balaenoptera
 musculus (Blue whale), 64
 physalus (Fin whale), 64
Belgrandia marginata, 127
Bembidion dauricum, 130
Betula (Birch), 120, 123
Birgus latro (Coconut crab), 17
Blarina brevicauda (Short-tail shrew), 138
Blennius pholis, 244

379

PHYSIOLOGICAL ECOLOGY

A Series of Monographs, Texts, and Treatises

EDITED BY

T. T. KOZLOWSKI

University of Wisconsin
Madison, Wisconsin

J. Levitt. Responses of Plants to Environmental Stresses, 2nd Edition.
Volume I: Chilling, Freezing, and High Temperature Stresses, 1980
Volume II: Water, Radiation, Salt, and Other Stresses, 1980

James A. Larsen. The Boreal Ecosystem, 1980

Sidney A. Gauthreaux, Jr. (Ed.), Animal Migration, Orientation, and
Navigation

In preparation

F. John Vernberg and Winona B. Vernberg (Eds.), Functional Adaptations of Marine Organisms